THE MYTH OF THE NEGRO PAST

D0850892

Melville J. Herskovits (September 10, 1895 – February 25, 1963) was on the faculty of Columbia University, Howard University, and Northwestern University where he was a Professor of Anthropology and Director of the University's Program of African Studies. He is the author of numerous books, the latest of which were published posthumously and include *Life in a Haitian Valley, Trinidad Village, The American Negro, Economic Anthropology,* and *New World Negro,* edited by Frances S. Herskovits.

THE MYTH
OF THE
NEGRO PAST

MELVILLE J. HERSKOVITS

BEACON PRESS   BEACON HILL   BOSTON

International Standard Book Number: 0–8070–5475–5
19   18   17   16   15   14   13   12   11

*To the men and women who, in Africa
and the New World, have helped me
understand their ways of life*

CONTENTS

FOREWORD

This volume is the first published result of a Study which was announced in the Annual Report of the President of Carnegie Corporation of New York for 1938 in the following terms:

The Corporation has for some time felt the need of a general study of the Negro in the United States, not only as a guide to its own activities, but for broader reasons. It appeared to be essential that such a study be made under the direction of a person who would be free from the presuppositions and emotional charges which we all share to a greater or less degree on this subject, and the Corporation, therefore, looked outside the United States for a distinguished student of the social sciences who would be available to organize and direct the project. It is a pleasure to announce that Dr. Karl Gunnar Myrdal has been granted a leave of absence from the University of Stockholm to enable him to accept the invitation of the Trustees to undertake this work.

Dr. Myrdal arrived in New York in September, 1938, and remained here until the European situation made necessary his return to Sweden in May, 1940. During this period he requested some twenty American students of the Negro to prepare memoranda on all the more important aspects of Negro life in America, and on numerous minor ones. Most of these memoranda were unfinished at the time of his departure, but the majority were completed by the following September. Uncertainty concerning the date of Dr. Myrdal's return demanded reconsideration of the original arrangement which provided that the right to use all materials collected in the course of the Study should be vested in the contributors after the main report was published. Because of the unavoidable delay in the completion of Dr. Myrdal's own work, it was decided to facilitate the publication of some of the memoranda in advance of the main report, and the undersigned Committee was appointed to advise in the selection of those contributions most nearly ready for publication. Dr. Samuel A. Stouffer of the University of Chicago, who acted as executive officer of the Study during Dr. Myrdal's absence, was invited by the Committee to serve as its secretary.

In general the memoranda were not designed for publication in the form written. Contributors' instructions were to prepare working memoranda rapidly and in a full and easy style which would make them most useful for Dr. Myrdal's purposes. Thus, by definition, they were not to be formal, balanced manuscripts ready for the printer and the public. The Committee found that every manuscript submitted offered significant contributions. In serving the purposes of the Study so well, the contributors necessarily subordinated their individual publication interests to the interests of the central project. This is evidence of unselfish team-play which deserves respect and commendation. The Committee, however, was pleasantly surprised to find an appreciable number of manuscripts so near the publication stage that it could proceed with plans for the prompt publication of a group of monographs in advance of the main report. It is possible that other contributors will later publish additional monographs and articles as a result of their association with the Study. Dr. Myrdal returned to the United States in March, 1941, and it is hoped that his own report may be released some time during 1942.

It is on the whole logical that the first monograph to be published as a result of this Study should be concerned with the African background of the American Negro. There is an understandable tendency in our civilization to order our thoughts with reference to sequence in time and to think in terms of origins. Obviously Negroes were not brought to the United States as culturally naked people, and the problem is to determine what of their African heritage has been retained to influence life in America today. We may concede that the greatest significance of the African heritage lies in the fact that most of it quickly and inevitably was lost before the ways of life of the dominant white man could be learned. Yet cultural differentials are so important in the social adjustment of different peoples to each other that the retention even of cultural fragments from Africa may introduce serious problems into Negro-white relations. On the positive side, the origin of the distinctive cultural contributions of the Negro to American life must not be overlooked. Furthermore, and entirely apart from immediate practical considerations, the social scientist can learn about the general nature of cultural development from the cultural history of the Negro in America.

Every major activity in the career of Dr. Melville J. Herskovits of Northwestern University as an anthropologist has contributed to his qualifications for the preparation of this monograph on *The Myth of the Negro Past*. He has studied the physical as well as the

cultural traits of the Negro, and has made outstanding contributions
to knowledge in both fields. His investigations have included work
not only in the United States but also in Africa, South America
and the West Indies. Before this book is in print he will be at work
in Brazil. His lifelong conviction that the study of the Negro in the
United States requires also comparable study of the Negro in Africa
and in all parts of the Western Hemisphere has made possible this
monograph, the first comprehensive analysis of the current beliefs
concerning the extent and significance of traits of African origin
persisting in the United States.

SHELBY M. HARRISON
WILLIAM F. OGBURN
DONALD YOUNG, *Chairman*

July 25, 1941

PREFACE

This work represents the documentation of an hypothesis, developed in the course of two decades of research. That the scientific study of the Negro and attempts to meliorate the interracial situation in the United States have been handicapped by a failure to consider adequately certain functioning aspects of Negro life has become increasingly apparent as this investigation has gone on. Problems in Negro research attacked without an assessment of historic depth, and a willingness to regard the historical past of an entire people as the equivalent of its written history, can clearly be seen to have made for confusion and error in interpretation, and misdirected judgment in evaluating practical ends.

The approach in the ensuing pages, though oriented toward the study of the Negro in the United States, takes into full account the West African, South American, and West Indian data, lacking which, I am convinced, true perspective on the values of Negro life in this country cannot be had, either by the student treating of the larger problems of cultural change or by the practical man seeking to lessen racial tensions. While it has been necessary to throw into relief the neglect by others of such background materials in favor of nonhistorical statistical analyses and *ad hoc* remedies, this is not because a full knowledge of existing conditions is held unimportant or that the urgent problems to be faced by Negroes and whites alike are unrecognized. On the contrary, this study has attempted to show that present-day situations are more complex in their underlying causes than has been grasped and that, whether in analyzing intellectual or practical problems, every consideration calls for insight into the influence of pre-American patterns.

A word may be said regarding the documentation itself. Only those works that have had the widest influence in the past and those that are the most cited today have been given extended treatment, for these are the sources upon which academic opinion, at least, is based. The citations that have been made are those which have the sharpest bearing on the problem as envisaged. Where antiquarian, quixotic, or speculative comments have been included, it is only because they have produced a literary lineage. It would have been possible, were polemics the aim of this discussion, to trace a consistent genealogy for many of the ideas that, with slight qualification, have come to us in recent "authoritative" presentations. It would have been equally possible, with the use of the materials from the field study conducted in Trinidad by Frances S. Herskovits and myself in 1939, to dissect current views on, let us say, the Negro family, in terms of attitudes and customs prevailing there that are directly comparable to those found in the United States. But since these materials are as yet unpublished and, moreover, since it was necessary to keep this study within manageable compass, this was not done.

I am deeply indebted to Frances S. Herskovits, whose many months devoted to exhaustive reading made available the materials for a control of the literature. Without this reading, which I myself was unable to undertake because of academic commitments, this study could not have been made. Nor could it have been delegated to another, for, since she has participated with me in all the ethnographic field studies in my research program, she was uniquely equipped to discern correspondences and to evaluate interpretations.

These pages have been written at the suggestion of Dr. Gunnar Myrdal for the Study of the Negro in America which was instituted by the Carnegie Corporation of New York. The survey character of the Study did not permit an intensive ethnological study of a southern community to be made, against which to project prevailing theories that expound Negro mores in the United States. It is gratifying that, though this analysis was undertaken with misgivings as to what the heterogeneous literature might yield, there did, in effect, emerge in clear outline a group of mythological categories that define the greater part of the literature in this field.

MELVILLE J. HERSKOVITS

Evanston, Illinois
August 24, 1940

PREFACE TO THE BEACON PRESS EDITION

This book first appeared during the month in which the United States entered the Second World War. Its major hypotheses were based on research into African and New World cultures that began seventeen years before writing of it was planned; this Preface is being written seventeen years after its outline was drawn up. I could not have foreseen at the time that the challenge set forth in it — a challenge to the stereotype set up, elaborated and defended by serious students of the Negro, and held to firmly by a vast majority of lay opinion — would be so caught up in the historical currents of the immediate years ahead as to force the profound revision of views concerning the potentialities and the role of Africans and their New World descendants that has actually occurred on the world scene.

The contributing factors in this fundamental and fundamentally significant change in the climate of opinion relating to the African and the New World Negro are many and complex. This is not the place to analyze these factors, but we must particularly bear in mind realignments of economic potential and political structure and, more than is ordinarily realized, the influence of the doctrine of the Four Freedoms, with its hold on the imagination of non-self-governing peoples and minority groups everywhere. We need only call the roll of the nations that have come into being in these post-war years to recognize the international repercussions — the Philippines, India, Pakistan, Indonesia, Burma, Ceylon, Israel, Syria, Lybia, Tunisia, Morocco, the Sudan, Ghana, Malaya, and the emerging Commonwealth of the West Indies. Internal changes are more subtle, but here, without doubt, the more far-reaching are those which have occurred in the position of minority groups in the United States, changes that have culminated in the dramatic drive toward Negro-white integration.

The fact that this book appeared at the moment it did was a fortuitous circumstance; but it was no accident that it was conceived and written, for this was in response to a long chain of developments in both the intellectual and the political spheres. That it may have made some contribution toward an understanding of human relations, on any level, and aided in bringing about a heightened awareness that the role played by cultural continuities in New World Negro life is no different from the one it plays in the lives of other peoples, is a cause for deep gratification.

Despite the far-reaching changes of the past decade and a half, there has been little reason to alter the form or the content of this book. Beyond its concern with the historic process, as this bears on the adjustment to a changed cultural milieu of that part of the population of the United States and of other parts of the New World which is wholly or partially of African descent, lie its implications for an understanding of the processes of human civilization as a whole. And these are not things of the moment. The reader will, I hope, find in the pages that follow information of immediate relevance for an understanding of the problems of interracial living which must be resolved if this country and the world are to achieve the adjustments essential to peaceful existence. He will, however, not find solutions to these problems there. It is the task of scholarship to provide the facts, ordered on a basis of scientific method and theory, which will permit the development of sound policy. But implementation moves to a different level, for here the questions posed can be resolved, in a democracy, at least, only by political debate and the exercise of the mandate of public opinion.

In terms of a broad classification of scientific effort, this book belongs to a field that has come to be known as the field of Afroamerican Studies. In 1941, when these materials were originally published, this field had scarcely more than taken form. Many new data have since been gathered, concepts have been rendered more precise, theory has been more clearly defined; but the basic orientation of research and analysis has remained stable. It is to be noted that in the lively debate that followed the publication of this book, the validity of these facts, as such, was never challenged; rather, discussion turned on questions of method and interpretation.

Thus, by consensus, the facts stand; the conceptual scheme has been sharpened, but not negated. This Preface, therefore, will supplement the original by indicating, in broad line, the findings of subsequent research in the field and the refinements in approach that have characterized its growth, thereby bringing the book into line with developments since its original publication.

2

How rapidly the field of Afroamerican Studies has grown becomes apparent when it is pointed out that the Appendix to the earlier edition, entitled "Directives for Future Research," is not reprinted here because so many of the studies named there have since been executed. By the same token, consultation of the supplementary bibliography which follows the list of titles originally given will disclose how much has been achieved that has bearing on the study of the place and significance of African cultures in the New World and the cultural adaptations that characterize them. It should be noted, moreover, that this supplementary list of works does not tell the whole story, since of necessity it cannot include reports of research as yet unpublished, to say nothing of investigations in process.

A decade and a half ago, the civilizations of the regions of Africa from which the African ancestors of the New World Negroes were derived were known to us in general outline, but there were substantial gaps, both as regards certain of its sub-areas and particular facets of its cultures. In the intervening period, the major portion of these gaps has been filled, an effort which, to an appreciable extent, is a reflection of the steadily growing importance of Africa in the world picture. Scholarly investigation has also been given stimulus and stability by the establishment of centers of higher learning and research institutes in Africa itself. In the relevant areas we find the Institut Français de l'Afrique Noire (IFAN) in Dakar, with its branches in various parts of French West Africa; the University Colleges of Ghana, and of Ibadan, in Nigeria; in the Belgian Congo, the Institut pour la Récherche Scientifique en Afrique Central (IRSAC), all with important subdivisions dealing with the human sciences; and the East African Institute for Social and Economic Research in Uganda. In addition, the many field researches carried out by the students and the staffs of universities in the United States, England, France and Belgium have materially added to our resources for studying the cultural background of the New World Negroes.

Indeed, in some respects, this phase of the work has gone well beyond anything that could have been envisaged in 1941. Researches that are giving new insights into historical relationships, especially between the peoples of West Africa, are being prosecuted with energy and imagination, many of them, significantly, by African scholars. In Dakar, at IFAN, a long-term study of the history of slaving operations from the island of Gorée, a major base in the early days of the trade, has been conducted by Dr. Ly ; Professor K. Onwuka Dike's volume on commerce and politics in the Niger Delta area during the nineteenth century has important implications for Afroamerican studies in their historical dimensions. Extensive investigations into the past of certain societies which contributed significantly to the peopling of the New World by Africans and their descendants are under way or projected, notably those having to do with Benin and with the Yoruba. The musicological and ethnographic study of Akan (Ghana) dirges by J. H. Nketia provides fresh comparative data from that country, whose cultures were so important in establishing characteristic Afroamerican patterns in many parts of the New World. Numerous linguistic researches have been and are being conducted, so that comparisons between New World Negro and African patterns of speech can be made with fuller documentation.

A similar tale is to be told for ethnographic studies in many parts of the New World. The researches carried out under the auspices of the Instituto Joaquim Nabuco, in Recife, Brazil, the University College of the West Indies, both by its Institute of Social Studies and its Extra-Mural Department, and the Haitian Institut d'Ethnologie, have added measurably to the data in hand. As with African research, to the work of organizations functioning *in situ* must be added the investigations that have been and are being made in the Caribbean and Latin America under the auspices of numerous institutions of higher learning elsewhere. In the aggregate, certainly as concerns those parts of the Negro New World that lie outside the United States, this body of research has met, and in certain instances exceeded, the schedule laid down in these earlier "Directives." Entirely new areas have also been opened for study, as is to be seen in the ethnohistorical research of Gonzalo Aguirre-Beltrán on the Negro population of Mexico, the work of Taylor and Coelho on the Black Caribs of Honduras, of Edouardo on the Afrobrazilians of Maranhão in northernmost Brazil, of myself and later of Bastide on those of the extreme south of the country, and of Arboleda and Price on the Negroes of Colombia. In addition, regions earlier studied have been subjected to more intensive re-analysis. Historical research in Jamaica, Haiti, Colombia, Brazil and the United States has added valuable insights to our understanding of the process of inter-group acculturation and the emergence of identifiably dominant cultural influences.

Studies in special fields have also been prosecuted with vigor. In linguistics we now have structural analyses of New World Negro creolized languages — *taki-taki* of Suriname, *créole* of Haiti, *papiamento* of Curaçao — which complement researches in Africa that are similarly oriented. Continuation of earlier work in the collection and systematic analysis of narratives, proverbs, riddles and other forms of the language arts, that sector of Afroamerican studies longest pursued by trained scholars, has materially extended our knowledge of this aspect of the field. Prewar interest in jazz has taken on an intellectualistic component marked by concern for its derivations and development. Its relation to other New World forms, such as the calypso of Trinidad, the rhumba of Cuba, or the samba of Brazil, is being explored, and its African roots analyzed through the use of the vast collections of music that have been recorded on both sides of the Atlantic. The potential of this development can be realized when it is pointed out that at the time this book was first published, the technique of recording on magnetic tape had not been developed, and electronic recording, on discs, then a novelty, required bulky equipment that presented endless problems in the field. Along with interest in Afroamerican music is an interest in the dance forms so closely associated with it. Here the methodological problems present many challenges, but these are being met, and analyses that will broaden our understanding of the operation of acculturative mechanisms in this important phase of New World Negro life should become increasingly available. Psychological studies, notably the research of Ribeiro into the Afrobrazilian *candomble* as a mechanism of adjustment, have materially enlarged our resources in this dimension of the field.

Perhaps the only geographical area where Afroamericanist field research has not been prosecuted is the United States. Numerous studies have, of course, been made here, but they have been primarily oriented toward the solution of critical aspects of interracial adjustment. They provide the Afroamericanist certain data that are of use to him, and that, indeed, were gathered from many earlier works in writing this book; but studies of this sort do not suffice, and the sentence written seventeen years ago remains as valid today as it was then: "In this country, the greatest need is for research in Negro communities wherein the life will be studied in all its phases and with all regard for the implications of those traditional values that . . . may be considered in the light of similarities in the African background."

A number of Negro communities, it is true, have been studied in the past decade and a half. But Afroamerican and "community" studies are by no means the same. The distinctive feature of Afroamericanist analysis lies in precisely what these other researches, which analyze the interaction between structure and function on the single time-plane of the present, fail to take into account. This reluctance to include ethnohistory as a methodological tool, and to make use of relevant comparative data, is the reason why, as far as the United States is concerned, systematic research within the field of Afroamerican studies still lies before us.

There is one other development that is essential if the potentialities of Afroamerican research are to be fully realized. If we compare the formulations of African and Afroamerican research, we find far more attention paid to the religious aspect of Afroamerican cultures than to their social or economic forms, while in African research emphasis is laid almost exclusively on social, political and economic institutions. As a result, the gears of researches that should mesh closely fail to come together, and developments of mutual advantage that might be anticipated do not eventuate.

In making this observation about the need for adequate study of the social and economic phases of Afroamerican life, or for more research into the religious aspects of African cultures, I am, however, in no sense urging the substitution of one emphasis for the other, in either direction. Research designs for particular aspects of a culture have obvious merits for the study of some problems, but such studies should be undertaken only where the total culture of which this aspect is a part has been studied. This caution becomes the more urgent when it is pointed out that the purely analytical, restricted approach too often has associated with it a neglect of the historic dimension, to the essential place of which, in the Afroamerican field, this entire work bears testimony.

3

It will be evident, from the initial chapter of this book, that its methodology was more sharply defined than the conceptual structure that shaped its hypotheses and guided the utilization of its data. In essence, its conceptual apparatus did not go beyond the general theory of culture and of culture change posited in it. The considerable contributions of Afroamerican research to cultural theory, and especially to cultural dynamics, that have been derived from the analysis of additional data had not as yet been formulated.

In essence, my earlier theoretical propositions held that culture is learned, not inborn, except insofar as the species *homo sapiens,* as a whole, derives its culture-building ability from being the kind of organism it is; that is, that the factor of race does not enter. I further held that culture, being learned, can be additively learned; that this additive learning is ever present in contact between peoples, and gives rise to new cultural forms. However, I noted that many of the pre-established modes of behavior are retained under contact, and that where cultural interchange occurs, those influence the form of what is taken over.

The emphasis in the book on method was a logical consequence of the particular state of development of the Afroamericanist field at the time, for this field was, in fact, an innovation in anthropological research, because it cut across established geographical frames for investigation. Thus, of necessity, stress was laid on ways and means of attacking the scientific problems which the conceptualization and delimitation of the new field had laid bare. It was important, also, to develop a sound critical apparatus with which to achieve the blending of historical and ethnographic data, a procedure that actually did eventuate in what was perhaps the first consistent use of the *ethnohistorical method.* And by no means least, in this catalogue of methodological problems that had to be met, were problems of field research. Because of the absence, in Afroamerican studies, of the conventional "tribal" setting where most anthropological research had been conducted up to that time, Afroamerican field-work necessitated a reappraisal and readjustment of existing field techniques.

One concept that had been used in the Afroamerican field was the concept of *syncretism*. Arthur Ramos, the Brazilian anthropologist, was one of the first to employ it in this context. Without exploring its implications for cultural theory, he used it in describing how, in the African cults that have been preserved in Bahia, cult members who are at the same time faithful Catholics identify their African deities with the saints of the Church. This striking phenomenon had also been reported from other parts of the New World, notably Catholic Haiti and Cuba. It was recognized that this process of identification represented a pattern of first importance in understanding the religious life of these New World Negro societies.

The very use of the term "syncretism" helped to sharpen my analyses, and led me to a more precise formulation of problem and of theory. As I continued the study of the accumulating data from Afroamerican field research, it became clear that this formulation had implications for the understanding of certain processes that had been overlooked in the study of cultural dynamics in this and other world regions. For, considered in the light of the theory of culture-change, it seemed to me that the syncretizing process really lay at one pole of a continuum that stretched from situations where items from two or more cultures in contact had been fully merged to those situations where there was the unchanged retention of pre-existing ones.

This conceptualization brought about a new orientation. It continued, of course, to be of paramount importance to establish African provenience of a given group as accurately as the documents permitted ; to analyze, wherever possible, on the basis of the historic facts, the elements which induced or retarded cultural change, and to utilize ethnographic field research to determine the Africanisms that were present. But to these established procedures a new dimension of research was added. It was now evident that if we accepted the proposition that culture-contact produces cultural change, and that cultures of multiple origin do not represent a cultural mosaic, but rather become newly reintegrated, then the next essential step was to ascertain the degree to which these reconciliations had actually been achieved, and where, on this acculturative continuum, a given manifestation of the process of reworking these elements might lie.

From this came the concepts of *retention* and *reinterpretation*. In terms of this approach, research was pointed not to the question of what Africanisms were carried over in unaltered form, but how, in the contact of Africans with Europeans and American Indians, cultural accommodation and cultural integration had been achieved. For the essence of the reinterpretative process lies in differentiating cultural form from cultural sanction. Under contact, a new form can be accorded a value that has a functioning role into which it can be readily fitted; or an old form can be assimilated to a new one, the most obvious example of this being the syncretisms that have been mentioned between African gods and Catholic saints.

Let us take an example from the United States, and see how these processes operate in a Protestant church. In most cities, the services of "shouting" Negro churches are to be heard over the radio, and I have for some years been following, recording and transcribing a sampling of the services of one such church, so that by analyzing the text of the sermons, the world-view in terms of the implicit theological propositions that underlie the observed ritualistic practices may be inductively derived. Many of these churches, including the one under discussion, are of the Baptist denomination, and are so named. The larger churches are in structure much like conventional white-Protestant-American churches. Those in attendance are indistinguishable from other Americans in their dress and bearing. Music is provided by organ or piano, or both. The Bible is cited as the authority for belief. The broadcasting is smoothly integrated with the rest of the service. Nothing could be further removed from the West African setting for polytheistic worship, in the open air, with the prominent placing of drums and other ritual paraphernalia.

Yet there are aspects of the ritual, and of the theology as expressed in the sermons, that at the very least must be regarded as wide deviations from the practices and beliefs of white Baptists. As the service progresses, spirit possession by the Holy Ghost takes place, with motor behavior that is not European, but African. The gospel hymns are sung by a trained choir. There is little singing by the congregation, its participation being the hand-clapping which maintains the rhythm, and the antiphonal responses of "Yes! Yes! Lord," "Yes, Jesus," "That's so," which punctuate the sermon. It is an interesting exercise in ethnomusicology to compare the printed music of these hymns, used by the choir, with transcriptions of them as actually sung, and to note how the scores are manipulated to create quite distinct patterns of harmony and, particularly, of rhythm. Nor is a sociological analysis of the structure of the church group needed to perceive the existence of differences in status and rôle, because these are immediately apparent in the differences in dress that characterize the special roles of certain members.

The manner in which the supernatural Being that rules the Universe is conceived in churches of this kind is a subject that has received little or no attention, for the sermons in which these theological concepts are expressed have all too often been dismissed as the illiterate gibberish of preachers who play on the emotions of their congregations and at best are but caricatures of their white counterparts. Yet careful study reveals the sermons to be structurally patterned and rich in imagery. When followed over a period of time, they present a consistent and logical point of view. God, Jesus, and the Holy Ghost are all concerned with the immediate fate of those who worship them. They are disturbed by the fact of segregation and are responsible for its diminution; the Holy Ghost visits with the minister, taking messages to God from those in need of help.

One specific instance can be given of the immediacy of God in this complex of religious belief and behavior. On one occasion, the broadcast of the service was replaced by music, and those who telephoned the radio station were informed that there had been mechanical difficulty. The minister took cognizance of this the following Sunday. "Last week," he said in a matter-of-fact voice, "we were off the air. We got seventeen hundred phone calls and letters asking why." His voice became deeper. "Each Sunday, we sing, we pray, we make you happy, we heal you. But do I get

seventeen hundred letters and phone calls? Not even one hundred."
Then he thundered: "Do you think God likes this? Do you think
He'll stand for it? Now let me have your messages, or God will
do again what He did last week!" The sermon went on to expound
the interdependence of God and His children, and what each must
do for the other.

Now, it is simple to find Africanisms in church services of this
kind, and their presence is attested by the comments of Africans,
as well as Europeans of long experience in Africa, in whose company I have witnessed these rites. But this is only the starting point
in the search to understand how Afroamerican groups, through the
exercise of that resilience discussed in the pages that follow, have
integrated old beliefs with new, reinterpreting both to fit a pattern
of sanction and value that functions effectively in meeting the
psychological needs of life.

4

The sharpening of the theoretical framework of Afroamerican
studies resulted in other formulations. It became apparent that in
all the New World retention manifested itself in a continuum from
pure African carry-overs to behavior indistinguishable from that
which characterizes the dominant culture of European derivation.
The data had also to be further analyzed by breaking them down
into units that were less than whole cultures and entire populations.
For the former purpose, the concept of the cultural aspect proved
useful, and each culture was analyzed in terms of the greatest degree of retention of Africanisms in its economic and political life,
its social organization, its religion and its magical practices, its art,
folklore, music and language. It would be tempting to generalize
from the findings and to say, perhaps, that under contact religion
is more tenacious than technology. But attempts to do this have
shown that generalizations of this order do not stand very long. Experience has taught that it is rather more sound to generalize in
terms of process; in this case, that is, to generalize on the immediacy of the relationship between the particular circumstances
under which a given contact has occurred and the degree of intensity
of retention, and the extent of reinterpretation in the several aspects
of culture.

Breaking down these bodies of Afroamerican custom in this way had two further results. The first was the demonstration that to place total cultures on a comparative continuum masked the fact of degree of actual difference; that a society which, on the basis of the essentially impressionistic technique of comparing entire modes of life, might be regarded as polar could, in some of its aspects, be in actuality at quite a different point. The second was the development of the concept of *cultural focus,* a concept whose full implications are now being explored in a number of researches carried on in various cultures. It was self-evident that the high degree of retention of African religious elements in pure or reinterpreted form, as empirically established for Afroamerican cultures everywhere, could not be due to chance. This conclusion was underscored by the pervasive role of the belief system in the life of Africans. It was thus generalized that every people tend at a given time in their history to lay stress on that aspect, or those aspects, of their culture which are of greatest interest to them; and that, in a situation of cultural contact where free choice is not allowed, they retain elements of the focal aspects, either in unchanged or reinterpreted form, more tenaciously than those of other aspects.

It also became evident, when considering the over-all picture of the particular series of acculturative situations with which we were dealing, that the impact of culturally new elements was by no means uniformly felt by all the members of a group exposed to them. Learning, we have long recognized, is a function of opportunity, and where prestige attaches to one culture in a contact situation, it follows that those who have the greatest opportunity to learn what is presented and which, because of the factor of prestige, is desired, will do this sooner and more effectively than those who have less access to the new culture. Herein lies the contribution of more recent West Indian studies which concentrate on social structure and class differences in societies where the Afroamerican component bulks so large. These studies are in line with the caution entered when the table of intensity of Africanisms was first drawn. For though this table was cast in terms of cultural aspects, the caveat was also entered that these extreme manifestations of African retentions characterized in the main those who belonged to the lower socio-economic strata of a given society; that retention was as much a function of social class as it was of such other factors as geographical isolation, or of historic circumstance.

Beyond all this, however, are the data from Afroamerican societies that indicate how systems of values derived from one cultural stream are integrated into a new system. We see how attitudes toward time, or the accumulation or use of wealth brought to the New World from Europe have been reinterpreted by the descendants of Africans, and how powerfully pre-established patterns of evaluation functioned in guiding the kind of reinterpretations that occurred — often reinterpretations of African values accepted by Europeans as well as of the values of Europe by Afroamericans.

A final point has to do with the lesson taught by the materials of this book and the studies that have been made since its publication concerning the tenaciousness of tradition. We have a tendency, understandable in the light of important patterns of thought, to emphasize change and take stability for granted. The findings of Afroamerican research, since its inception, have demonstrated how essential it is to take equal cognizance of both. Important as this approach is, methodologically, for achieving balance in scientific analysis, it goes still further, as this book has sought to demonstrate, and points the need to hold this balanced view in attacking practical issues.

Today a new implication of these researches enters. This has to do with Africa itself, where Africans are increasingly faced with the need to make their adjustments to cultural innovations that belong to the same stream of tradition as that to which, in generations past, Africans and their descendants in the New World had also to adjust. For those who hold that African cultures give way before pressures brought about by induced change, the lesson is clear. For it would seem that in Africa, far more than could have been possible anywhere in the New World, the principle of cultural tenacity must hold; especially since this is a principle which has been found to be operative in all societies experiencing cultural change. The conclusion that we reach is that in Africa, as in the New World, the cultural processes that will be operative will be those of addition and synthesis to achieve congruence with older forms, rather than of subtraction and substitution, with their resulting fragmentation.

5

This book, when first published, discussed and documented a position that at the time was less than congenial to the considerable number of intellectuals who accommodated their thinking to the position of an important and established group of social scientists and of students of language and literature, a position which was challenged by its conclusions. Its point of view, moreover, was unpalatable to economic determinists. It is thus not surprising that some of its reviewers questioned or criticized the method used and the interpretations drawn. Most of those who commented on it, however, welcomed the book for what it was — an attempt to lay bare assumptions underlying interracial conflict, and thus to help place programs of action on a foundation of fact rather than of presupposition.

Today, the receptivity to the point of view the book documents can be attributed to several factors. To an appreciable degree, it has resulted from changing attitudes toward race and race differences. These have made possible the mature consideration of an approach that, in the heat of conflict, could be dismissed as remote from day-to-day realities, where it was not held to be undesirable in its practical implications.

This newer point of view has been the result of a far better understanding of African culture in the United States. For generations, America was indeed the dark continent in so far as concerns any sound knowledge of Africa. But with increased interest in Africa, greater opportunity to meet and to know Africans, especially those who came to our institutions of higher learning, and, above all with the realization that the nationalist movements of Africa were directed by men of maturity and competence, it was inevitable that attitudes should change. For the Negroes in the United States, Ghana has become a symbol; in the face of the achievements of Africans, the distortions in the caricature of the Africans and their ways of life, as given in the early chapters of this book, no longer carry conviction to serious students. On occasion, it is true, we still find Negroes in the United States or the West Indies who reject their past. But the number of those who do this is steadily diminishing, as is the number of their white fellow-citizens who hold to the earlier point of view. And the American Negro, in discovering that he has a past, has added assurance that he will have a future.

Melville J. Herskovits

1958

THE MYTH OF THE NEGRO PAST

Chapter I

THE SIGNIFICANCE OF AFRICANISMS

The myth of the Negro past is one of the principal supports of race prejudice in this country. Unrecognized in its efficacy, it rationalizes discrimination in everyday contact between Negroes and whites, influences the shaping of policy where Negroes are concerned, and affects the trends of research by scholars whose theoretical approach, methods, and systems of thought presented to students are in harmony with it. Where all its elements are not accepted, no conflict ensues even when, as in popular belief, certain tenets run contrary to some of its component parts, since its acceptance is so little subject to question that contradictions are not likely to be scrutinized too closely. The system is thus to be regarded as mythological in the technical sense of the term, for, as will be made apparent, it provides the sanction for deep-seated belief which gives coherence to behavior.

This myth of the Negro past, which validates the concept of Negro inferiority, may be outlined as follows:

1. Negroes are naturally of a childlike character, and adjust easily to the most unsatisfactory social situations, which they accept readily and even happily, in contrast to the American Indians, who preferred extinction to slavery;

2. Only the poorer stock of Africa was enslaved, the more intelligent members of the African communities raided having been clever enough to elude the slavers' nets;

3. Since the Negroes were brought from all parts of the African continent, spoke diverse languages, represented greatly differing bodies of custom, and, as a matter of policy, were distributed in the New World so as to lose tribal identity, no least common denominator of understanding or behavior could have possibly been worked out by them;

I

4. Even granting enough Negroes of a given tribe had the opportunity to live together, and that they had the will and ability to continue their customary modes of behavior, the cultures of Africa were so savage and relatively so low in the scale of human civilization that the apparent superiority of European customs as observed in the behavior of their masters, would have caused and actually did cause them to give up such aboriginal traditions as they may otherwise have desired to preserve;

5. The Negro is thus a man without a past.

Naturally, there have been reactions against this point of view, and in such works as Carter Woodson's *The African Background Outlined* and W. E. B. Du Bois' *Black Folk, Then and Now* serious attempts have been made to comprehend the entire picture of the Negro, African and New World, in its historical and functional setting. In still another category of those who disagree with this system are writers whose reactions, presented customarily with little valid documentation, center attention on Africa principally to prove that "Negro culture" can take its place among the "higher" civilizations of mankind. Scientific thought has for some time abjured attempts at the comparative evaluation of cultures, so that these works are significant more as manifestations of the psychology of interracial conflict than as contributions to serious thought. They are in essence a part of the literature of polemics, and as such need be given little attention here.

It must also be recognized that not every writer who has made statements of the type oulined above has accepted or, if he has accepted, has stressed all the elements in the system; and that popular opinion often underscores the African character of certain aspects of the behavior of Negroes, emphasizing the savage and exotic nature of the presumed carry-overs. Yet on the intellectual level, a long line of trained specialists have reiterated, in whole or in part, the assumptions concerning the Negro past that have been sketched. As a consequence, diverse as are the contributions of these writers in approach, method, and materials, they have, with but few exceptions, contributed to the perpetuation of the legend concerning the quality of Negro aboriginal endowment and its lack of stamina under contact. We may best begin our documentation of this system with a series of citations concerning the final, culminating element, leaving to later pages excerpts which demonstrate the tenaciousness of the other propositions that lead up to this last point.

2

Though the historical relationship between the present-day Negroes of the United States and Africa admits of no debate, there is little scientific knowledge of what has happened to this African cultural heritage in the New World. Statements bearing on the absence or the retention of Africanisms, even though these are drawn out of differing degrees of familiarity with the patterns of Negro life in this country, share one character in common. That is, their authors, whether lay or scholarly, not only are unencumbered by firsthand experience with the African civilizations involved, but the majority of them know or, at all events, utilize but few, if any, of the works wherein these cultures are described; while such works as are cited in documentation are commonly the older sources, which today are of little scientific value.

Scholarly opinion presents a fairly homogeneous conception as to African survivals in the United States. On the whole, specialists tend to accept and stress the view that Africanisms have disappeared as a result of the pressures exerted by the experience of slavery on all aboriginal modes of thought or behavior. As a starting point for subsequent analysis, a few examples of this body of thought may here be given to make available its major assumptions. Representative of this point of view is the following statement of R. E. Park, who in these terms summarizes a position he has held consistently over the years:

My own impression is that the amount of African tradition which the Negro brought to the United States was very small. In fact, there is every reason to believe, it seems to me, that the Negro, when he landed in the United States, left behind him almost everything but his dark complexion and his tropical temperament. It is very difficult to find in the South today anything that can be traced directly back to Africa.[1]

E. F. Frazier, in his study of the Negro family, stressed this position in a passage where, speaking of the "scraps of memories, which form only an insignificant part of the growing body of traditions in Negro families" and which "are what remains of the African heritage," he says:

Probably never before in history has a people been so nearly completely stripped of its social heritage as the Negroes who were brought

---

[1] For all notes see References beginning p. 300.

to America. Other conquered races have continued to worship their household gods within the intimate circle of their kinsmen. But American slavery destroyed household gods and dissolved the bonds of sympathy and affection between men of the same blood and household. Old men and women might have brooded over memories of their African homeland, but they could not change the world about them. Through force of circumstances, they had to acquire a new language, adopt new habits of labor, and take over, however imperfectly, the folkways of the American environment. Their children, who knew only the American environment, soon forgot the few memories that had been passed on to them and developed motivations and modes of behavior in harmony with the New World. Their children's children have often recalled with skepticism the fragments of stories concerning Africa which have been preserved in their families. But, of the habits and customs as well as the hopes and fears that characterized the life of their forebearers in Africa, nothing remains.[2]

Another student of the American Negro, E. B. Reuter, reviewing Frazier's work, gives unconditional assent to the point of view expressed in the preceding passage, when he writes:

The . . . Negro people . . . were brought to America in small consignments from many parts of the African continent and over a long period of time. In the course of capture, importation, and enslavement they lost every vestige of the African culture. The native languages disappeared immediately and so completely that scarcely a word of African origin found its way into English, owing to the dispersion, to the accidental or intentional separation of tribal stocks, and to the suppression of religious exercises. The supernatural beliefs and practices completely disappeared; the native forms of family life and the codes and customs of sex control were destroyed by the circumstances of slave life; and procreation and the relations of the sexes were reduced to a simple and primitive level, so with every element of the social heritage.[3]

In Charles S. Johnson's analysis of present-day Negro plantation life, the comment on background similarly follows the accepted position:

The Negro of the plantation came into the picture with a completely broken cultural heritage. He came directly from Africa or indirectly from Africa through the West Indies. There had been for him no preparation for, and no organized exposure to, the dominant and approved patterns of American culture. What he knew of life was what he could learn from other slaves or from the examples set by the white planters themselves. In the towns where this contact was close there was some effect, such as has many times been noted in the cultural differences between the early Negro house servants and the plantation hands.

On the plantation, however, their contact was a distant one, regulated by the strict "etiquette" of slavery and the code of the plantation.[4]

The same point of view concerning the retention of the African background is expressed by nonacademic specialists in the field, and by those professional students whose concern is with particular segments of culture wherein Africanisms may have persisted. Embree, who translates the physical homogeneity of the mixed Negroes of the United States into the concept of a "brown American" type, and expresses the opinion that, "it is astonishing how completely the Negro has been cut off from his African home,"[5] explains the process in these terms:

Torn from their previous environment, Africans found themselves grouped in the homesteads and plantations of America with fellow blacks from divergent tribes whose customs differed widely, whose languages even they could not understand. A new life had to be formed and was formed in the pattern of the New World. The old African tribal society was completely destroyed. From membership in their primitive social units, Negroes were forced into the organization required by the plantation and by the demands of the particular American families to which they were attached. The only folkways that had elements in common for all the slaves were those they found about them in America. The Africans began to take hold of life where they could. They began to speak English, to take up the Christian religion, to fall into the labor pattern demanded by American needs and customs, to fit themselves as best they could into all the mores of the New World.[6]

Cleanth Brooks, directing the techniques of his special field, phonetics, to the study of Negro dialect, concludes that, "in almost every case, the specifically negro forms turn out to be older English forms which the negro must have taken originally from the white man, and which he has retained after the white man has begun to lose them." And from this he derives the methodological principle on which his research is based: "For the purpose of this study the speech of the negro and of the white will be considered as one." It is clear that no African element is granted in the background of the unified mode of speech, which is assumed to be solely European in origin, for no single reference appears in his work to any African phonetic system.[7] Concerning the speech of the Sea Island Negroes, Guy B. Johnson writes in not dissimilar terms:

Gullah has been called the most African of any of our dialects, yet it can be traced in practically every detail to English dialect speech. There has been a popular belief in this country to the effect that Southern white speech is what it is because of the Negro. This idea needs to be reversed, for both the Negro and the white man in the South speak English as they learned it from the latter's ancestors.[8]

Or, to take another instance in a still different field, we find that Doyle, in his detailed study of etiquette as it functions in determining the pattern of Negro-white relations in the South, maintains the conventional position by implication when he begins his historical analysis with the period of slavery, and says nothing even as to the possible existence of codes of etiquette in Africa itself.[9]

3

If in this discussion assumptions are made that diverge from the point of view of those cited, and of the many other students holding this same position left uncited, these are to be regarded as deriving directly from facts discovered over an extended period of research. Whatever disagreement exists in basic approach is, indeed, the result of opportunity to investigate at first hand certain New World Negro societies outside the United States. It was investigation on this broader base, wherein the problem of Africanisms in present-day Negro behavior was only incidental, that forced revision of an hypothesis which, in the initial stages of research, there was no tendency to question.[10]

The nature of this experience may be sketched here, to make more explicit how research findings repeatedly forced revision of prevailing hypotheses. The citations given in the preceding note represent a point of view deriving from studies oriented toward the analysis of racial crossing in the United States; that is to say, they are based on observations made during investigations wherein the major issues lay outside the relevant sociological field. In studying race-crossing, however, it became apparent that without comparable measurements from ancestral African populations, the findings must have less value than were such data available. Consequently, ethnological researches, aimed at discovering the precise localities from which these African ancestral populations had been derived, were instituted.

Out of this program has come firsthand field study of New World Negroes in Dutch Guiana, in Haiti, and in Trinidad. Extended research has also been carried on into the history of slaving, and close contact has been maintained with specialists in Negro studies in the countries of South America and with those devoted ethnological *amateurs* who, in several of the colonies of the Caribbean, have been impelled by a desire to know more of the folk about them to contribute to the store of data on New World Negro cultures. From the need to trace African origins has come research in Africa itself where, in Nigeria, the Gold Coast, and more especially in Dahomey it has been possible to study at first hand the important ancestral civilizations. Through this continually widening experience has grown recognition of the need for scientific reinvestigation of the problem of the retention of Africanisms in the New World—in the United States itself. It was through this same experience, also, that a sense of the practical significance of the conclusions to be drawn from such investigations for the interracial situation as it exists at the present time was developed, especially as such conclusions can give to those concerned with everyday issues a sense of the historical depth in which these issues are lodged, and the assumptions under-lying the thinking of most Americans, white or Negro, regarding the values involved.

At this point it must be again emphasized that exact knowledge touching survivals of African traditions and beliefs in the behavior of present-day Negroes in the United States and elsewhere in the New World, or of the effect of these survivals on the daily life of their carriers, is not at hand. Materials are scattered and fragment-ary, where they are not altogether lacking; but the controversy aroused when the very problem is broached attests its vitality and its importance.[11]

The study has progressed far enough, however, to indicate some of the main lines of approach. We know today that the analysis of African survivals among the Negroes of the United States involves far more than the commonly attempted correlation of traits of Negro behavior in this country with aboriginal tradition in Africa itself. On the contrary, such an analysis, to be adequate, requires a series of intermediate steps. A knowledge of the tribal origins of the Negroes of this country is indispensable if the variation in custom found among the tribes from which the African ancestry was drawn is to be properly evaluated; and this is the more to be desired since almost all those who write of the Negro make a capital point of this variation—variation in terms of the African continent as a whole, however, rather than of that relatively restricted area from which the slaves were predominantly derived. An analysis of the slave trade as revealed in contemporary documents and in African tradi-tions, to give us a knowledge of any selection it may have exercised, and the reaction of the slaves to their status, is similarly essential. The mechanisms of adjusting the newly arrived Africans to their situation as slaves, and the extent to which these operated to permit the retention of old habits, or to force the taking over of new modes of behavior, or to make for a mingling of old patterns and newly experienced alternatives, must be understood as thoroughly as the data will permit.

Nor may any investigation on these lines confine itself to the United States alone. For if any methodological caution has emerged from exploratory research, it is that a knowledge of the Negro cultures of the Caribbean islands and of Latin America is indispensable. The matter has been well put by Phillips:

As regards negro slavery the history of the West Indies is insepa-rable from that of North America. In them the plantation system origi-nated and reached its greatest scale, and from them the institution of slavery was extended to the continent. The industrial system on the islands, and particularly on those occupied by the British, is accord-ingly instructive as an introduction and a parallel to the continental regime.[12]

From the point of view of the study of Africanisms, also, it is as important to know the variation in Negro customary behavior, tra-ditions, and beliefs over the entire New World as it is to understand the variation in the ancestral cultures of Africa itself, for only against such a background can the student project a clear picture of what has resulted from the differing historical experiences that con-stitute the essential control in the research procedure. And only with this background mastered are those mores of Negro life in the United States which deviate from majority sanctions to be realis-tically analyzed.

The discussion in these pages will therefore be oriented in accord-ance with these principles. Our initial concern will be the African background, the processes of enslavement, and the reaction of the Negro to slavery. The accommodation of Negroes to their New World setting and the resultant variation in the degree of accultura-tion over the entire area where slavery existed will then be indicated, while the aspects of Negro culture where Africanisms have been most retained and those where the least of aboriginal endowment is manifest, and the reasons for these differentials, will be pointed to show the complexity of what in general has hitherto been consid-ered a single problem. Finally, further steps in research will be out-lined to the end of attaining a better understanding of the processes of culture as a whole, and of an attack on the social issues presented by the Negro in the United States, in so far as the elements of conflict in the interracial situation are sharpened by beliefs concerning the quality of the cultural background of the Negro.

4

Before turning to an analysis of available materials, let us consider the theoretical problems and practical issues on which our broad approach can throw some light, indicating at the outset the questions of most concern to students of culture for which the data of our investigation have relevance.

The problems whose answers are to be sought in the study of data from many civilizations fall under several general heads. The organization of human civilization as a whole, and the interrelation and integration of the several aspects of culture when combined into a given body of traditions, technologies and beliefs, are the most fundamental points at issue. The manner of cultural borrowing and, where possible, the circumstances under which an interchange of tradition takes place are similarly important, as is the related problem of the degree to which any culture represents inventions originating from within or taken over from foreign sources. The relation between culture and its human carriers, focused especially on the manner in which the cultural setting of an individual conditions not only his general mode of life but the organization of his personality and the character of his motor habits, has in recent years come to the fore as a significant problem. Finally, the question of the degree to which the individual, admittedly in large measure the creature of his culture, can influence it while adapting himself to its patterns brings up the essential question of the various forces making for cultural change and cultural conservatism.

The comparative study of culture, like intensive analyses of individual civilizations, has in the past attempted to base its hypotheses on data from the nonhistoric peoples—those nonliterate folk termed "primitive"—who are relatively but little disturbed by European influence. Until recent times, students have been reluctant to include in their programs of investigation the consideration of changes which have occurred, and are taking place as a result of the contact of these nonhistoric peoples with the historic cultures under European colonial expansion and the westerly march of the American frontier. Yet for the study of problems of cultural dynamics and of social integration, of objective patternings and of psychological interrelations, the contact situations have much of value to offer. For here the conclusions from the study of relatively undisturbed and more static societies may be taken into the laboratory of observable change. Diffusion in process, the forces that make for cultural stability or instability, the reactions of individuals to new situations, the development of new orientations, the rise of new meanings and new values in life—all these may be observed where a people are in continuous contact with modes of life other than their own. What is accepted and what rejected, the influence of force as against mere exposure or verbal persuasion, and the effect on human personalities of living under a dual, nonintegrated system of directives may be analyzed under ideal conditions for observing and recording the pertinent data.

Social studies of this type have in recent years come to be designated as acculturation studies, and it is as studies in acculturation that research into the problem of African survivals in the behavior of New World Negroes may be looked to to make their greatest contribution to the understanding of the nature and processes of culture as a whole. Acculturation has been most comprehensively defined as the study of "those phenomena which result when groups of individuals having different cultures come into continuous first-hand contact, with subsequent changes in the original cultural patterns of either or both groups."[13] It is unnecessary here to examine the implications of this definition or the methods of studying these contact situations, as these matters have been treated elsewhere.[14] At this point it need only be recalled that by taking his cultural data into the laboratory provided by the historical situation, the social scientist may test his hypotheses in reference to conditions subject to historic validation; to obtain, that is, something of the control that is the essence of scientific method.

That the Negro peoples in the New World offer unusual opportunities for research has been remarked by several students. Sir Harry H. Johnston, in the volume wherein he reports a visit to the West Indies and the United States in 1910,[15] shows with clarity how rich a yield can be provided by knowledge of the ancestral continent when directed toward the New World scene. Despite the shortness of his stay, and the undisciplined observation and analysis of data that here, as in his other works, characterize the writings of this soldier, writer, and artist, his book is illuminating. For it demonstrates how much in the way of aboriginal tradition exists in West Indian and South American regions where, in disregard of even its surface manifestations, it has been overlooked by those students from the United States who, without grounding in African cultures and equipped only with the hypothesis of the disappearance of African customs as a frame of reference, have tended to minimize African retentions. W. M. Macmillan, a South African, also realized the closeness of affiliation of the West Indies with Africa—though his concern was with the special socio-economic problems of the British possessions—and he indicates the importance of service for colonial officials in the islands before their tour of duty in Africa itself.[16]

Perhaps the earliest student in the United States to point out the importance of research in the West Indies was U. G. Weatherly. He stressed the significance of "social groups in an insular environment," particularly where, as here, an historical record is at hand to aid in determining the experience of the people, and where contact with the outside world and other factors such as "internal revolutions" or "radical shifts in control from the outside" have made for "something more than rectilineal development." The "smaller West Indian Islands, extending from St. Thomas to the South American coast," according to him, "possess many of these characteristics." Here the present culture is the result of contact between Africans and Europeans of many nationalities, and their historical experience has been that of transfer from one of these European powers to another, with consequent historically known changes in cultural impulses. In addition, with the "European population as a fluctuating and diminishing element, there remain as major factors the Negro and the East Indian." The method and value of study in these islands is then indicated in the following terms:

Systematic research on the problems here outlined would of necessity be a cooperative undertaking. It would call for specialists in social technology, ethnology, culture, history, agricultural economics, psychology and education. The most obvious appeal of such a study would be that of practical problems: and yet it is possible that the most valuable results might come from the opportunity of working out some of the principles of pure social science. These communities, by reason of their isolation and peculiar cultural status, offer a nearer approach to social experimentation than cosmopolitan groups of the continental areas and are no doubt better adapted to the elaboration of a special methodology for the social sciences. The units are sufficiently small and detached to be easily dealt with, and the social forces at work are less muddled than in the complex environment of larger groups.[17]

Park likewise has called attention to the research possibilities presented by the Negro, though he does not in this place envisage the problem as falling outside the limits of the United States:

For a study of the acculturation process, there are probably no materials more complete and accessible than those offered by the history of the American Negro. No other representatives of a primitive race have had so prolonged and so intimate an association with European civilization, and still preserved their racial identity. Among no other people is it possible to find so many stages of culture existing contemporaneously.[18]

In a later paper[19] Park considers the resources of the West Indies for such research. Reuter, who phrases his conception of the problem in these words, also holds comparative study to be of importance:

For this scientific study the Negroes in America are valuable above most other social groups. They represent various stages of cultural development. In the group are men and women highly and fully educated and defined, persons who have thoroughly assimilated the European cultural heritage and have in some respects added to it. At the opposite extreme are persons but slightly removed from the African culture level. There are other groups of longer time in America but whose residence in the isolated regions of the hinterland has so retarded the assimilative process that they are still, in many respects, outside the modern culture. There are Negroes in America who speak dialects hardly intelligible to outsiders. . . . In the group it is possible to study the evolution of human and social institutions in process. Almost every stage in cultural evolution may be seen in coincident process of becoming. What must usually be studied by an historic method may here be studied by an observational and scientific procedure.[20]

From the point of view of the methods, objectives, and achievement to be discussed in ensuing pages, these earlier proposals, as we shall see, must be regarded as eddies in the principal stream of interest of the authors and their colleagues in the social sciences. Certainly these formulations have stimulated no exploration of the problems sketched; and it is of some importance to examine into the cause or causes that have determined this.

If we refer once again to the assumption of American students that Africanisms have failed to survive under contact with European civilization, we at once come upon a valuable clue. For, with this approach, the question in West Indian research becomes not, "What has happened to the aboriginal cultural endowments of those concerned in the contact of Africans and Europeans?" but rather, "Since African culture has given way before European contact, to what extent does the resulting adjustment indicate inherent aptitudes for specific forms of tradition, and what light can research throw on the innate ability of Negroes to handle European civilization?" The answer to this latter query is patent. It needs no training in scientific method to discover that Negroes in the New World have mastered European culture where opportunity has permitted; or that, where their modes of behavior diverge most strikingly from those of the majority, the reasons for this can be phrased in such terms as "isolation," "discrimination," and the like.

In the minds of these students, however, the major problem has been the obvious one of racial aptitudes and limitations, as, for example, is to be seen in the following:

Now the Negro belongs perhaps to the most docile and modifiable of all races. He readily takes the tone and color of his social environment, assimilating to the dominant culture with little resistance. Further, he is ordinarily, though not quite correctly, assumed to have brought with him from Africa little cultural equipment of his own. If culture is diffused only through contact, there is here a means of following, in the experience of an especially susceptible people, the processes of transformation which different types of association have generated. If the racial theory is sound, race traits ought here to have persisted; or at least definitely modified the new influences with which the dominant European peoples have brought the Negro in contact.[21]

Stating the matter somewhat differently, Park envisages the matter on the same level when, in his discussion of the problem of Negro studies previously cited, he writes:

I have sought in this brief sketch to indicate the modifications, changes and fortune which a racial temperament has undergone as a result of encounters with an alien life and culture. This temperament, as I conceive it, consists in a few elementary but distinctive characteristics, determined by physical organization and transmitted biologically. These characteristics manifest themselves in a genial, sunny and social disposition, in an interest and attachment to external, physical things rather than to subjective states and objects of introspection; in a disposition for expression rather than enterprise and action. The changes which have taken place in the manifestations of this temperament have been actuated by an inherent and natural impulse, characteristic of all living things, to persist and maintain themselves in a changed environment. Such changes have occurred as are likely to take place in any organism in its struggle to live and to use its environment to further and complete its own existence.[22]

Now, it is not to be denied that the problem of the relationship between innate endowment and cultural aptitudes is important, but the deeper this problem is probed, the larger the methodological difficulties it presents. When, therefore, the materials on the New

World Negro are attacked in terms of this problem alone—or as, in the case of Park and Reuter, of a concept of "social evolution" which must inevitably involve an attempt to trace "stages of cultural development"—the data are too complex, too cumbersome to see in workable perspective, and in consequence, the suggested researches die stillborn. But if an assumption of the vitality of African cultural traits is accepted as a working hypothesis to be tested, and the geographical area for study is conceived to include the United States, the Caribbean, Latin America, and the relevant regions of Africa itself, attainable directives are made possible and research is encouraged.

It is recognized, to be sure, that no matter how the problem is formulated, it brings the student extremely close to the fundamental quality of the relationship between the biological and cultural potentialities of human groups. Yet by posing the question on the cultural level, this issue does not become paramount. The analysis is consistently held to the plane of learned behavior, so that whatever role innate endowment may play, it is not permitted to confuse the issues of the research. The problem thus becomes one of accounting for the presence or absence of cultural survivals in all of the New World, assessing the intensity of such survivals, discovering how they have changed their form or the way they have assumed new meaning in terms of the historic experience of the peoples concerned, and indicating the extent to which there has been a mutual interchange between all groups party to the contact, whether European, Indian, or African. Should certain constants be discovered in the behavior of all Negroes in the New World and of their Old World relatives as well, then, and then only, need the question be faced of the degree to which we are dealing with a deeply set traditional factor or with inherent drives; or the cognate problem of the degree to which, under the racial crossing that has occurred everywhere in the New World, such inherent tendencies have persisted.

The study of the results of race-crossing is important, and there are aspects of such research where, as in the matter of social selection, the mores must be taken into account. But the reverse is not true, and had the social scientists who have indicated the research potentialities in the study of New World Negroes been more concerned with their major field of interest, and less with the relationship between race and culture; if, above all, had they not assumed that the Negro presented a cultural *tabula rasa* on which to receive this New World experience, their suggestions might have stimulated the studies whose usefulness they recognized.

5

A plan to profit by the research potentialities of the New World historical "laboratory" through a coordinated attack on Negro cultures in Africa and the Western Hemisphere was first suggested in 1930.[23] It envisaged study in West Africa to establish the cultural base line from which the differing traditions of the dominant New World Negro peoples might be assessed, and concomitant study of the life of Negroes in the West Indies and South America, where acculturation to European patterns has proceeded less rapidly than in the United States. Negro communities of the United States were to be held as later research objectives, since it was recognized that only on the basis of the broadest background could an adequate investigation of the presence of Africanisms and their functioning in such groups be achieved.

This plan, which outlined a reconsideration of the problem of Africanisms sketched in preceding pages, resulted essentially from findings of field research among the coastal and Bush Negroes of Dutch Guiana. It was evident, for example, even on initial acquaintance, that many ancestral African customs were to be found among the Negro tribes of the Bush, who because of their long isolation had experienced a minimum of contact with Europeans. But to one expecting a modicum of Africanisms in the Bush, and an absence of them in the coastal city of Paramaribo, where the Negroes have had close and continuous contact not alone with Europeans, but with Caribs, Javanese, British Indians, and Chinese as well, the results of close study were startling. In the interior, a full-blown African religious system, a smoothly functioning African clan organization, African place and personal names, African elements in economic life, a style of wood carving that could be traced to African sources showed what might be looked for in the institutions of any isolated culture that is a going concern. In the coastal region, however, underneath such Europeanisms as the use of European clothing and money, or baptismal certificates and literacy, numerous African institutions, beliefs, and canons of behavior were likewise encountered.[24] The question thus posed itself: If this obtains in Guiana, might not Negro behavior elsewhere in the New World be profitably reinvestigated with the lessons of this research in mind? An intensive review of published sources made an affirmative answer inescapable, and culminated in this outline of a comprehensive approach.

The conceptual tool which represented the widest departure from earlier usage was described in the following passage:

It is quite possible on the basis of our present knowledge to make a kind of chart indicating the extent to which the descendants of Africans brought to the New World have retained Africanisms in their cultural behavior. If we consider the intensity of African cultural elements in the various regions north of Brazil (which I do not include because there are so few data on which to base judgment), we may say that after Africa itself it is the Bush Negroes of Suriname who exhibit a civilization which is the most African. . . . Next to them, on our scale, would be placed their Negro neighbors on the coastal plains of the Guianas who, in spite of centuries of close association with the whites, have retained an amazing amount of their aboriginal African traditions, many of which are combined in curious fashion with the traditions of the dominant group. Next on our scale we should undoubtedly place the peasants of Haiti . . . and associated with them, although in a lesser degree, would come the inhabitants of neighboring Santo Domingo. From this point, when we come to the islands of the British, Dutch, and (sometime) Danish West Indies, the proportion of African cultural elements drops perceptibly, . . . though . . . we realize that all of African culture has not by any means been lost to them. Next on our table we should place such isolated groups living in the United States as the Negroes of the Savannahs of southern Georgia, or those of the Gullah islands off the Carolina coast, where African elements of culture are still more tenuous, and then the vast mass of Negroes of all degrees of racial mixture living in the South of the United States. Finally, we should come to a group where, to all intents and purposes, there is nothing of the African tradition left, and which consists of people of varying degrees of Negroid physical type, who only differ from their white neighbors in the fact that they have more pigmentation in their skins.[25]

Revisions of detail in this outline, necessitated by the work of a decade, must obviously be made in charting the intensity of Africanisms in the various areas of the New World when this conceptual tool is employed today; but the technique itself has richly proved its usefulness.[26] Thus, more concentrated research has been done on the African forms of religious life of the Negro in Brazil during the past decade than in any other part of the New World, and this has made available materials of first importance not at hand ten years ago. Studies in race relations and, most recently, in nonreligious patterns of the African survivals in the Negro culture of that country have also been initiated.[27] Haitian peasant life is far better known,[28] and the importance of the syncretisms which mark the reconciliation between African and European custom in many places of this culture has been pointed out.[29] These discussions bring into bolder relief the corresponding syntheses in the field of religion which likewise exist among the Catholic "fetishist" Negroes of Brazil and Cuba,[30] and indicate aspects of New World Negro acculturation that have far wider meaning for an understanding of the results of cultural contact than its significance for this particular problem. With fuller comparative materials at hand, it has also been possible to utilize more effectively older sources from these regions and from Jamaica, the customs of whose Negro population were described in one of the pioneer ethnographic works from the entire New World Negro area.[31]

More recent research in West Africa has also emphasized the complexity of the cultures of that part of the continent from which so large a number of Negro slaves were captured. This work affords us leads toward the solution of the riddle of how the Negroes, coming from different tribes and speaking different languages, have by a hitherto unrecognized least common denominator in tradition and speech found it possible to preserve elements of their heritage.[32] Research in the Virgin Islands,[33] and Trinidad,[34] also, has similarly made revision of the scale of intensity of Africanisms necessary, for this field work has demonstrated the principle that the acculturative process in each locality is to be analyzed in terms of the peculiarities of its own historic past and its own socio-economic present.

Ten years ago it would not have seemed possible that the survivals to be found in the southern United States could be comparable to those discernible in any of the West Indian islands. Yet variation in intensity of Africanisms in the Antilles, while undoubtedly greater than in the South on the side of its African elements, runs the full range toward the most complete acculturation to European patterns that might be encountered not alone in the South, but also in northern states. Certainly it is not commonly understood that the socio-economic situation in such an island as Trinidad presents aspects that have meaning when compared with that of the Negroes of the United States—sharecropping, the presence of the industrial worker, and the like. Impressive parallels, however, are to be found —and, it may be hoped, will be saved from the neglect which students of these phases of the Negro problem, in their reluctance to make comparative analyses, have so unanimously accorded them.

6

A further tool, which has been of increasing use as research has proceeded, is the concept of an Old World cultural province. As one comes to know the cultures of the entire African continent, one becomes cognizant of numerous cultural correspondences between African, European, and Asiatic civilizations. As will be indicated later in our discussion, this is most apparent in the field of folklore, where, for example, animal tales of the Uncle Remus type are found in the Reynard cycle of medieval Europe, in the fables of Aesop, in the Panchatantra of India, in the Jataka tales of China, and in the animal stories of Indonesia. Certain aspects of the use of magic, of ordeals, of the role and forms of divination, of conceptions of the universe (especially the organization of deities in relationship groups), of games, of the use of proverbs and aphorisms, are also widely distributed over the Old World.

All these, and many others not possible to detail here, have bearing on the study of the survival of Africanisms in the New World. For it is here we must turn for an explanation of the seemingly baffling fact, so often encountered, that given traits of New World Negro, and especially of American Negro behavior, are ascribable equally to European and African origin. This may well be viewed as but a reflection of the fact that deep beneath the differences between these varied civilizations of the Old World lie common aspects which, in generalized form, might be expected to emerge in situations of close contact between peoples, such as Europeans and Africans, whose specialized cultural endowments are comprehended within the larger unity.

7

It is apparent that research into the problem of African survivals in the United States, when set in its proper context, carries the student into areas of importance for an understanding of the nature and processes of human civilization. It is apparent, further, that the problem cannot be studied most profitably except under terms of such a broad approach; and that, above all, if it is to realize its potentialities as a means of scientific comprehension, there must be an end to unsupported assumption concerning the disappearance of the traditions brought by the slaves from their homeland. As will later be indicated, African culture, instead of being weak under contact, is strong but resilient, with a resiliency that itself has sanction in aboriginal tradition. For the African holds it is pointless not to seek an adaptation of outer form, where this can "in a manner" be achieved. Before this point can be discussed at length, however, we must consider the significance of our analysis of cultural tenacity and resilience for those issues of practical importance, suggested in the opening section of this chapter, which no student of the Negro, however detached his approach, may disregard.

We turn again, therefore, to the phenomenon of race prejudice, the factor that provides the rationalization for many of the inter-racial strains that are the essence of concern to the practical man. Racial prejudice, when analyzed, is found to rest on the operation of two closely interrelated factors, one socio-economic, the other historical and psychological. These social and economic factors are well recognized; certainly, it is with these that both practical and academic studies of Negro life have been primarily, and often exclusively, concerned. The reason for this is clear. Stresses lodging in this area are immediate, and call so compellingly for solution that the impulse to render first aid is difficult to resist. Moreover, on the surface, at least, these stresses can be referred to the situation of slavery; and their accentuation during the slave regime and since its suppression can thus be readily and satisfactorily explained. Finally, in programs of action, many of these difficulties are of a kind encountered in analogous form elsewhere in the socio-economic configurations of this country, and can thus reasonably be regarded as susceptible of effective attack through the operation of short-time ameliorative projects.

The effect of this approach has been to relegate to the background the psychological basis of the race problem, and its less immediate historical aspects, when not entirely ignoring them. Again, this is understandable, for phenomena of this order cannot be studied, much less evaluated, without long and sustained analysis, such as has been already sketched. And this, too often, gives these problems an air of remoteness which militates against their appeal to those seeking the immediate solution of pressing needs. Yet these factors are as deeply intrenched in the interracial situation as are those other elements which lie on the social and economic level, and they are far more insidious. In the light of current thinking about racial differences in general, they are the most effective cause in perpetuating all shades of superiority-inferiority ranking given whites and Negroes by members of both groups. For here we are dealing with points of view that have received their directive force through generations of reiteration of cultural values, of comparative worth, of historic dignity. It is, therefore, at this point that the entire historical setting, which includes the problem of Africanisms in American Negro behavior, becomes crucial, since the question of social endowment enters intimately into the determination of the assumptions on which attitudes regarding Negro inferiority rest. And it is these attitudes, as validated by the series of conceptions grouped under the heading of the myth concerning the Negro past, which rationalize and justify the handicaps that, perpetuated from one generation to the next, cause current unrest among the Negroes who suffer under them and make for a diffused, all-pervading sense of malaise and even guilt among those who impose them.

8

Whatever position concerning Africanisms is being considered, assumed functional relationships between various forces—innate endowment and natural environment, on the one hand, and overt behavior, on the other—must be taken into account. In works of the earliest students, especially of Nott and Glidden,[35] an inescapable biological inferiority was advanced as the explanation for those traits of Negro customary behavior that were held undesirable. This point of view has been well summarized by L. C. Copeland:

The South's dependence on the Negro is further obscured by the belief in the complete dependence of the black race upon the white race for moral as well as for economic support. The Negro is thought of as a child race, the ward of the civilized white man. We are told: "The savage and uncivilized black man lacks the ability to organize his social life on the level of the white community. He is unrestrained and requires the constant control of white people to keep him in check. Without the presence of the white police force Negroes would turn upon themselves and destroy each other. The white man is the only authority he knows."[36]

This is the end result of constant repetition of the inferiority of Negroes and of African culture, indicated in the same discussion in these terms:

In commenting on the books current in his youth Booker T. Washington was struck by the manner in which they "put the pictures of Africa and African life in an unnecessarily cruel contrast with the pictures of the civilized and highly cultured Europeans and Americans." In one book a picture of George Washington was "placed side by side with a naked African, having a ring in his nose and a dagger in his hand. Here, as elsewhere, in order to put the lofty position to which the white race has attained in sharper contrast with the lowly condition of a more primitive people, the best among the white people was contrasted with the worst among the black." Washington related that he unconsciously took over the prevalent feeling that there must be something wrong and degraded about any person who was different from the customary.[37]

Leaving aside for the moment the significant comment of Booker T. Washington that the contrast between African and American or European life was made *unnecessarily* cruel, it is understandable how Negroes and unbiased whites, intent on analyzing the interracial situation, came to feel that all strategic considerations made desirable as emphatic a denial of African influence on present-day Negro custom as possible.

Some of the statements bearing on the point, found in works that have had wide circulation and considerable influence, and which this negative attitude was designed to combat, may be cited. Dowd, in his much-quoted *Negro in American Life*, makes the flat assertion, "Nowhere in Africa have the Negroes evolved a civilization," qualifying this only with the statement:

. . . but they have shown capacity to assimilate it. In the region of the Fellatah Empire, before the arrival of the Europeans, the natives had learned to read and write in Arabic, and had established several notable educational centers.[38]

Whenever this view of the inferior creative ability of the Negro is brought forward, it is customarily coupled with an observation on his imitative gift, which in turn becomes an additional reason for a policy of rigid control of Negroes by whites, of the type already indicated. And still further implications are drawn from it:

The characteristic of self-abasement, involving as it does a lack of self-respect, explains the Negro's extraordinary imitativeness. "This slavish imitation of the white," says Mecklin, "even to the attempted obliteration of physical characteristics, such as wooly hair, is almost pathetic, and exceedingly significant as indicating the absence of feeling of race pride or race integrity. Any imitation of one race by another, of such a wholesale and servile kind as to involve complete self-abnegation, must be disastrous to all concerned."[39]

Imitativeness is only one phase of what is written of as the Negro's childlike mind and his accompanying cheerfulness. Though comparable statements are not often met in the more recent works, the idea is sufficiently current to deserve its place in the mythological system, as presented. One example of this point of view continues the Dowd citations:

The mind of the Negro can best be understood by likening it to that of a child. For instance, the Negro lives in the present, his interests are objective, and his actions are governed by his emotions. . . . William H. Thomas, himself a Negro, also noted the childish traits of his race: "The Negro lives only in the present, and though at times doleful in language and frantic in grief, he is, like a child, readily soothed by trifles and easily diverted by persuasive speech." . . . If cheerfulness is characteristic of children and of the Negro mind, so also are impulsiveness and fits of anger. The Negro, like a child, is easily irritated, and prone to quarrel and to fight. When angered he becomes a "raving Amazon, as it were, apparently beyond control, growing madder and madder each moment, eyes rolling, lips protruding, feet stamping, pawing, gesticulating."[40]

A study that has attracted a great deal of attention as an authoritative work—the excerpt from it in the preceding passage is merely one instance of the many times it is referred to—is an early contribution by Odum. The present position of this author follows that of the more liberal group of students of the Negro; yet the paragraph presented below must be quoted if only to give an example of the position still taken by many concerning the Negro's mentality and the relative worth of his African background. It must be regarded as especially significant, indeed, because of the place the work from which it is taken has assumed in the history of Negro research:

Back of the child, and affecting him both directly and indirectly, are the characteristics of the race. The Negro has little home conscience or love of home. . . . He has no pride of ancestry . . . has few ideals . . . little conception of the meaning of virtue, truth, honor, manhood, integrity. He is shiftless, untidy, and indolent . . . the migratory or roving tendency seems to be a natural one to him. . . . The Negro shirks details and difficult tasks. . . . He does not know the value of his word or the meaning of words in general. . . . The Negro is improvident and extravagant; . . . he lacks initiative; he is often dishonest and untruthful. He is over-religious and superstitious . . . his mind does not conceive of faith in humanity—he does not comprehend it.[41]

Thus Odum conceived Negro mentality in 1910; his concept of the form of education best suited to Negro children may also be indicated as cogent:

. . . Let the influences upon the Negro child, at least as far as the school is able to effect this end, lead him toward the unquestioning acceptance of the fact that his is a different race from the white, and properly so; that it always has been and always will be; that it is not a discredit not to be able to do as the whites, and that it is not necessarily a credit to imitate the life of the white man. . . .[42]

And in developing this thesis, an expression of his conception of the comparative worth of the Negro's African past is indicated: "He may learn from reading stories of Africa how much better off he is than his cousins."

John Daniels, who has written a history of the Negroes of Boston that is frequently cited, evaluates their aboriginal heritage in these terms:

It is of course undeniable that the precedent conditions out of which the Negro population of Boston is derived, have, from the earliest period down to the present, been of a peculiarly inferior kind. The first members of this race to appear in that city were brought, by way of the Bahamas, from their native African jungle, where from time immemorial their ancestors had lived in a stage of primitive savagery. They were savages themselves, utterly ignorant of civilization, having no religion above a fear-born superstition, and lacking all conception of a reasoned morality.[43]

One further example, which shows how this position persists in unexpected nooks of the world in thought, is the following passage from a psychoanalytic journal, recently cited by J. Dollard:

Leaving out of question the anthropometric tests which correspond closely to those of the native African, we find a number of qualities indicative of the relationship. The precocity of the children, the early onset of puberty, the failure to grasp subjective ideas, the strong sexual and herd instincts with the few inhibitions, the simple dream life, the easy reversion to savagery when deprived of the restraining influence of the whites (as in Haiti and Liberia), the tendency to seek expression in such rhythmic means as music and dancing, the low resistance to such toxins as syphilis and alcohol, the sway of superstition, all these and many other things betray the savage heart beneath the civilized exterior. Because he wears a Palm Beach suit instead of a string of cowries, carries a gold-headed cane instead of a spear, uses the telephone instead of beating the drum from hill to hill and for the jungle path has substituted the pay-as-you-enter street car his psychology is no less that of the African.[44]

This quotation is especially meaningful because of the manner in which it indicates how, employing psychoanalytic phraseology and concepts, this older view of Negro endowment and the Negro past may be rationalized. That it is to be hoped that this writer's evaluation of the Negro of the United States is somewhat more accurate than his exposition of African civilization is aside from the point, which has to do with the light the quotation throws on the way a concept, once developed, may present itself in new form.

It must be stressed that we are at the moment concerned only with those evaluations of the Negro past which served to establish the denial to African culture traits of any vitality in the scheme of American Negro life. A few more of these evaluations may be reviewed. Hoffman, in stating that the materials in his study of the Negro constitute "a most severe condemnation of modern attempts of superior races to lift inferior races to their own elevated position" shows the attitude taken in another of these works.[45] Tillinghast's influence, which has been very great, is especially apparent in references to the supposed effect of the African climate on the Negro's cultural endowment. Two quotations may be given to illustrate this approach:

The direct influence of African climate is adverse to persistent effort. Where high temperatures, and low humidity prevail, the rapid evaporation from the body cools it, and permits considerable exertion, as is the case in Egypt. Great humidity, combined with a low temperature, as in the British Isles, has no bad effect. But West Africa enjoys neither of these advantages, it swelters under a torrid heat combined with excessive humidity. Such conditions deaden industrial effort. The white man, whose capacity for energetic and prolonged labor in most circumstances is so great, whose wants are so numerous and insatiable, finds himself irresistibly overcome. Rich rewards await those who can put forth a little effort, yet as Ellis says, so intense is the disinclination to work, that even the strongest wills can rarely combat it. In fact, this very will itself seems to become inert.

We are now prepared to appreciate the workings of the vitally important factor of natural selection. It is obvious that in West Africa natural selection could not have tended to evolve great industrial capacity and aptitude, simply because these were not necessary to survival. Where a cold climate and poor natural productiveness threaten constant destruction to those who cannot or will not put forth persistent effort, selection operates to eliminate them and preserve the efficient. In torrid and bountiful West Africa, however, the conditions of existence have for ages been too easy to select the industrially efficient and reject the inefficient.[46]

That such statements are of themselves not to be taken seriously need scarcely be mentioned at the present time, since Tillinghast's references to Spencer adequately date his conception of the role of natural selection in determining traits of social behavior. Nevertheless, in the light of the number of times this work has been quoted, it must be taken fully into account.

U. B. Phillips, the best-known historian of slavery, phrases his estimate of the African scene in the following manner:

The climate in fact not only discourages but prohibits mental effort of severe or sustained character, and the negroes have submitted to that prohibition as to many others, through countless generations, with excellent grace. . . . It can hardly be maintained that savage life is idyllic. Yet the question remains, and may long remain, whether the manner in which the negroes were brought into touch with civilization resulted in the greater blessing or the greater curse. That manner was determined in part at least by the nature of the typical negroes themselves. Impulsive and inconstant, sociable and amorous, voluble, dilatory and negligent, but robust, amiable, obedient and contented, they have been the world's premium slaves.[47]

N. N. Puckett has likewise expressed himself concerning African characteristics, references to inborn and learned traits of behavior, as in previous instances, being intermingled:

Impulsiveness is another African trait which in the United States is gradually being laid aside in favor of greater self-restraint. . . . Laziness is found both in Africa and in America; in Africa being enhanced by the enervating tropical environment. . . . While a well regulated sex life is in part a result of cultural background, yet the sexual indulgence of the Negro, so open in Africa and in many parts of the rural South, may conceivably be a racial characteristic developed by natural selection in West Africa as a result of the frightful mortality. . . . Despotism in West Africa seems to win loyalty, pride and popularity, possibly because a strong-minded master has spirit enough to resent aggression and self-reliance enough to protect his followers from outside annoyance. . . . Shortsightedness, indifference and disregard of the future are traits common not only to Africans and many Negroes, but to almost all undisciplined primitive peoples.[48]

In Weatherford's *The Negro from Africa to America*, the evaluations of the African background, the importance of knowing which is stressed, follow the same pattern, as is shown in this passage from the introductory remarks:

We believe that much of the present response of the Negro to social environment is influenced by the social heritage, not only from slavery but from the far African past. This is in no way an intimation that the Negro has not progressed far beyond that past. Indeed no one can read the story of his marvelous progress without great amazement, and the amazement is all the more heightened when one sees the humble beginnings of the race. On the other hand there can be little doubt that there are vestigial remains of a far social heritage in the present social reactions of the American Negro, which if not understood will vitiate all of our judgments concerning him.[49]

The cultural handicap of the Negro is explained in these terms:

Like the mountain peoples of Tennessee, Kentucky and North Carolina, who for two centuries or more have been held in isolation by their mountain fastnesses and hence have fallen two centuries behind the procession of civilization, so the African peoples, shut in by the natural barriers of their own continent for thousands of years, have dropped many centuries behind the progress of civilization, not altogether because of less capacity, but mainly, at least, because of less contacts.[50]

The call for tolerance is consistent with the position taken:

The student of the American Negro today must therefore come to his task with a knowledge of the Negro's past if he is to really understand him at the present. He must be willing to judge him as to the distance he has traveled since he left his African home, rather than compared with the white man who had thousands of years the start. He must recognize the traits built into a race during long centuries cannot be bred out in a few years or even a few decades, and that the political and economic life of the present American Negro in the light of his background, is nothing less than amazing.[51]

Ten years after the publication of the volume in which the above passages appeared, the author obtained a collaborator and saw a new light. Whether the collaborator was responsible for the change of opinion, or further consideration on his own part brought this about, cannot be said, but a far different point of view is expressed in the collaborative work, which, for the sake of perspective, can also be quoted:

Since the culture of Africa therefore is quite different from our own, we are apt to conclude it is inferior. For have we not felled the forest, dug up the ores from the earth, fashioned powerful machines, annihilated space, subdued nature to our bidding? The African had done none of these things in such marked degree as have we; hence we are inclined to say his culture is inferior and the Africans as a people are inferior. Aptitudes of people may be proven far more by ability to adapt their culture to the environment in which they live than by ability to borrow culture from all the rest of the world, and on this basis we may find upon further study that the African peoples have much to commend them to our respect.[52]

9

It is not strange that the extremes to which the statements quoted above go should have brought the conviction that, since the African past constituted a serious handicap, the best thing to do was to disregard it wherever possible; from which the rationalization that nothing of this African handicap remains was but a short step, especially since, as has been indicated, the degree of acculturation to the patterns of the white majority actually manifested by Negroes in the United States is so considerable. The quotations that have been given are, of course, from the pens of men accustomed to writing with restraint; scholars, whose words were calculated to promote the search for knowledge rather than to lead to action. It is unnecessary to do more than call attention to the innumerable statements and repetitions of statements that came from the journalists, the clergy, the politicians, and those others who took for granted the inferiority of the Negro, and were able to enforce their conviction by a control of the power that, after emancipation no less than before, they used to keep the Negro in that place in the social scene he was deemed fit to occupy by reason of his inferiority.

This is not the point in our discussion to inquire into the extent that the evaluations of African culture are tenable.[53] Here we may but suggest that for those concerned with the best interests of the Negro, there was ample reason to conclude that strategy demanded a refutation of the claim that the Negro always has been, and always must be, the bearer of an inferior tradition, which, since he can never shake it off, must doom him to a perpetual status of inferiority. That they may have overshot the mark in looking to change of emphasis rather than the erasing of misconception is beside the point; the reasons why they took the position they did take are, granting their point of view, unassailable.

Yet it is not permitted us to conclude that this is the sole reason why the presence of Africanisms has been denied, both in the United States and in the New World generally. An ethnocentric point of view is congenial to any people and, in our society, this has become stabilized in terms of the idea that no nonliterate folk can withstand contact with Euro-American culture. It is assumed, for example, that primitive cultures—customarily denoted by nonscientists as "inferior" or "simple"—are everywhere dying out because of this contact. In this country, the disappearance of many Indian tribes did much to strengthen this ethnocentrism; applied to the transplanted African, whose pliability and subservience were used to explain his physical survival under slavery, it took the form of matter-of-fact acceptance of the disappearance of African tradition in the face of association with the whites. A case in point can be cited, the more interesting since it does not concern the Negroes of the United States and has to do with no theory of relative cultural merit. In 1888, William W. Newell, in discussing the voodoo cult of Haiti, says:

> Although all the writers who have alluded to these superstitions have assumed that they are an inheritance from Africa, I shall be able to make it appear: first, that the name Vaudoux, or Voodoo, is derived from a European source; secondly, that the beliefs which the word denotes are equally imported from Europe; thirdly, that the alleged sect and its supposed rites have, in all probability, no real existence, but are a product of popular imagination.[54]

With the last claim we are not concerned, but in "establishing" his other propositions, Newell curiously adumbrates the method of those who, in various fields, have derived the peculiarities of Negro custom from European sources. For he conceives the almost purely African rites of the *vodun* cult as mere misunderstanding, by Haitian Negroes, of the French heretical sect of the Waldenses, whose name was mispronounced, as its ritual was distorted by the blacks when they adopted this cult in favor of their aboriginal religious beliefs.

The quotations given in the opening pages of this discussion may be taken as typifying the point of view of most present-day students of the Negro. The African past may perhaps be considered as appearing fragmentarily in a few aspects of contemporary Negro life of the United States, but such survivals are to be studied only in the most complete antiquarian sense. African culture may be conceded a greater degree of comparative respectability than was earlier the case—though, as has been seen, the point of view which Hoffman, Tillinghast, Dowd, Mecklin, the earlier manifestations of Odum and Weatherford, and others put forward concerning the low caliber of African modes of life has by no means lost its vitality. But African culture, in any event, is held unimportant and hence can be disregarded in studying Negro life in this country today. This is the point of view expressed in Powdermaker's study of Indianola:

> The Negro did not come here culturally naked, but the conditions of slavery were such that a large part of his aboriginal culture was of

necessity lost. He was separated from fellow-tribesmen, taught a new language, and inducted both subtly and forcibly into the culture of the white masters. Beyond doubt, there are some survivals of African culture, but to determine exactly what those are would require a very different type of research. Historical elements enter into this point of view only as they make themselves felt in current processes and attitudes.[55]

Other examples, too numerous to be quoted here, could be drawn from studies whose concern is with a single phase of the problem. Thus Krapp[56] argues that all Negro speech can be traced to English dialect, though this is a proposition in accord with his general position that, "so far as pronunciation is concerned, it is doubtful if in a single instance the pronunciation of normal American English has been modified by the influence of a foreign language," and that, in the matter of vocabulary, "American English has been very slow to borrow new words from other languages."[57] The more familiar assumption, however, is not absent from his writings:

The native African dialects have been completely lost. That this should have happened is not surprising, for it is a linguistic axiom that when two groups of people with different languages come into contact, the one on a relatively high, the other on a relatively low cultural level, the latter adapts itself freely to the speech of the former, whereas the group on the higher cultural plane borrows little or nothing from that on the lower.[58]

If "dominant" and "minority" groups might be substituted for "high" and "low" as cultural designations, there would be at least a logical presumption that such a process might occur—though whether in enough instances to give rise to a "linguistic axiom" cannot be said. But the statement as phrased, in so far as it touches on the matter of our concern, merely is another example of the conventional pattern of expression regarding Negro aboriginal endowment.

It follows logically, then, that men of good faith might well conclude that the less said of Africanisms the better. But what if the estimate of Africanisms is not correct? What if the cultures of Africa from which the New World Negroes were derived, when described in terms of the findings of modern scientific method, are found to be vastly different from the current stereotype? What if these cultures impressed themselves on their carriers, and the descendants of their carriers, too deeply to be eradicated any more than were the cultural endowments of the various groups of European immigrants? More than this, what if the aboriginal African endow-

ment were found, in certain respects, even to have been transmitted
to the whites, thus making the result of contact an exchange of culture—as it was in the case of other groups—rather than the endowment of an inferior people with habits of a superior group? Let us
suppose, in short, it could be shown that the Negro is a man with
a past and a reputable past; that in time the concept could be spread
that the civilizations of Africa, like those of Europe, have contributed to American culture as we know it today; and that this idea
might eventually be taken over into the canons of general thought.
Would this not, as a practical measure, tend to undermine the assumptions that bolster racial prejudice?

There are other, more immediate, ways in which truer perspective concerning Africanisms might be helpful from a practical point
of view:

Granting that current social and economic forces are predominant in
shaping race relations, it must never be forgotten that psychological
imponderables are also of first importance in sanctioning action on any
level. And it is such imponderables . . . that . . . are now being
strengthened by the findings of studies that ignore the only valid point
of departure in social investigation—the historical background of the
phenomenon being studied and those factors which make for its existence
and perpetuation. When, for instance, one sees vast programs of Negro
education undertaken without the slightest consideration being given to
even the possibility of some retention of African habits of thought and
speech that might influence the Negroes' reception of the instruction
thus offered, one cannot but ask how we hope to reach the desired objectives. When we are confronted with psychological studies of race
relations made in utter ignorance of characteristic African patterns of
motivation and behavior, or with sociological analyses of Negro family
life which make not the slightest attempt to take into account even the
chance that the phenomenon being studied might in some way have been
influenced by the carry-over of certain African traditions; when we
contemplate accounts of the history of slavery which make of plantation
life a kind of paradise by ignoring or distorting the essential fact that
the institution persisted only through constant precautions taken against
slave uprisings, we can but wonder about the value of such work.[59]

Yet such studies are being undertaken, and in them the ordinary
procedure of the scientist, whereby he attempts to take into account
all possible factors, is invariably neglected in favor of uncritical
repetition of statements touching the aboriginal Negro past. And
this, it is submitted, achieves one result with a sureness that would
shock those who are the cause of this distortion of scholarship. For
though it has often been pointed out that the skin-color of the Negro

makes him an all too visible mark for prejudice, it is not so well
realized that the accepted opinion of the nature of the Negro's cultural heritage is what makes him the only element in the peopling
of the United States that has no operative past except in bondage.

There is still another point of practical importance that should
not be overlooked in appraising the implications of proper study of
Negro backgrounds and of the retention of Africanisms. And this
is the effect of the present-day representatives of this race without
a past, of the deprivation they suffer in bearing no pride of tradition. For no group in the population of this country has been more
completely convinced of the inferior nature of the African background than have the Negroes. Woodson has phrased the point in
these terms:

Negroes themselves accept as a compliment the theory of a complete
break with Africa, for above all things they do not care to be known as
resembling in any way these "terrible Africans." On the other hand, the
whites prate considerably about what they have preserved of the ancient
cultures of the "Teutons" or "Anglo-Saxons," emphasizing especially
the good and saying nothing about the undesirable practices. If you tell
a white man that his institution of lynching is the result of the custom
of raising the "hue and cry" among his tribal ancestors in Germany or
that his custom of dealing unceremoniously with both foreigners and
Negro citizens regardless of statutory prohibition is the vestigial hark-
ing back to the Teuton's practice of the "personality of the law," he
becomes enraged. And so do Negroes when you inform them that their
religious practices differ from those of their white neighbors chiefly to
the extent that they have combined the European with the African
superstition. These differences, of course, render the Negroes undesirable to those otherwise religious-minded. The Jews boldly adhere to
their old practices while the Negroes, who enjoy their old customs just
as much, are ashamed of them because they are not popular among
"Teutons."[60]

No better documentation of Woodson's point could be made than
the following comment of a Negro scholar on the cultural endowment of his own group:

The tradition and culture of the American Negro have grown out of
his experience in America and have derived their meaning and significance from the same source. Through the study of the Negro family one
is able to see the process by which these experiences have become a part
of the traditions and culture of the Negro group. To be sure, when one
undertakes the study of the Negro he discovers a great poverty of
traditions and patterns of behavior that exercise any real influence on
the formation of the Negro's personality and conduct. If, as Keyserling

remarks, the most striking thing about the Chinese is their deep culture, the most conspicuous thing about the Negro is his lack of a culture.[61]

It is little wonder that to mention Africa to a Negro audience sets up tensions in the same manner as would have resulted from the singing of spirituals, the "mark of slavery," to similar groups a generation ago. Africa is a badge of shame; it is the reminder of a savage past not sufficiently remote, as is that of European savagery, to have become hallowed. Yet without a conviction of the worth of one's own group, this is inevitable. A people that denies its past cannot escape being a prey to doubt of its value today and of its potentialities for the future.

To give the Negro an appreciation of his past is to endow him with the confidence in his own position in this country and in the world which he must have, and which he can best attain when he has available a foundation of scientific fact concerning the ancestral cultures of Africa and the survivals of Africanisms in the New World. And it must again be emphasized that when such a body of fact, solidly grounded, is established, a ferment must follow which, when this information is diffused over the population as a whole, will influence opinion in general concerning Negro abilities and potentialities, and thus contribute to a lessening of interracial tensions.

Chapter II

THE SEARCH FOR TRIBAL ORIGINS

Fundamental to any discussion of the presence or absence of Africanisms in Negro custom in the New World is the establishment of a "base line" from which change may be judged. Two elements enter into this; it is necessary to discover, as precisely as possible, the tribal origins of the slaves brought to the New World, and on the basis of these facts to obtain as full and accurate knowledge as we can of the cultures of these folk.

In this chapter, only the historical materials employed to determine tribal provenience will be presented. Yet historical analysis, of itself, has not given and cannot give the needed information. Only in coordination with the ethnological phase of the dual attack implicit in the ethno-historical method can documentation, otherwise meaningless, realize its greatest significance. Historical scholars have for years considered the problem of the African origins of the slaves, but without knowledge of the cultures of the regions toward which the materials in the documents pointed, they were unable to validate their hypotheses, even when, as was not always the case, they had adequate acquaintance with the local geography of Africa and could thus make effective use of the place names which recur in the contemporary literature.

Here, indeed, the appropriate elements in our "mythology" played their part well. Historians, with the principle that the slaves were derived from most of the African continent and the convention of the "thousand-mile march" of the slave coffles in mind, were understandably reluctant to draw the conclusions their data indicated—that the greater number of Negroes imported into New World countries came from a far more restricted area than had been thought the case. It is generally recognized that the ports from which slaves were shipped were preponderantly those along the western coast, and that south and east African points of shipment were rare. Yet coastal shipping points do not necessarily mean that the goods exported from them have not been gathered from a hinterland—in this case by distant native chieftains whose avarice was stimulated by the rewards held out to them by slavers.

Only when the scrutiny of the documents was complemented by acquaintance with the ethnography of Old and New World Negro communities, and the traits of the cultures of these groups were correlated with data from Africa to discover correspondences, did the question of African origins become susceptible of attack. A base line for the study of cultural change among New World Negroes— and, from the point of the focus of attention in this discussion, of the Negroes of the United States—has been established by means of this cross-disciplinary technique. We may thus turn to the results obtained from analyzing the historical documentation, leaving a consideration of survivals of tribal custom in New World Negro societies for a more appropriate later point.

2

The pattern for the prevailing conception of slaving operations was set by early writers, and this has been reinforced by the tendency to interpret African commercial relations in terms of European methods of trading, and a lack of knowledge of the density of population in West Africa. For it is not difficult to reason that, with a demand for slaves such as the American markets created, word of the commercial advantages in trading to the coast would spread to the hinterland, and captives would be brought to the factories of the European companies. It is said that slaves in some numbers were traded from tribe to tribe across the entire bulk of Central Africa, so that members of East African communities found themselves at Congo ports awaiting shipment to the New World! Yet this disregards not only the vast distances involved— some 3,000 miles—but also the dangers attendant upon such journeys in terms of the hostility between many of the tribes over the area and the absence of adequate lines of communication, to say nothing of the slight economic gain from such hazardous commerce even were the highest prices to be paid for such slaves.

The earlier writers give astonishingly little justification, in their own works, for their statements on the extensiveness of the slave traffic. For example, though Bryan Edwards writes of the "immense distance to the sea-coast" traveled by slaves, yet of the cases he cites of Jamaican slaves he questioned about their African homes, only for "Adam, a Congo boy," is his assertion borne out, though even

Adam is merely claimed to have come "from a vast distance inland." The four Negroes from the Gold Coast could not have lived far from the sea, the Ebo village Edwards speaks of was "about one day's journey from the sea-coast," while the fifth, a Chamba from Sierra Leone, also came from relatively near the coast.[1]

Mungo Park has given us our only firsthand account of the adventures of a slave coffle, but his description gives little support to those who emphasize the "thousand-mile" journeys made to the coast.[2] On April 19, 1797, Mr. Park departed from Kamalia, in the Bambara country, with slaves for the American market. The Gambia was reached on June 1, at a point some hundred miles from the sea where seagoing ships could come, and fifteen days later an American ship, bound for South Carolina, took slaves as cargo and Mr. Park as passenger. The distance traversed by the caravan was about 500 miles, or, for the 44 days in transit, an average of between 11 and 12 miles per day. The question may well be raised whether the translation of time into space, of the slow progress from sources of supply to the coastal factories, might not have been an important factor in giving the weight of logic to the conception of the slave trade as reaching far into the interior.

A source of information on the provenience of slaves which has remained almost entirely unexploited is the tradition about slaving in Africa itself. In Dahomey, for example, a kingdom on the Guinea coast which extended from its port, Whydah, some 150 miles into the interior, the annual "war" operated to supply the slave dealers. There are no traditions among these people that they acted as middlemen for traders farther inland; they were, in fact, avoided by the merchant folk, such as the Hausa, since the stranger in their kingdom was himself fair game. The peoples raided by the Dahomeans lived no farther from the coast than 200 miles, while most of their victims came from much nearer. Tribes to the east and west, rather than to the north, were the easiest prey, and hence the Nago (Yoruba) of Nigeria and the people of the present Togoland are found to figure most prominently in native lists of the annual campaigns.[3]

The coastal area of the Gold Coast was occupied by the Fanti tribes, who, because of their control of this strategic region, acted as middlemen for tribes to the north. Yet all evidence from recognizable survivals such as the many Ashanti-Akan-Fanti place names, names of deities, and day names in the New World are evidence that the sources of the slaves exported by the Fanti were in greatest proportion within the present boundaries of the Gold Coast colony. This is quite in accordance with the population resources of the coastal belt. The numerous villages, and the presence of cities of considerable size all through the area, suggest that the conception of an Africa depopulated by the slave trade, without the numbers necessary to support a drain that is to be figured in the millions, stands in need of drastic revision.

Senegal and the Guinea coast are two of the four principal areas mentioned in contemporary writings. The regions about the mouth of the Niger, named Bonny and Calabar in the documents, and the Congo are the other two, and it may be profitable to outline the situation with respect to the potentialities of the hinterland which was exploited for slaves there, and the nature of such historical facts as are available about the slave trade in these areas. The Niger Delta is a teeming hive, its low marshy plains densely inhabited by groups which, like those in the country lying behind it, are small and autonomous and thus, politically, in contrast to those larger entities, which we term kingdoms and empires, of the other parts of the slaving area. In this hinterland it was group against group, and kidnaping was probably more customary than anywhere else, though the oral tradition of the care taken by mothers to guard their children from unsupervised contact with strangers heard everywhere in West Africa and the use of folk tales to impress children with the danger of leaving the familial compound are eloquent of the fears engendered by the slavers in all the vast region.

Large numbers of slaves were shipped from the Niger Delta region, as indicated by the manifests of ships loaded at Calabar and Bonny, the principal ports. These were mainly Ibo slaves representing a people which today inhabits a large portion of this region. Their tendency to despondency, noted in many parts of the New World, and a tradition of suicide as a way out of difficulties has often been remarked, as, for example, in Haiti where the old saying "*Ibos pend' cor' a yo*—The Ibo hang themselves"[4] is still current. That this attitude toward life is still well recognized among the Ibo in Africa was corroborated in the field recently by Dr. J. S. Harris.[5] The same tendency was noticed among the "Calabar" Negroes— another generic name for Ibos among the slaves—in the United States, as is indicated by the remark of the biographer of Henry Laurens, that in South Carolina "the frequent suicides among Calabar slaves indicate the different degrees of sensitive and independent spirit among the various Negro tribes."[6]

To the east of the Cross River lie the Cameroons and Gaboon, which figured little in the slave trade. The worth of these Negroes was held to be slight, as the following quotation indicates:

The "kingdom of Gaboon," which straddled the equator, was the worst reputed of all. "From thence a good negro was scarcely ever brought. They are purchased so cheaply on the coast as to tempt many captains to freight with them; but they generally die either on the passage or soon after their arrival in the islands. The debility of their constitutions is astonishing." From this it would appear that most of the so-called Gaboons must have been in reality Pygmies caught in the inland equatorial forests, for Bosman, who traded among the Gaboons, merely inveighed against their garrulity, their indecision, their gullibility and their fondness for strong drink, while as to their physique he observed: "they are mostly large, robust well-shaped men."[7]

Raiding to the north or northeast of the Calabar area was rendered unlikely by the existence of better-organized political units, and consequently this locus of the trade fed on itself. This it could do, for with its dense peopling it could export the numbers it did without significant recourse to its neighboring territories, so that here again, where provenience is concerned, research must look to tribes well within a belt stretching, as a maximum, not more than two or three hundred miles from the coast as the area meriting closest attention.

For the Congo, relatively little information is available. We know from the data concerning two New World centers, Brazil and the Sea Islands off the Carolina Coast, that the peoples of Angola figured largely in the trade to these areas. Many traits of Congo religion and song have been recorded from Brazil[8] while linguistic survivals in the Gulla Islands, as studied by Dr. Turner,[9] likewise show a substantial proportion of words from this region. The works of Père Dieudonne Rinchon have dealt more carefully with the Congo slave trade than those of any other student of the history of slaving, and his testimony indicates that in the Congo, as elsewhere, the slavers were not compelled to range far from the coast to obtain their supply. The following passage, indeed, reinforces this hypothesis by the manner in which the case described in the last three sentences is singled out for special attention:

Les esclaves exportés sont principalement des Ambundus, des gens de Mbamba et de Mbata, et pour le reste des Nègres du Haut-Congo achetés par les Bamfumgunu et les Bateke du Pool. Quelques-uns de ces esclaves viennent de fort loin dans l'intérieur. Le capitaine négrier Degrandpré achète à Cabinda une Négresse qui lui paraît assez familière avec les Blancs, ou du moins qui ne témoigne à leur vue ni surprise, ni frayeur;

frappé de cette sécurité peu ordinaire, le negrier lui en demande la cause. Elle répond qu'elle a vu précédement des Blancs dans une autre terre, où le soleil se lève dans l'eau, et non comme au Congo où il se cache dans la mer ; et elle ajoute en montrant le levant *monizi monambu,* j'ai vu le bord de la mer ; elle a été en chemin, *gonda cacata,* beaucoup de lune. Ce récit semble confirmer les dires de Dapper que parfois des esclaves du Mozambique sont vendus au Congo.[10]

A passage with such definite information is rare in the literature. It can most easily be utilized for our purpose if the excellent tribal maps of the Belgian Congo now available are consulted.[11] The Bamfumungu and the Bateke, living in the region of Stanley Pool on the upper Congo, are a bare 200 miles in the interior. The Bambata (the Mbata of the citation) are found about 100 miles from the coast; the Mbamba, given by Rinchon as living between 7° 5' South and 14° East, are today included in the Portuguese colony of Angola, and hence not recorded on the Belgian maps. Aside from Rinchon's list, another clue to Congo origins is had in the name of a people mentioned often in Haiti, the Mayombe. This tribe lives directly behind Cabinda, which was the principal slaving port, and their most easterly extension is not more than 50 or 75 miles from the coast.

Material gained during field work further reinforces our hypothesis that the coastal area of West Africa furnished the greater proportion of the slaves. This information was recorded in the Hausa city of Kano, in northern Nigeria, where by the kindness of the Emir, it was possible to query four old men who themselves, like their forebears for many generations, had been important slavers— "merchants," as they insisted on being called—trading with the Gold Coast. The route they traveled was some 1,800 miles long, and is still traversed today; and every point on it could be checked with standard maps. The distance involved was largely a matter of east-and-west travel rather than southwardly to the coast (Kano lies between 500 and 600 miles inland), since it was necessary, were they to remain in friendly Hausa and allied territory, to strike far to the west before moving southward—otherwise they would have encountered the hostile peoples of Dahomey and what is today western Nigeria. It is not necessary to repeat here information already available,[12] except to indicate briefly the operations of these men as they describe something of the numbers of slaves brought from this relatively deep point in the interior. That is, slaves were taken to the Gold Coast for trade with the Ashanti only as an incident to the major purpose; it was goods, not human beings, that were the object of their attention, and the slaves in the caravans were burden-bearers, and hence only for sale on the most advantageous terms. They were never, as far as these men knew, traded directly to the whites. The matter has been summarized in the following terms:

. . . though perhaps six or seven thousand slaves left Kano every year for the Gold Coast, two-thirds or three-fourths of that number returned north as carriers, the capacity in which they had acted during the south-ward journey. And though we may suppose that more than five caravans departed from Kano each year when the slave trade was at its height, and that a smaller proportion of slaves than that named were returned as carriers of merchandise, even then the number who arrived at the coastal factories could constitute but a fraction of the enormous num-bers of slaves whom the record tells us were shipped from Gold Coast ports.[13]

Let us permit the question of provenience, as it bears on the dis-tance in the interior from which slaves were brought, to rest at this point. The present status of our knowledge permits us only to indi-cate that a reinvestigation of prevailing hypotheses in the matter of slaves brought to America deriving from vast distances inland is necessary. A qualification must, however, be made explicit and emphatic, for, as in other points at which research tends to con-travene the accepted "mythology," an exception taken is interpreted as an assertion made. Hence it is necessary here to state unequiv-ocally that in positing derivation from coastal tribes for the major portion of the slaves, it is not intended to convey that no slaves came from distant points inland or even from East Africa. Some undoubtedly traveled great distances—the case cited by Rinchon is to the point—since the demand, particularly in the later days of the slave trade, must have been so great that a certain number of cap-tives were brought from the deep interior. The point at issue is not whether any slaves were derived from far inland, but whether enough of them could have been brought from these localities to the New World to place the stamp of their tribal customs on New World behavior; whether, that is, these extensions of the trade in-land are significant for the study of African survivals in the United States and elsewhere in the New World. To answer this question we shall need to anticipate ethnographic materials to be given later; but it may here be stated that survivals of the known customs of such interior peoples are practically nonexistent in the Americas even where, and especially where, the most precisely localized African traits have been carried over. This means that, as we turn again to the documents, we may evaluate our findings without misgivings about whether or not points of shipment may be taken to be signifi-cant of points of tribal origin.

3

Before considering fresh data, let us examine some of the state-ments made on slave derivations. The late U. B. Phillips, outstand-ing as an historian of slavery, demonstrates handicaps under which the historian labors when he is not in a position to control the eth-nography and geography of Africa as well as he does his documents. To illustrate this, an incident hitherto unrecorded may be set down. In discussing the problem of African tribal distribution, Phillips, a few years ago, posed the following question: "Have you, in your field work in West Africa, ever encountered a people often men-tioned in the literature, the Fantynes?" So varied has been the man-ner of writing tribal names that the meticulous technique of this historian did not permit him to deduce that the Fantynes of the documents and the great Fanti tribe of the present-day Gold Coast were identical, and that these people, who number considerably more than a million souls, were indeed the people meant in the slavers' reports. When the answer was given, the reaction of this scholar was primarily one of pleasure at having finally arrived at the solution to a problem that, as he stated at the time, had long troubled him.

Notwithstanding this, Phillips was essentially correct in his as-sumption on provenience, even though he felt compelled to endorse the patterned conception of the wide range of operations in Africa itself. For if this conception caused him to write:

The coffles came from distances ranging to a thousand miles or more, on rivers and paths whose shore ends the European traders could see but did not find inviting. . . . The swarm of their ships was particularly great in the Gulf of Guinea upon whose shores the vast fan-shaped hinterland poured its exiles among converging lines . . .[14]

he also reported many names of coastal tribes in giving the evalua-tions of the planters on various types of slaves they distinguished, and properly described the principal area of shipment in stating that:

The markets most frequented by the English and American separate traders lay on the great middle stretches of the coast—Sierra Leone, the Grain Coast (Liberia), the Ivory, Gold and Slave Coasts, the Oil Rivers as the Niger Delta was then called, Cameroon, Gaboon and Loango.[15]

Acting on a basic principle of his research—a principle that is of the soundest—that for purposes of the study of slavery the West Indies and the United States must be considered a unit, he details the categories of slaves recognized in Jamaica and elsewhere in the area. The names of these slave types are tribal or place names, all of which are to be found on present-day maps of Africa if one knows where to look and has sufficient detailed knowledge of the geography of Africa to facilitate equating the nomenclature of the period— Senegalese, Coromatees, Whydahs, Nagoes, Pawpaws, Eboes, Mocoes, Congoes, Angolas, Gambia (Mandingoes), Calabar—with that now used. Phillips indicated the types of Negroes advertised as fugitives from the Jamaica workhouse, in the following passage:

It would appear that the Congoes, Angolas and Eboes were especially prone to run away, or perhaps particularly easy to capture when fugitive, for among the 1046 native Africans advertised as runaways in the Jamaica workhouses in 1803 there were 284 Eboes and Mocoes, 185 Congos and 259 Angolas as compared with 101 Mandingoes, 60 Chambas (from Sierra Leone), 70 Coromantees, 57 Nagoes and Pawpaws, and 30 scattering, along with a total of 488 American-born negroes and mulattoes, and 187 unclassified.[16]

Other citations which indicate better knowledge than might be thought of the locale of slaving may also be quoted. Puckett—with proper qualification in regard to the "thousand-mile" hypothesis— under the heading "Sources of American Slaves" gives the following:

Roughly speaking, the six to twelve million Negro slaves brought to America came from that part of the West Coast of Africa between the Senegal and the Congo rivers. True enough these West Coast slave markets did in turn obtain some slaves from far in the interior of the continent, but the principal markets were about the mouths of the Senegal, Gambia, Niger and Congo, and the majority of the blacks were obtained from this West Coast region.[17]

The absence in this passage of any mention of the Gold Coast, "Slave" Coast (Dahomey and Togo), or western Nigeria is noteworthy, yet despite these omissions he reasons cogently regarding the population resources of this area, and the economic advantages of operation from it, though again he accepts explanations for the density in terms which, to say the least, are open to debate:

Here was the locality closest to America, the one with the densest population (more than half the total population of Africa was located in this western equatorial zone), with the inhabitants consisting largely of the more passive inland people driven to the coast by inland tribes expanding towards the sea. This mild and pacific disposition was enhanced by the tropical climate and excessive humidity of the coast.[18]

Reuter gives the conventional statement:

The Negroes brought to America were in the main of West African descent. For the most part they were bought or captured along the West Coast and the Guinea Negroes were by far the more numerous, constituting well over fifty per cent of the total importation. But the slaves secured along the Guinea Coast were by no means all of local origin. There were representatives of many different tribal stocks from many parts of the continent. The slave trails extended far into the interior of the continent and the slave coffles came by river and forest path sometimes for a thousand and more miles to the markets on the coast.[19]

Park likewise accepts the customary point of view:

The great markets for slaves in Africa were on the West Coast, but the old slave trails ran back from the coast far into the interior of the continent, and all peoples of Central Africa contributed to the stream of enforced emigration to the New World.[20]

Weatherford and Johnson, who discuss the problem of Negro origins on the basis of at least one contemporary source,[21] take a position somewhat more in accord with the facts. The conception of the "thousand-mile" trek to the coast here appears in somewhat reduced form—700 miles:

The slaves brought to America came almost exclusively from the west coast. The English brought captives from the Senegal and Gambia rivers, from the Gold Coast, slave coast, and even as far south as Angola. The Dutch had forts on the Gold Coast, and in 1640 captured the Portuguese forts at Angola, where they gathered many slaves. The French had Fort Louis at the mouth of the Senegal river, and other forts scattered down the west coast. Anthony Benezet, who made a careful study of the slave trade, said that the slaves were regularly shipped from all points from the Senegal to Angola, a coast of nearly 4,000 miles. The heart of the trade was the slave coast and the Gold Coast, and behind this a territory extending into the interior for 700 miles or more. From this territory Senegalese Negroes, Mandingoes, Ibos, Efikes, Ibonis, Karamantis, Wydyas, Jolofs, Fulis, together with representatives of many of the interior Bantus were brought to America.[22]

It is not possible here to reproduce the detail in which Sir Harry H. Johnston indicates tribal origins, but the very richness of his suggestions, arising out of his acquaintance with the African scene itself, reinforces the point already made of the importance of a proper background in studying provenience.[23] Of interest here is his statement that, of the language of the Sea Island Negroes off the Carolina and Georgia coast, "the few words I have seen in print appear to be of Yoruban stock or from the Niger Delta."[24]

4

Recent data furnished by students of slaving and slavery give the historian far better information on all aspects of the trade than has hitherto been available. This is not to say that the resources, whether archival or of the earlier published works, have been exploited to the degree possible; for example, the works of French and Belgian students of the slave traffic, particularly as regards the trade out of the port of Nantes, seem to have been entirely ignored. Of these, one can cite the volumes of Rinchon, already mentioned in another connection, or of Gaston-Martin,[25] wherein many aspects of slavery, ordinarily subject to speculation, are treated with a wealth of fresh information. The importance of Rinchon's earlier work has already been indicated; his most recent volume presents, among other matters, the most precise data that have as yet been brought to light on the proportion of slaves shipped to the New World who lost their lives in the hazards of the "middle passage." In giving abstracts of the manifests of shipping out of Nantes, this author notes a startlingly low percentage of losses, as compared with previous estimates.[26] For he shows that, between 1748 and 1782, 541 slavers bought 146,799 slaves, and disposed of 127,133. The difference, 19,666, or 13 per cent, would indicate that the losses from all causes during shipment—and it by no means follows that these were deaths —were much smaller than has been thought.[27]

This failure to go to sources is especially difficult to understand in the case of the West Indies—a region that, as has long been recognized, took its slaves from the same localities of Africa as did the United States, and that, indeed, for many slaves served as an acclimatizing ground for Negroes resold to the mainland. The most precise information as to the sources of slaves in the Virgin Islands, between 1772 and 1775, is contained in a report of the inspector general of the Moravian Missions, C. G. A. Oldendorp.[28] A man who lived before the science of ethnology was known, or such a subject as applied anthropology was dreamed of, he produced a model report which goes far in enabling the student of today to understand why Africanisms have been forced so deeply underground in the life of these islands, when their inhabitants are compared to the Negroes in other parts of the Antilles. For in this rarely cited work, we find what a man could discover when he queried "salt-water" slaves—those born in Africa—and asked them their places of birth, the names of their tribes and the peoples bordering on the areas their own groups occupied, the names of their rulers, their gods, and various words from their vocabularies. The harvest for the student of New World Negro origins who uses this book is, as might be imagined, a rich one—so rich that only mere reference can here be made to its contents. So accurate is the reporting that almost every tribal name Oldendorp gives can be found on present-day maps of Africa—an accuracy that is doubly assured when we find that he correctly distinguishes such confusing tribal designations as Mandingo (Senegal) and Mondongo (Congo basin).[29] Hartsinck and Stedman, writing of Dutch Guiana,[30] or Moreau de St. Méry and Charlevoix and Père Labat,[31] reporting on the French West Indies, or Monk Lewis and Bryan Edwards for Jamaica,[32] have likewise been far too little employed. Of the many significant works written by those active in the slave trade, Bosman and Snelgrave[33] are almost exclusively encountered. Brantz Mayer's *Captain Canot*,[34] popularized through a recent reprint, is sometimes used, but without any apparent realization that the case is an abnormal one, since Canot was a slaver who operated during the last years of the trade, when all the accentuated viciousness of an outlawed traffic would be expected to appear.

The most precise information on the African sources of slaves brought to the United States is to be found in the documents published and analyzed by Miss Elizabeth Donnan.[35] Here, in convenient compass, has been assembled information of special significance to the students of the trade as it affected the United States; a sampling so extensive that it is doubtful whether data from other collections, English or otherwise—these volumes deal exclusively with British slaving operations—can do more than fill in details of the picture she outlines. Especially important are the abstracts of manifests of slaving vessels landing cargoes in ports of continental United States. Since only raw materials are given, it is necessary for one who uses this work to compute totals and analyze the data statistically, but once this is done, a remarkably clear idea is had of the degree to which the various parts of Africa were drawn upon for human materials. Such an analysis discloses that those portions of West Africa named in this chapter as the regions where the forerunners of survivals are to be sought are mentioned in greatest prominence.

These documents indicate the large difference in immediate sources of slaves brought to the northern colonies in the earlier days of the trade, on the one hand, and of those imported into the southern states. Phillips has remarked that the majority of the Negroes brought to the north were imported from the West Indies:

> In the Northern colonies at large the slaves imported were more generally drawn from the West Indies than directly from Africa. The reasons were several. Small parcels, better suited to the retail demand, might be brought more profitably from the sugar islands whither New England, New York and Pennsylvania ships were frequently plying than from Guinea whence special voyages must be made. Familiarity with the English language and the rudiments of civilization at the outset were more essential to petty masters than to the owners of plantation gangs who had means of breaking in fresh Africans by deputy. But most important of all, a sojourn in the West Indies would lessen the shock of acclimatization, severe enough under the best circumstances. The number of negroes who died from it was probably not small, and of those who survived some were incapacitated and bedridden with each recurrence of winter.[36]

This statement is entirely borne out by Miss Donnan's figures, for of the 4,551 slaves received in New York and New Jersey ports between 1715 and 1765, only 930 were native Africans.[37] The small numbers of slaves in each cargo—the "retail" aspect of the trade to the North—is likewise shown in the few credited to each ship, more than ten per vessel being the exception, and the large majority of manifests listing but two or three. This "retail" nature of slaving operations to the North, furthermore, is a factor of some consequence in assessing differential rates of acculturation to European patterns as between northern and southern Negroes, since the opportunities for learning European ways were far greater for these northern Negroes than for slaves sent to the plantations of the South, or for those in the West Indies who lived even more remote from white contact.

From the point of view of the African provenience of northern Negroes, the manifests tell us little even where a shipment came direct, since the entry, "coast of Affrica," is the one most frequently set down; something which, indeed, contrasts interestingly with the specific names of the West Indian ports where the slaves were procured. What is important in the documents from the northern states are the letters of the slavers, who in reports to their owners mention the ports and the tribes which have been indicated as the source of New World slaves. Thus Captain Peleg Clarke, writing from "Cape Coast Roade" in the Gold Coast to John Fletcher under date of 6 July, 1776, says:

D'r Sir, In my former letters Pr Capt'ns Bold, Smith and others, I fully informed you in what manner I had disposed of my Cargo, and time agree'd on for payment was the middle of Augt. And Expected to be at Barbadoes in Novr and I Should purchase About 275 Slaves. I now add that Trade has been entirely Stop'd for this 6 Weeks pass'd, Owing to the Chief of the people a going back against the Asshantees to Warr, and are not yet returned, but there is no likelyhoods of comeing to battle as the Asshantees is returned back to their Country again, and it is not likely their will be any great matter of trade for this some time again, (as the Chiefest of our trade comes through the Asshantees).[38]

There is full documentation to prove the unity of sources for the New World Negroes in these accounts, particularly in instructions to the captains of slaving vessels as to ports of call to dispose of their cargoes, and in the reports of these men telling of the ports at which they were to call or actually had called. Thus Samuel Waldo, of Boston, on March 12, 1734, handed these instructions to Captain Samuel Rhodes, in command of Waldo's sloop *Affrica*:

You will be a Judge of what may be most for my Intrist, so I shall entirely confide that You'll act accordingly in the Purchase of Negros, Gold Dust or any other the produce of that Country with which You'll as soon as possible make your Return to me either by way of the West Indies or Virginia where You'll sell Your Slaves either for Gold Silver or good Bills of Exch'e. . . . If Your coming from off the Coast with Slaves will bring it towards Winter or late in the Fall before You can reach Virginia it will be best to go for the West Indies where by trying more or less the Islands You may probably do better than by selling att Virginia, . . .[39]

The abstracts of ships' manifests which account for slaves imported into the southern colonies give much exact information about the African regions where their cargoes were procured. As is to be anticipated from the comparison of direct trading between Africa and northern and southern ports, the proportion of slaves reimported from the West Indies drops sharply. A tabulation of the raw data found in the manifests recorded from Virginia between the years 1710 and 1769 gives the following results:[40]

Source of origin given as "Africa"...................... 20,564
Gambia (including Senegal and Goree)................. 3,652
"Guinea" (from sources indicated as Gold Coast, Cabo-
corso Castle, Bande, Bance Island, and Windward Coast) 6,777
Calabar (Old Calabar, New Calabar and Bonny)......... 9,224
Angola.. 3,860
Madagascar... 1,011

Slaves brought directly from Africa...................... 45,088
Slaves imported from the West Indies.................... 7,046
Slaves from other North American ports................. 370

52,504

It is to be observed that in addition to ships indicated as arriving from "Africa"—which gives no clue at all as to provenience except in so far as we wish to compare direct importation from that continent with the West Indian trade—the regions that figure most prominently are "Guinea," which means the west coast of Africa from the Ivory Coast to western Nigeria, Calabar, which represents the Niger Delta region, Angola, or the area about the lower Congo, and the Gambia.

The shipments from Madagascar are of some interest, if only because they indicate what small proportion of the slaves were drawn from ports other than those lying in the regions given as the principal centers of slaving operations. These 1,011 slaves out of 52,504 brought to Virginia—and, as will be seen shortly, the 473 listed as coming from Mozambique and East Africa out of 67,769 imported into South Carolina—merely underscore the points made as to provenience and indicate how relatively slight the exceptions were. More importantly, such figures show how little basis exists for the widespread idea that New World Negroes represent a sampling of the population of the entire African continent. Various other documents make this point—one of the most striking is the following decision handed down by the general court of Maryland:

Negro Mary v. The Vestry of William and Mary's Parish.
Oct. 1, 1796   3 Har. & M'Hen. 501

Petition for freedom. It was admitted the petitioner was descended from negro Mary, imported many years ago into this country from Madagascar, and the question was, whether she was entitled to her freedom.

It was contended that Madagascar was not a place from which slaves were brought, and that the act of 1715 related only to slaves brought in the usual course of the trade. On the other side, it was contended, that the petty provinces of Madagascar make war upon each other for slaves and plunder; and they carry on the slave trade with Europeans.

Per. Cur. Madagascar being a country where the slave trade is practiced, and this being a country where it is tolerated, it is incumbent on the petition to show her ancestor was free in her own country to entitle her to freedom.[41]

In view of all available figures, it is understandable how Negro Mary came to base her hope of freedom on the fact that she, of Malagasy descent, was to be exempted from bondage because her enslaved ancestress had been taken from a country outside "the usual course of the trade."

The slaves imported into South Carolina between 1733 and 1785 as listed by Miss Donnan, when tabulated, show them to have been derived from the following African sources:[42]

Origin given as "Africa"	4,146
From the Gambia to Sierra Leone	12,441
Sierra Leone	3,906
Liberia and the Ivory Coast (Rice and Grain Coasts)	3,851
"Guinea Coast" (Gold Coast to Calabar)	18,240
Angola	11,485
Congo	10,924
Mozambique	243
East Africa	230
Imported from Africa	65,466
Imported from the West Indies	2,303
	67,769

It would be of interest further to document the origins of the various groups of New World Negroes with comparable figures from the West Indies and South America, but far less data for these regions have been made available than for the United States. Such as do exist, however, support the assumption of essential unity stated in these pages. Rinchon's materials for the French West Indies name sources of origin of cargoes only in terms of "Senegal," "Guinea," "Angola," and "Mozambique," though figures even in such categories do offer supporting data in showing that only 17 out of the 1,313 cargoes listed for the years 1713 to 1792 came from East Africa. One interesting point is the small number of Senegalese shipments; which means that as far as the West Indian receivers in Haiti, Martinique, Cayenne, Trinidad, Cuba, Puerto Rico, and Suriname were concerned, the vast majority of the slaves they bought, from ships owned and operated out of the French port of Nantes, were from the region lying between the Gold Coast and Angola.[43] Mr. J. G. Cruickshank, archivist of British Guiana, has studied the materials to be found in the files of the *Essequibo and Demerary Gazette* for the years 1803-1807.[44] These materials consist mainly of advertisements of "new" Negroes, designated as to African type. When classified according to the regions given in the tables of figures calculated from Miss Donnan's work, they indicate the same points of origin:[45]

Windward Coast	3,014
Gold Coast	3,593
Evo (Calabar)	820
Angola	1,051
Others	1,029

In 1789, Stephen Fuller, agent for Jamaica in London, published (by order of the House of Assembly of that colony) two reports for the Committee of the House which had been appointed to examine into the slave trade and the treatment of Negroes. Bryan Edwards, known for his *History of the British West Indies*, who gathered the materials, reproduced the accounts of five brokerage firms, giving records of the Negroes imported from Africa and sold by them. Four of these gave the sources of origin of their slaves in such fashion as to make tabulation possible. Combined in the following table, the four lists represent shipments for the years 1764-1774 for the first firm, 1782-1788 for the second, 1779-1788 for the third, and 1786-1788 for the last; the data themselves, when tabulated, support the position taken here:[46]

Gambia		673
Windward Coast		2,669
Gold Coast		14,312
Anamaboe	8,488	
"Gold Coast"	5,824	
Togo and Dahomey		3,912
Pawpaw	131	
Whydah	3,781	
Niger Delta		10,305
Benin	1,039	
Bonny	3,052	
Calabar and Old Calabar	6,214	
Gaboon		155
Angola		1,984
	Total	34,010

With the principal areas of slaving established, and direct comparability in terms of the cultural background common to Negroes in all the New World proved by the documentary evidence, the final step in discovering the most significant tribal origins is greatly simplified. For we need merely turn to those works, written by men and women who surveyed the scene of slavery while it was at its height in the West Indies, and utilize the many tribal names con-

tained therein—names which, when located in Africa, are found to
lie within the regions indicated as those where the most intense slaving was carried on. The Ashanti and Fanti of the Gold Coast, the
former most frequently termed Coromantes after a place name of
their homeland, are mentioned most often by those who wrote of
the British possessions, continental as well as insular. The Dahomean
and allied peoples, at times called Whydahs, after the major sea-
coast town of Dahomey, or Pawpaws, from Popo, a town not far to
the west, are especially prominent in the French writings. The
French planters had little liking for the Gold Coast slaves, and these
scarcely figure in Moreau de St. Mèry's listing of tribes represented
in Haiti; in similar manner, the Dahomeans, who were the favorite
slaves of the French, were not fancied by the English, as is to be seen
when we contrast the 14,312 Gold Coast Negroes in our list of
Jamaica imports with the 3,912 from Dahomey.

Another type of slave frequently mentioned is the Nago. This
term is used for the Yoruba of western Nigeria, whose language is
called by that name. Historical records for those parts of Latin
America where present-day Negro customs have been studied, Cuba
and Brazil, are not available. In the case of the latter, they were
burned to wipe out every trace of slavery when the Negroes were
emancipated in that country; and if they exist for Cuba, they have
not been published. But such data as we have establish that the Nago
slaves were favorites of the Spanish and Portuguese planters; from
which it follows that it is logical to find Yoruban customs pre-
ponderant in the African survivals reported from these countries.
For the rest of the slaving area, evidences of Africanisms are fragmentary. The Mandingo, Senegalese, and Hausa of the subdesert
area to the north have left traces of their presence, principally in
Brazil. The vast masses of Congo slaves that we know were imported have made their influence felt disproportionately little, though
a few tribal names, a few tribal deities, some linguistic survivals,
and more often the word "Congo" itself are encountered.

The mechanism that determined survival of customs and nomenclature of some African tribes in the New World, and not others,
may probably be connected with the geographical spread of the slave
trade itself. In the earliest days, before the trade became a major
industry, Senegal was most important. Yet though in the aggregate
many slaves were brought from this region, not enough from any
one group came at this earlier period to make possible the establishment of their common customs in the new home. As the demand for
slaves surpassed the human resources of this less densely populated

region, the traders came more and more to seek their goods along the Guinea coast, and here most of the slaving was carried on during the last half of the eighteenth century. With the nineteenth century, the weight of the abolitionist movement began to make itself felt, and when the trade was outlawed, the captains of slave vessels had to cruise more widely than before. They found the Congo ports, under Portuguese control, most hospitable; and this is reflected in Miss Donnan's work, which on analysis makes it apparent that slaves were shipped from the Congo in increasing numbers toward the latter days of the traffic.[47]

The fact that the slave captains ranged more widely as time passed, perhaps because the difficulties of supplying their needs on the Guinea coast increased as the demands for slaves became greater, is likewise shown if a recapitulation be made of the ships sailing from Nantes, in terms of the African ports where they obtained their cargoes:[48]

Year	Region of Origin of Cargo			
	Senegal	Guinea	Angola	Mozambique
1748...............	—	6	4	—
1749...............	—	24	9	—
1750...............	1	14	7	—
1751...............	1	11	5	—
1752...............	2	20	8	—
1753...............	3	17	8	—
1754...............	1	13	7	—
1755...............	1	5	5	—
* * *				
1763...............	—	24	9	—
1764...............	2	20	8	—
1765...............	—	18	12	—
1766...............	3	11	11	—
1767...............	—	15	6	—
1768...............	1	13	4	—
1769...............	—	18	6	—
1770...............	2	10	8	—
1771...............	1	13	9	—
1772...............	2	8	6	—
1773...............	1	13	10	—
1774...............	2	6	10	—
1775...............	—	11	6	—
1776...............	1	12	5	—
1777...............	—	11	2	—
* * *				
1783...............	8	11	18	—
1784...............	3	8	9	—
1785...............	11	11	14	2

Year	Region of Origin of Cargo			
	Senegal	Guinea	Angola	Mozambique
1786...................	6	*19*	15	3
1787...................	—	15	17	1
1788...................	—	16	16	2
1789...................	—	19	22	—
1790...................	—	18	22	2
1791...................	1	9	27	2
1792...................	2	8	*18*	3

Not until 1783, except for one year, does the Congo traffic exceed that from the Guinea coast; but after this time the French traders also, because of economic and political reasons that need not be gone into here, turned increasingly southward. And while not a single Mozambique cargo appears until 1785, after that year the demand seems to have been great enough to cause a few ships to be sent there annually. It must be observed, however, that the number is too slight to influence appreciably the demography of the New World Negro population.

Let us consider another facet of the problem. It is not difficult to see that the slaves who came late to the New World had to accommodate themselves to patterns of Negro behavior established earlier on the basis of the customs of the tribes represented during the middle period of slaving. In Haiti, Congo slaves are said to have been more complacent than those from other parts of Africa, and were held in contempt by those Negroes who refused to accept the slave status with equanimity. Tradition has it that when the blacks rose in revolt, these Congo slaves were killed in large numbers, since it was felt they could not be trusted. Mr. Cruickshank has advanced a cogent suggestion:

. . . from what I have learned from old Negroes . . . it would appear that the three or four African Nations who were brought here in predominant numbers imposed their language, beliefs, etc., gradually on the others. In course of time there were not enough of the minority tribes on an estate to take part in customs, dances and the like, or even to carry on the language. There was nobody left to talk to! Children growing up heard another African language far oftener than their own; they were even laughed at when they said some of their mother's words —when they "cut country," as it was said—and so the language of the minority tribes, and much else—though probably never all—died out.[49]

Though African survivals in the United States are far fewer than in British Guiana, nonetheless, a similar process may well have operated. It might also be hazarded that, in the instance of early Senegalese arrivals, whatever was retained of aboriginal custom was overshadowed by the traditions of the more numerous Guinea coast Negroes; while as for late-comers such as the Congo Negroes, the slaves they found were numerous enough, and well enough established, to have translated their modes of behavior—always in so far as Africanisms are concerned, and without reference to the degree of acculturation to European habits—into community patterns.

The indisputable survivals in those parts of the New World where a considerable degree of African culture is found today in pure form are to be traced to a relatively restricted region of the area where slaving was carried on; this simplifies the problems we must face in drawing up our base line for the study of deviation from African tradition, leading to the two tasks which constitute the next step in our analysis. The cultures of the tribes of the area must first be described as an aid to direct comprehension of the New World data, and this description must be compared with published accounts of the Negro's cultural heritage. We must then determine whether more generalized aspects of West African culture are to be discovered. For if such aspects are held in common both by the dominant tribes and by all the other folk of the entire area from which slaves were brought, we will be afforded further insight into those more subtle survivals which, because they represent the deepest seated aspects of African tradition, have persisted even where overt forms of African behavior and African custom have completely disappeared.

Chapter III

THE AFRICAN CULTURAL HERITAGE

Judged by references in the literature, the writers who in the United States have most influenced concepts of the Negro's African heritage are Tillinghast,[1] Dowd[2] and Weatherford.[3] But since all these went to the same sources for their African materials, where they did not draw on the works of each other, and none had firsthand contact with any of the native peoples he mentions, their substantial agreement in describing and, what is more significant, in evaluating the civilizations of Guinea is not surprising. The unanimity of their findings is important for the support it has afforded the concepts of aboriginal cultural endowment of the Negro presented by any one of them.

It is of some interest to outline briefly the materials which they employed. Most frequent are references to what were but secondary sources even when they were first made available. Especially useful to them were the several compendia that were written to give ready access to the various forms of primitive civilizations known at the time they were written, a feat impossible today, with the development of scientific ethnology and its rich and numerous field studies. Citations to such works as A. H. Keane's *Man: Past and Present*, Ratzel's *History of Mankind*, Waitz' *Anthropologie der Naturvölker*, D. G. Brinton's *Races and Peoples*, and Elisée Reclus' *Universal Geography*, appear again and again. Even granting the contemporary usefulness of these works, it is questionable whether there ever was justification for the student of the Negro in the United States, concerned with the problem of African background, to base his analysis of the aboriginal cultures on "sources" such as these. Yet the tradition lingers on, and the failure of more recent scholars to employ the modern data and the critical tools at their disposal is lamentable. If the plea is entered that these recent scientific analyses of West African cultures are difficult to use, this is

54

but a confession of an inadequacy which speaks for itself when conclusions are evaluated.

How persistent is the tradition of being content with well-worn obiter dicta may be seen in the following quotation from the most recent edition of Reuter's textbook—a work wherein African background takes a prominent place in the argument:

> The African Negroes, representing as they do many separate tribal groups, have a variety of sex mores and marriage and family customs differing widely from one another. The reliable data are still fragmentary; dependence must be had in some part upon the reports of missionaries and officials and upon the impressionistic accounts of travelers. These accounts are of course prone to a considerable degree of biased error. The scientific and dependable studies are mainly local and of somewhat limited tribal application. A further difficulty to the understanding of the African Negro family organization results from the fact that the present native family structure is in many cases highly disorganized through tribal intermixture and as a result of foreign contacts and missionary activities. General statements are in consequence difficult and subject to numerous individual and tribal exceptions.[4]

Most of this passage is sheer nonsense. Denial of most of its assertions can be documented by reference to the relevant passages in the following section, or to the monographs on West Africa previously cited.[5] The variety of "sex mores and marriage and family customs" in "Africa" (West Africa?) can be considered unusually great only if we are unaware of the underlying similarities which support local variations. Present native family structure is not at all "highly disorganized" but, as a matter of fact, has been scarcely touched by European contact which, because of the debilitating effect of the climate on whites, is for most individual tribesmen casual in West Africa. It is surely unnecessary today to rely upon the "impressionistic accounts of travelers" for information concerning African customs. That the "scientific and dependable studies are mainly local and of somewhat limited tribal application" should be no obstacle to a scholar who wishes to make use of them. This passage may thus be taken as a complaint that no recent summary is at hand for those who are bewildered by technical descriptions of complex institutions in societies whose simplicity has long been uncritically taken for granted.

Aside from the compendia mentioned, those who, like Tillinghast, Dowd, Weatherford, and others to whom we are at the moment giving our attention, write of Africa also lean heavily on the works of A. B. Ellis.[6] Though Ellis had actual experience on the Gold

Coast and at least visited the other areas of which he wrote, he is notorious among Africanists for his uncritical borrowing from other authors. The vogue of his volumes has perhaps come from their logical organization, facile style of writing, and congeniality of interpretation. Today specialists in the field hold him outmoded and of but negligible value—the dates of publication of his books are eloquent on this point; the deficiencies of his best work, on the Gold Coast tribes, have been repeatedly pointed out by such an authority as Rattray. Among students of the Negro in the United States who are not willing to do the reading necessary to take advantage of scientific studies made with modern field methods, his authority continues undiminished. The older travelers' accounts also are drawn on in the American attempts to describe West African culture, with the writings of Bosman, Barbot, Norris, Proyart, and Snelgrave[7] figuring repeatedly in the references. Some of the best of these observers are almost entirely neglected—Bowdich for the Gold Coast, for example, or Burton for Dahomey.[8] The works of Miss Mary Kingsley, a Victorian lady of much courage and an excellent observer, but whose evaluations were far too often influenced by the period in which she lived, are also favored, as are the writings of Nassau, a missionary whose biases are patent.[9]

While in the case of Tillinghast, at least, most of the available sources of his time were drawn on, no attempt was made by him to test his conclusions by reference to the textual consistency of the data themselves. An acquaintance with the writings which he and others like him cite need not be extensive to show that the estimates of African culture found in these books by no means always flow from the facts as presented. Using him as an example, then, his assertions may be sampled to determine whether his descriptions of West African culture are valid in the light of modern findings. We are told that the West African:

. . . lives under conditions adverse to the growth of industrial efficiency; . . . so abundant is nature's provision for food and other wants, that with little effort they obtain what is needed. . . . In the case of cultivated produce, the fertility of the soil and the climatic advantages are such that very large returns are yielded to slight labor.[10]

Actually, the climate of West Africa is like all tropical climates at low altitudes. It permits a rich yield if crops are undisturbed, but crops are so rarely undisturbed that the hazards of agriculture are far greater than in the temperate zone. The conception of the native as one pampered by nature is thus entirely fallacious. The African's ability for sustained toil, his need to work and work hard if he is to extract a living from the soil, have been remarked by all those who have made serious firsthand studies of the labor required to maintain life in the region. Should precise testimony be desired awaiting the appearance of Harris' analyses of the actual number of hours spent in work by the Ibo,[11] reference may be made to the study by Forde,[12] wherein the effort and planning involved in carrying on agriculture among the Yakö of the region which lies at the bend of the west coast are made plain.

Again, Tillinghast informs us:

Previous to the appearance of Europeans, the extreme west coast of Africa was completely isolated from the outside world; its inhabitants lived in scattered villages buried in the forest, and remained in dense ignorance of any other desirable objects than the necessities of their own savage life. Among the forces which have helped to civilize other peoples has been the stimulus to effort arising from newly conceived wants, quickened into being at the discovery of commodities, first brought by strangers. The appearance of Europeans with new and attractive commodities, produced a great effect. To get them in exchange for native products, thousands of negroes were moved to unwonted exertions, while foreigners taught them new and better methods of production. All this, however, has been comparatively recent, and for ages the negroes were without such incitements to industry.[13]

Once more, misstatements are found in almost every line. The philosophy underlying most of the assertions, a kind of naïve laissez-faire economics which holds progress to be in some way related to a constant accretion in the range of wants, is immediately apparent. That isolation in terms of lack of contact by sea might be replaced with land-borne commerce across the Sahara never seems to have occurred to this writer, as it seldom occurs to others who speak of the "isolation" of Africa. Yet from the earliest times the Ashanti, for example, acquired silk cloths from Tunis and Morocco, which they unraveled, redyed and rewove into great chiefs' silk cloths. The "isolated villages" spoken of are in many cases population centers of considerable size—Ibadan, Nigeria, has some 325,000 inhabitants—while the dense forests are in many parts nonexistent, since the land is required to support this population.

"Division of labor has proceeded but a very little way,"[14] we are told—this perhaps being written with Adam Smith's statement regarding the importance of this factor in making for economic advancement in mind.

The number of handicraftsmen in any given tribe is small, and their special skill is jealously withheld from the common herd. . . . These simple folk exist somehow on an incredibly meagre supply of implements and weapons. Even in the manual arts women are compelled to do all the drudgery of collecting raw material, etc. All these facts reveal how the great mass of male population escapes distasteful toil.[15]

Here also we are confronted with assertions that are directly contravened by the facts. As will be seen, the large number of specialized crafts are indicative of a corresponding degree of division of labor. The popular assumption of the savage male as lazy, allowing his women to carry on the work necessary for subsistence, is far removed from the actuality of the sex division of labor, which invades all fields. That the women do agricultural labor[16] is but an expression of the forms of sex division of labor universal in human societies, literate or not; in Africa the arrangement makes the men responsible for the really heavy work of preparing the fields, and leaves to the women the lighter tasks of caring for the growing plants, harvesting the crops, and preparing food. As a matter of fact, the economic position of women in West Africa is high. It is based on the fact that the women are traders quite as much as agricultural workers, and on recognition that what they earn is their own. They do none of the ironworking or wood carving or house-building or weaving or carrying of burdens or other heavy labor. This is reserved for the men. They unquestionably contribute their share to the support of the household and the community; but they are not the exploited creatures undisciplined fancy would have them.

It is not possible here to detail all the misconceptions which characterize Tillinghast's descriptions of West African life, among them statements expounding a presumed inadequacy of West African technology, simplicity of the system of trade, and absence of social morality in the religious concepts of the people—a fact refuted by the widely spread incidence of belief that the gods punish antisocial behavior, which, needless to say, is an important moral sanction. It could be shown how Tillinghast agrees that wives are "bought" and that cannibalism was "once practised universally."[17] Political development is indicated as being "on a par with the low stage attained in all other directions"[18]—specific reference being made to the Ashanti and to Dahomey, where "vanquished tribes are extinguished by slaughter or held as slaves." Or we learn of "customs regulating property and personal relations after a crude fashion,"[19] another error the more glaring in the light of general recognition of the African's "legal genius." All these misconceptions are evaluated with a wealth of adjectival embroidery which makes it impossible for the reader to conceive of the civilizations of the region as anything but outstanding examples of a low state of savagery—a savagery that, as the author surmises, is the source of the Negro's assumed insufficiency in mastering white culture in the United States.

Similarly, it is not possible, even were it necessary, to cite from the works of those others who have perpetuated these misinterpretations of African culture. Excerpts from the writings of Dowd, or Weatherford, or others would be repetitious, but to illustrate the tenaciousness of the point of view, quotations from two volumes will be given. The first of these books, by Mecklin, was published in 1914, and, like the others, is found in most bibliographies of books and articles dealing with the Negro in the United States:

The most striking feature of the African negro is the low forms of social organization, the lack of industrial and political cooperation, and consequently the almost entire absence of social and national self-consciousness. This rather than intellectual inferiority explains the lack of social sympathy, the presence of such barbarous institutions as cannibalism and slavery, the low position of woman, inefficiency in the industrial and mechanical arts, the low type of group morals, rudimentary art-sense, lack of race pride and self-assertiveness, and an intellectual and religious life largely synonymous with fetichism and sorcery.[20]

It is scarcely necessary to point out once more that almost every assertion in this statement is incorrect; indeed, it is rare, even in works on the Negro, to come upon a paragraph with such a high concentration of error as this. The most glaring of these misconceptions, viewed from the perspective of the last three decades of art history, is the statement concerning the "rudimentary art-sense" of the Africans. For an outstanding development of modern art has been the steady growth of interest in African—West African— wood carving and other art forms, and the influence of these forms on many of the painters of the present day.

Reuter, whose textbook has already been quoted as an example of the manner in which this approach and point of view still lives in standard works dealing with the Negro population of the United States, will give our series its most recent instance. The excerpt is from the second edition, which, appearing in 1938, can be taken to represent the present position of its author. Social life in West Africa is dismissed in this edition (the earlier one[21] went into some detail concerning African family life in terms typical of what has been cited in the way of misconception), with the statement that, "The family institution [was] never highly developed among the

West African tribes." No qualification is given this statement, as the author proceeds to explain how such a weak institution could not but give way under slavery in the United States, when it encountered the presumably stronger European type of family.

More revealing for our purpose at the moment, however, is the description of the patterns of West African religious life. As will later be seen, this is unusually complex, and represents one of the most sophisticated aspects of the cultures in the region. Because the following is taken as authoritative by the large numbers of those it has reached, it may be cited at some length, to permit a realization of its total effect:

The religion of the African was, basically, a crude and simple demonology. It began and ended in a belief in spirits and in the practices designed to court their favor and to avoid the consequences of their displeasure. There was a lack of unity and system resulting from the decentralization and absence of unity in the political and social life. . . . Fear was the basic element in the religious complex of the Negroes. In the conditions of primitive existence in the African environment it could not well have been otherwise. The life of the native was never safe. Personal danger was the universal fact of life. There was an almost complete lack of control of natural forces. The forests and rivers were full of dangerous animals, and dangerous human enemies were always close at hand. The insect pests and the tropical diseases made the conditions of life hard and its duration brief. To the real dangers were added an abundance of malignant spirits. An ever present fear of the natural and supernatural enemies was the normal condition of daily life and protection was the ever present need. These facts everywhere found expression in the religious and magical beliefs and practices. The state of religious development varied considerably with tribal groups. In some tribes nature worship was elaborated to the point where definite supernatural powers had been differentiated to preside over definite spheres of life. In other groups the basic fetichism was modified by and combined with a worship of nature. In certain of the politically more advanced groups ancestor worship was an important element in the religious complex. But everywhere the practices were directly designed to placate or coerce the malignant and insure the co-operation of the beneficent powers. Since it was the nature of the latter to aid, the cultus procedure in their case was less important and was quite commonly neglected. Magic, both sympathetic and imitative, was practiced by private individuals as well as by professional magicians. Sickness, accident, injury, death, and other misfortunes were attributed to evil influences exercised by or through some person, and the effort to find the persons guilty of exercising evil influence lay at the basis of the witch trials and the other bloody religious sacrifices of the African peoples.[22]

A brief analysis of this passage is called for, since only in this way may its misstatements be set forth. Much of what is said here, if divested of comment and evaluative adjectives, might be regarded as true in a generalized sense, as, for example, that there are good as well as evil forces or that nature worship obtains. Yet words like "crude" and "simple," or the emphasis on naïveté, and, above all, the picture of the fear-ridden native; the conception of the dangers of the environment, which, it may be said, gives an excellent glimpse into the imaginings of the armchair traveler as he dreams of the tropical jungle and its denizens, are far from the truth. Equally fallacious are the presumed neglect of the beneficent powers, whose existence in the system is mentioned because of the author's belief in the preoccupation of the native with the forces of evil, and the manner in which the role of magic is conceived. All these leave a residue of impression calculated to prepare the reader for the incapability of the Negro, with a background of tradition such as this, to grasp higher and more restrained aspects of belief and ceremonialism such as he presumably encountered in the New World.

2

Today, as in the days of the great traffic in slaves, the tribes living in the heart of the slaving area are the Akan-Ashanti folk of the Gold Coast, the Dahomeans, the Yoruba of western Nigeria, and the Bini of eastern Nigeria. Composites of many smaller groups, welded through a long process of conquest into more or less homogeneous kingdoms, they share many traits in common. Their numbers are large as primitive societies go, and consequently many problems of economic, social, and political organization must be met if smooth functioning is to be achieved. It follows that complex institutions in those fields are the rule. The ensuing discussion will touch upon those aspects of the cultures of these kingdoms which, germane to their functioning, have been impinged upon but little by the circumstances of European political domination.

The economic life, adapted to the support of large populations, is far more intricate than is customarily expected or, indeed, found among nonliterate folk. Essentially agricultural, all these societies manifest a considerable degree of specialization, from which are derived the arrangements for the exchange of goods that take the form principally of stated markets, wherein operations are carried on with the aid of a monetary system which, in pre-European days, was based on the cowry shell to facilitate the expression of values.

The economic system permits the production of a substantial surplus over the needs of subsistence, and the support of rulers, priests, and their subordinates. As a result, a class structure has been erected on this economic base that has tended to encourage that disciplined behavior which marks every phase of life. In the field of production, this discipline takes the form of a pattern of cooperative labor under responsible direction, and such mutual self-help is found not only in agricultural work, but in the craft guilds, characteristically organized on the basis of kinship. This genius for organization also manifests itself in the distributive processes. Here the women play an important part. Women, who are for the most part the sellers in the market, retain their gains for themselves, often becoming independently wealthy. With their high economic status, they have likewise perfected disciplined organizations to protect their interests in the markets. These organizations comprise one of the primary price-fixing agencies, prices being set on the basis of supply and demand, with due consideration for the cost of transporting goods to market.

Slavery has long existed in the entire region, and in at least one of its kingdoms, Dahomey, a kind of plantation system was found under which an absentee ownership, with the ruler as principal, demanded the utmost return from the estates, and thus created conditions of labor resembling the regime the slaves were to encounter in the New World. Whether this system was the exception rather than the rule cannot be said, for this aspect of the economic order, as the first suppressed under European rule, is not easy to document satisfactorily. On the whole, slaveholding was of the household variety, with large numbers of slaves the property of the chief, and important either as export goods (to enable the rulers to obtain guns, gunpowder, European cloths, and other commodities) or as ritual goods (for the sacrifices, required almost exclusively of royalty, in the worship of their powerful ancestors).

The economic base of the social structure is most apparent when the role of the relationship group in the production and distribution of wealth is considered. Essentially, this structure comprises as its principal elements the polygynous family; legal recognition of kinship through one line, with the nonrelated side of varying importance—ranging from the noninstitutionalized sentimental relationship with the mother's family in patrilineal Dahomey and among the Yoruba to the Ashanti system wherein an individual inherits his position in society on the maternal side and his spiritual affiliations from the father; the "extended family," a well-recognized institution which affords a more restricted relationship grouping than the sib (clan); and finally the sib itself, comprising large numbers of persons whose face-to-face contact with each other may be intimate or casual or nonexistent. Guild organization tends, in the majority of cases, to follow the lines of these kinship groupings. Since the principal occupation is agriculture, landholdings are conceived in terms of family rather than individual rights; and while, as in all primitive societies, a man has the exclusive ownership of the produce of whatever land he works, the land itself is not his. As a member of a relationship group of considerable size, however, he has an assurance of support in time of need. This has contributed largely to the stability of these societies, since the economic aspect reinforces the social one in a peculiarly intimate manner, and causes the relationship group to hold added significance for its members.

The fundamental sanction of the kinship system is the ancestral cult, which, in turn, is a closely knit component of the prevailing world view. The power of a man does not end with death, for the dead are so integral a part of life that differences in power of the living are carried on into the next world. Just as among the living individuals of royal or chiefly blood are the most powerful, so the royal or chiefly ancestors are conceived as the most potent of all the dead. The dead in Dahomey and among the Yoruba, at least, are deified; among the Ashanti, this remains to be studied. The relationship between the ancestors and the gods is close, but the origin of this collaboration is obscure and extremely difficult to establish. Evidence adduced by Bascom[23] indicates that at least in the Nigerian city of Ife, the spiritual center of Yoruba religious life, the beings conceived elsewhere as gods are there regarded as ancestors. The sib mythologies collected in Dahomey would also seem to indicate something of a similar order, certain sibs being considered as descended from various gods, though there is no sib without its "oldest ancestor," who figures importantly in the daily life of each member of the group.

The elaborateness of funeral rites in the area is cast in terms of the role of the ancestors in the lives of their descendants, and because it is important to have the assurance of the ancestral good will, the dead are honored with extended and costly rituals. In all this region, in fact, the funeral is the true climax of life, and no belief drives deeper into the traditions of West African thought. For the problem of New World survivals this is of paramount importance, for whatever else has been lost of aboriginal custom, the attitudes toward the dead as manifested in meticulous rituals cast in the mold of West African patterns have survived.

As in most primitive societies, the sib functions in regulating marriage, since mating between sib mates or other relatives of legally established affiliation and, among the Ashanti, on the side of "spiritual" affiliation, is forbidden as incestuous. This is not as much a handicap in finding a mate as in smaller groups having similar prohibitions, since with dense populations such as are found in West Africa there is no lack of eligible mates outside the sib. The major problem, where a marriage is contemplated, thus merely involves the tracing of descent lines to ensure that no common affiliation stands in the way. Far more important, as a matter of fact, is the assurance that the suitor has resources and substantial family support to make of him a responsible husband, and that the young woman has had the training to make a competent wife.

Qualifications are carefully scrutinized by both families, for, as in so many primitive societies, marriage is a matter of family alliance. This is not to be construed that the common dictum, that affection does not enter, is valid. In contradiction to this may be cited the frequency of runaway marriages—recognized by the Dahomeans as one of the principal forms of marriage—as an expedient of the young people to circumvent unwelcome matings arranged by those of their social group who have legal control over their behavior. In all this region the obligations of the man to the parents of his bride are paramount, not only before but after marriage. Yet the characterization of African marriage as "bride purchase" is no more valid here than elsewhere. As a matter of fact, in this region what the husband gives his parents-in-law is regarded essentially as a form of collateral for good behavior, though the social worth to a man's sib of prospective issue does figure psychologically.

The widespread character of polygyny gives rise to a number of important research problems. In so far as New World Negro life is concerned, the deep-seated nature of the pattern of plural marriage aids greatly in accounting for some of the aberrant types of family organization to be found. Of outstanding significance in this connection is the relationship between father and children as against mother and children. For where a man has plural wives, the offspring of any one woman must share their father with those of other women, while they share their mother with none but other children by her. This psychological fact is reinforced by the physical setting of family life in this area, as well as by the principles of inheritance of wealth, which obtain at least among the Yoruba and in Dahomey. The family is typically housed in a compound, which is a group of structures surrounded by a wall or a hedge, to give the total complex a physical unity. The head of the household, the eldest male, and all other adult males, married or unmarried (for in some parts of the area, young married sons, or younger married brothers and their children, may live in a father or elder brother's compound), have individual huts of their own, to which their wives come in turn to live with them and, for a stated period, to care for their needs. Each wife has her own dwelling, however, where she lives with her children. Once she conceives, she drops out of the routine of visits —a factor in restricting the number of children a woman may bear, and well recognized in Dahomey, at least, as a hygienic measure— to resume it only when she has weaned her child. Naturally, not every household is polygynous, though the degree to which even those not among upper-class groups have more than one wife gives rise to a problem, as yet unsolved, of a possible differential in sex ratio.

Among the Yoruba and Dahomeans, a chosen son succeeds to the wealth of his father, and here again, as in matters of personal jealousies, conflict among wives in terms of jockeying for position to obtain advantageous consideration for a son makes for closeness of relationship between mother and children as against father and children. Among the Ashanti, wealth, like position, is inherited from a maternal uncle, hence this particular economic factor does not obtain in attitudes toward the father, but takes form in rivalries for the uncle's favor. But even where questions of succession do not enter, the very nature of the life in any polygynous household is such that it gives the psychological generalization validity. In Dahomey the explicit recognition of the difference is emphatic. Phrased in terms of inheritance, while there is always bitter dispute over the apportionment of the wealth of a father, such quarreling, it is asserted, is unthinkable even when the property of a wealthy woman is to be distributed. "They are children of the same mother," would seem for these people, as for various New World Negro folk, to be an explanation that needs no clarification.

The political organization of the tribes of our region has two distinct aspects, historically of great importance. It is simplest to think of each of the three aggregates we are describing as political entities, since in each we find the kings, courts, and subchiefs that mark them as units. Yet once we probe more deeply, it becomes apparent that for the people themselves the unit is smaller. Where, as among the Ashanti, in Dahomey, in Benin, and among the Oyo of Nigeria, powerful states were in existence during the time of the slave traffic, and until European conquest, they were actually but glosses on an underlying pattern of local autonomy and local loyalties. One of the most confusing aspects of the study of New World Negro origins based on the documents is a semantic one, which arises out of the difference between the conception held by the early writers of a king and a kingdom, and the ethnological reality of African concept. To this day, in a small village, one may be introduced to its "king" by a loyal follower of this petty, powerless potentate; and the village will likewise be designated a "kingdom." It is this, as much as anything else, that has misled students in attempting to understand the importance of the political units named in the slavers' accounts. If we take, for example, the oft-mentioned kingdom of Pawpaw—the Popo of present-day Togoland—we find it to be a village whose ruler commanded a "kingdom" of perhaps not more than 250 square miles at its greatest! Yet the identity of this "kingdom" has persisted under the French as it persisted in the face of Dahomean conquest. In exactly the same way we encounter the local loyalties of the inhabitants of Kumawo as against Mampong among the Ashanti of the Gold Coast, of Allada as against Abomey in the interior of Dahomey, of Ife as against Oyo among the Yoruba. These reflect identifications which, earlier, were to independent states whose inhabitants, after their absorption, never attained complete identification with the larger kingdom. From this it follows that the realms found in our area had to exercise control over the local chiefs; while, in addition, in the interstices between their fluid boundaries local communities could, and did, persist without giving up their autonomies.

The larger aggregates were no less significant political realities, nor did they function any the less efficiently because of these local loyalties, for their organization was remarkable in the light of conceptions generally held of the simple nature of the primitive political institutions. Given cultures without writing, and with local traditions as strong as those existing in West Africa, the rulers accomplished their ends with an expeditiousness that can only be realized by studying the writings of firsthand observers who visited the courts of the Ashanti, Dahomean, and Yoruban potentates. It is thus particularly ironical that, in this field, the simplicity and crudity of primitive life attributed to the African should have been permitted to loom as so important a trait of African culture. Stable dynasties were the rule, not the exception. Courts and related institutions ensured the operation of orderly processes of law, while specialists in warfare saw to it that the territory of the ruler was not only defended in case of attack, but that he could extend his dominion as opportunity offered.

In outlining the ordering of life in this area, there is no intention of picturing the West African as a kind of natural man living in a golden age. For if rulers were efficient, they were also exacting and ruthless; if they ensured orderly processes of law in the courts, they were also given to pecuniary persuasion that helped to dim the identity of law with justice. In terms of native standards, their way of life was lavish, and they did not scruple to tax heavily in order to maintain their status. In war, all males were liable to service, and any member of an enemy people who came within reach of their armies was fair game; men, women, and children were taken captive, and the category of noncombatant was unknown. The institution of polygyny reached fantastic proportions, for any woman who took the fancy of the ruler was liable to be claimed for his harem. In Dahomey, also, where centralization of authority and the despotic exercise of power were most developed, battalions of women warriors were kept as nominal wives of the ruler, and hence unapproachable by another man. Many women were thus not permitted normal life, which from the point of view of population policy prevented the kingdom from reproducing the numbers needed to support the expense, in human life and wealth, of its expansionist policy, and eventually contributed to its downfall.

Yet within these despotisms, life went on with a degree of regularity and security rarely envisaged when African polity is thought of. Authority was divided and redivided in terms of a precept under which the delegation of power was accompanied by sharply defined responsibility. The head of the family group, for example, was responsible to the village chief or, in the more populous centers, to the head of his "ward" or "quarter." The local chief was responsible to a district chief, and he to the head of a larger area, who in turn had to account for the administration of his "province" to the king himself or to one of the highest ministers of state. These chiefs sat with their elders and passed judgment when disputes arose among their people. Various devices were employed in the courts. In some cases testimony was taken, in others an ordeal was administered; on occasion a chief might point the way to informal amicable compromise of a dispute. Appeals to a higher court might be taken by plaintiff or defendant. Such crimes as theft were rare, but when the culprit was apprehended his punishment was severe.

The cost of these central governments was met by the taxes levied on the population at large. As reported for the Ashanti and for

Dahomey, tax programs were administered so as to exact from the people the greatest possible return. Taxes in Ashanti took two principal forms, death dues and levies on goods in transit. The Ashanti traded with the tribes to the north and with coastal folk to the south, and caravans going in either direction were liable for imposts according to the nature of the goods they carried. Commodities which were seasonal might not be traded in by commoners until royalty had had the opportunity to profit by the high prices for the early crop. Ashanti death dues were indirect but heavy. A proportion of each estate became the property of the local ruler, to go to the next higher officer on the death of this official, until it finally reached the royal treasury.

In Dahomey, indirection was the rule. Everything, including the population, was counted, and all commodities were taxed; but the people were not told when they were being counted, and taxable goods were often enumerated by subterfuge. In some of the methods the priests collaborated; in others, it was merely a matter of subtracting a balance on hand from an observed rate of production. As an indication of the ingenuousness of some of these indirections, the case of pepper, a prized commodity, may be indicated. To prohibit its general cultivation would have made for discontent, hence each man was permitted to raise enough plants to give him a small bag for his own use. But this was far from sufficient, and he had to buy the rest in the market, which was supplied by plantations in remote parts of the country. Even then, no direct tax was levied on the sale of pepper, but since all roads had tollgates at which porters' taxes were collected, and since all the pepper sold had to be brought from these plantations far removed from centers of population, the tolls paid by those who brought this commodity to market came to a substantial sum which was, in effect, a tax on pepper. Death dues in Dahomey were more directly assessed than among the Ashanti. The movable goods of the dead were brought to the local administrative center; what was returned to the heir was given as a gracious gift of the king. That the portion of the estate returned never equaled what had gone into the royal enclosure was no excuse for the recipient to fail in a show of gratitude.

One point must be emphasized concerning the political, social, and economic institutions of this part of West Africa which, it may again be recalled, was the heart of the slaving belt, and from which came the people who have left the most definite traces of their culture in the New World. Despite wars that were at times of some magnitude and the serious inroads on population made by the slave

trade, so well integrated were the cultures that little or no demoralization seems to have resulted. Today, in all this area, despite the
fact of European control which has changed the role of the native
ruler where it has not obliterated him; which, in the economic
sphere, has been responsible for the introduction of stable currency
and the raising of cash crops that make the native dependent on the
vagaries of the world market; and which, in the realm of social institutions has subjected such a deep-lying pattern as that of the polygynous family to the impact of Christian conceptions of morality, these
cultures continue with all vitality. Even in the field of technology,
European influence has had but relatively slight effect.

In coastal cities, it is true, certain indications of deculturization
are to be perceived, especially in those centers where Africans from
all parts of the coast have been indiscriminately thrown together.
But once away from these seaports, the aboriginal culture is found
functioning much as it must have functioned during the days before
European control; and even in the coastal cities, far more of aboriginal pattern persists than is apparent on first sight. This resilience,
when manifested in those aspects of culture most susceptible to outside influence, argues a high degree of tenaciousness for the cultures
of this part of West Africa, and, if this is true, is a significant point
for New World Negro studies.

3

The religion of West Africa, as described by most of those who
have written of it, is customarily encompassed in the word "fetish."
Without defining the term, it is broadly held to refer to magical
practices of some sort or other, which characteristically are represented as so preoccupying the minds of the people that they live in
a state of abject fear. How loosely the word has been used has been
demonstrated by Rattray who, in one passage, indicts the practice
as something "the indiscriminate use of which, I believe, has done
infinite harm." In the specific case of the Ashanti, after showing
how it is applicable only to charms of various sorts from any "category of non-human spirits," he continues:

The native pastor and the European missionary alike found a word
already in universal use, i.e., "fetish." They were possibly quite ready
to welcome a designation which obviated any necessity for using a term
which, even when written with a small initial letter, they considered
much too good to apply to these "false gods" about whom we really still

know so little. Thus West Africa became "the Land of Fetish" and its religion "Fetishism." It would be as logical to speak in these terms of the religion of ancient Greece and Rome, pulling down from their high places the Olympian Deities and . . . Daemons— (those which were the souls of men who lived in the Golden Age, and those which were never incarnate in human form, but were gods created by the Supreme God), and branding all indiscriminately as "fetishes," and the great thinkers of old, e.g., Plato and Socrates, as fetish worshippers. "I owe a cock to Aesculapius," said the latter almost with his last breath, and this pious injunction to his friend would be understood by every old Ashanti today.[24]

In so far as the complex concepts that mark the world view of the Ashanti, the Dahomeans, and the Yoruba are given systematic expression, their religion may be analyzed into several major subdivisions. As has been said, the ancestral cult sanctions and stabilizes kinship groupings, and there is reason to believe that in some cases these sanctions are to be traced even back to the major deities. For the Ashanti, Rattray was convinced that the ultimate force of the universe is lodged in the Great God Nyame, as befits another widespread conception of the African world view, in terms of which the universe, created by an all-powerful deity, has been so left to itself by the creator that he need not be worshiped. This is not the place to discuss whether this is in fact a valid concept of the African's belief; it may be indicated, however, that on the basis of field studies, of comparative analyses, and of the internal evidence in Rattray's own works there is reason to believe that this hypothesis will ultimately be revised. In Dahomey and among the Yoruba, in any event, the Great Gods are envisaged as a series of family groupings, who represent the forces of nature and function as agencies for the enforcement of right living as conceived in terms of conformity to the patterns of morality and probity. That is, the gods, in Dahomey fully, and in a manner not entirely clear among the Yoruba, are grouped in pantheons, which follow the organization of the social units among men, each member having specific names, titles, functions, and worshipers. The cult groups are organized in honor of these deities, and the outstanding religious festivals are held for them.

Closely associated with the gods, yet not included in the pantheons, are certain other deities or forces. The cult of Fate, with its specialized divining technique, is particularly important in Dahomey and among the Yoruba. Here divination is principally based on a complex system of combinations and permutations arrived at by throwing a set number of seeds, and ties in with a whole body of mythology, interpreted by the diviner in the light of the particular situation involved, and the relation to this body of lore of the particular tale that is called for by a given throw. This means that the training which the diviner must have is quite comparable to that of specialists in our own culture; the very period of study required to become a diviner, between five and ten years, suggests an analogy with the doctorate of philosophy or medicine among ourselves. In the Gold Coast, where, as has been indicated, divination is less formal, training of this kind has not been recorded.

In Dahomey, and among the Yoruba, the philosophical implications of the divining system are impressive. Though the universe is held to be ruled by Fate and the destiny of each man worked out according to a predetermined scheme, there are ways of escape through invoking the good will of the god, the youngest child of the principal deity, who speaks the differing languages of the various divine "families," and as interpreter carries to them the messages which ensure that a man experience whatever is in store for him. For this divinity, the trickster among the gods, can be induced to change the orders he carries, and does so on occasion; so that if an unpleasant fate is in store for an assiduous worshiper, it is believed a simple matter for him to aid such a person by substituting a good for an evil destiny. Yet as a philosophical conception, this deification of Accident in a universe where predetermination is the rule is evidence of the sophistication of the prevailing world concept. For our special problem, it has a further significance. For it gives insight into deep-rooted patterns of thought under which a man refuses to accept any situation as inescapable, and thus reflects the diplomacy of the New World Negro in approaching human situations that is quite comparable to the manner in which the decrees of Fate itself are in West Africa not accepted as final.

Thus far we have seen that the West African's world view comprehends Great Gods (who may be remote deified ancestors), other deities and forces, such as Fate and the divine trickster, and the ancestors who, in the other world, look after the concerns of their descendants moving on the plane of the living. Other phases of African religion will be considered shortly, but one aspect of this polytheistic system which likewise concerns the flexibility of Negro thought patterns must be discussed at this point. This has to do with the lack of interest the Africans manifest in proselytizing; which, in obverse, means that they have no zeal for their own gods so great as to exclude the acceptance of new deities. In this area

they themselves recognize this fact, and will readily give an affirmative answer to direct questions concerning the tradition of accepting new gods or, more convincingly, will of their own volition
designate certain gods as theirs and indicate other deities they worship as adopted from outside the tribe.

This tendency to adopt new gods is to be referred to the conception of the deities as forces which function intimately in the daily
life of the people. For a supernatural power, if he is to be accepted,
must justify his existence (and merit the offerings of his worshipers) by accomplishing what his devotees ask of him. He need
not be completely effective, for errors in cult practice can always be
referred to in explaining why on occasion the prayers of worshipers
are not fulfilled. But the gods must as a minimum care for the well-
being of their people, and protect them not only from the forces of
nature but also from human enemies. If one tribe is conquered by
another, it therefore follows that the gods of the conquerors are
more powerful than those of the conquered, and all considerations
dictate that the deities of this folk be added to the less powerful gods
already worshiped. Yet this is not the entire story, for an autoch-
thonous god, if not propitiated, may still turn his considerable powers
against the conquerors and do them harm. Therefore political ferment in West Africa was something correlated with religious ferment, and brought about an interchange of deities which tended to
give to the tribes in this part of Africa the gods of their neighbors.

The relevance of this for the situation to be met with in New
World Negro cultures is apparent, for it sanctions a conception
of the relationship between comparative power of gods and the
strength of those who worship them. In these terms, the importance
of the European's God to people enslaved by those who worshiped
Him must have been self-evident. That this was actually the case
is to be seen in those parts of the New World where opportunities
have presented themselves to retain African gods despite contact
with Europeans; it will be seen how in such countries, especially
where Catholicism prevails, the resulting syncretisms furnish one
of the most arresting aspects of Negro religious life. In Protestant
countries, especially the United States, where retention of the African gods was made difficult if not impossible, this attitude likewise
goes far toward explaining the readiness of the Negroes to take
over the conceptions of the universe held by the white man; and
this points the way, also, to an understanding how, though forms
of worship may have been accepted, not all of African world view
or ritual practice was lost.

Magic is extremely important in all our area; as has been seen, the ubiquity of the magic charm is such that the term "fetishism" has come to be applied to all West African religion, with its other resources ignored in favor of this most immediate—and most apparent—technique of coping with the supernatural. Magic is easy to understand; it is not foreign to European belief, and, in its African form, is so specific in its operation that it can be readily explained by the native to an untrained inquirer. That its underlying philosophy is not so simple, and its relationship to the other forces of the universe still more obscure, is another matter. Its outward manifestations, to be encountered everywhere, are the charms people wear on arm or leg or about the neck, or that they suspend from their houses or insert in carved figurines or in the very shrines of their gods. The principle of "like to like" operates here as elsewhere, and the knowledge of how to manipulate the specific powers that reside in specific charms is widespread. There are, of course, specialists who deal in charms, but many laymen also know enough about these matters to make charms that are entirely adequate for a given purpose.

As has been indicated, the outstanding trait of the charm is its specific reference. Characteristically, a charm has certain taboos which its owner or wearer must observe lest it lose its power, while its ownership entails certain definite prescribed actions which must be carried out if it is to retain its force. Charms help in meeting every situation in life and magic has its place even in the worship of the gods themselves. It is customary to classify magic as good and bad, but whatever dichotomy obtains is not the kind ordinarily thought of, for good and bad are conceived as but the two sides of a single shield. A charm, that is, which protects its owner can bring harm to an attacker; thus a charm to cure smallpox turned out to be a virulent instrument of black magic which could kill—by giving a man the same disease.

From this fact we gain further insight into African patterns of thought, for here we encounter a refinement of concept in terms of a hardheaded realism that is as far removed as can be from that simplicity held to mark the mentality of "savages." For it is realistic, not naïve, to refuse to evaluate life in those terms of good and bad, white and black, desirable and undesirable that the European is so prone to employ in responding to an equally deep-seated pattern of his own manner of thinking. The African, rather, recognizes the fact that in reality there is no absolute good and no absolute evil, but that nothing can exert an influence for good without at the very least causing inconvenience elsewhere; that nothing is so evil that it cannot be found to have worked benefit to someone. The concepts of good and bad thus become relative, not absolute, and in understanding the magic of West Africa from which New World Negro magic has derived, we can the better understand why, of all Africanisms, this element of belief has most persisted in the mores of Negro life everywhere in the New World.

What of the fear so often indicated as the outstanding aspect of the Negro's reaction to the universe—especially his fear of the magic forces he must constantly contend against? Such an assertion runs quite contrary to the findings of students who have succeeded in peering beneath the surface of West African life. Religion is close to the everyday experience of the West African. Supernatural forces are potentially dangerous, it is true, but so are wild animals or illness. An analogy can be drawn in terms of our own reaction to electricity and automobiles. For those who work with either or benefit from the use of either, the potential dangers—of shock or of accident—are considerable. Yet, if we are normal, we do not set up phobias which preoccupy our waking moments and torment our sleep with nightmares concerning electricity and automobiles. For if these are dangerous, they are also helpful; if they can harm us when not handled properly, their proper use is beneficial. So with the West African's gods, and so with his magic. What can potentially harm, if not handled properly, can also be of the greatest aid; and just as we have specialists who see to it that our electrical devices are properly insulated and our automobiles are in proper working order, so in West Africa priests and diviners and dealers in magic charms are likewise on hand to exercise the proper controls.

Religion, in short, is important in the life of West Africa because it is an intimate part of that life. If it is difficult for us to comprehend such a point of view, this merely means that the institution in our culture which we label by the term "religion" does not, in the case of vast numbers of our people, enter into considerations of everyday living. It thus follows that what we designate by the term is not the same reality as what is similarly designated in the case of these folk. Just because the supernatural does function intimately in the daily life of West Africa, because the powers of the universe are of passionate interest to these West Africans, it does not follow that they have no time for other thoughts or that their emotional life is centered about fear of a universe which is held by outsiders to be far more hostile to them than they themselves regard it.

As might be expected among people whose world view is so complex, a rich mythology is encountered. These myths, however, are only one part of the literary repertory of the folk living in this core of the slaving belt, since "historical" tales, stories for children, and other types are likewise of great number. The popularity of the Uncle Remus stories in this country, and the circulation of Joel Chandler Harris's volumes of Negro tales[25] over the entire world, have caused these American Negro stories to be regarded as the characteristic form of African folk tales. In Africa, however, even where animal tales are told, they are neither naïve nor necessarily for children. Many elements of the Uncle Remus stories are encountered in the sacred myths, and these elements, even where the animal personnel has been retained, are handled in a subtle and sophisticated manner. They often exhibit a *double-entendre* that permits them to be employed as moralizing tales for children or as stories enjoyed by adults for their obscenity. In addition to the tales are numerous proverbs and riddles, the former in particular being used at every possible opportunity to make a point in an argument, or to document an assertion, or to drive home an admonition. Poetry is likewise not lacking, though poetic quality derives principally from a rich imagery; the association of poetry with song, moreover, is so intimate that it is not found as an independent form.

Aesthetic expression is profuse in other fields. The outstanding musical form of these folk is the song, though musical instruments are found—the ubiquitous drum in its many forms, the gong, rattles, and types of zithers and flutes. The musical bow, the sanza, or "African piano," and other instruments that have a distribution elsewhere on the continent are absent from the part of the west coast with which we are at present concerned. Though only one collection of songs of any size has been made in this region,[26] the four hundred and more recordings not only indicate that many different kinds of songs are to be encountered, but that an equally wide range of singing styles exists. If it does nothing else, indeed, this collection shows the impossibility of comprehending "African" music under a single rubric or even of considering the songs of one tribal group as constituting a single describable type. The significance of this fact for the problem of New World Negro music will be probed later in our discussion; here it is sufficient to indicate the complexity of West African musical forms with respect to scale, rhythm, and general organization, and to mention the many varieties of songs—ranging from lullabies through work songs, and songs of derision, and social dance songs, to sacred melodies as varied as are the individual deities to whom they are directed—that are found not alone in this region as a whole, but in the musical resources of any one of its tribal units.

Nor is it possible here to do more than make mention of the dance, also a fundamental element in aesthetic expression everywhere in Africa. Dancing takes multitudinous forms, and all who have had firsthand contact with the area of our special interest speak of the many varieties of dances found there. These may be ritual or recreational, religious or secular, fixed or improvised, and the dance itself has in characteristic form carried over into the New World to a greater degree than almost any other trait of African culture. To attempt verbal descriptions of dance types requires a technique as yet scarcely developed;[27] since analysis must also await the utilization of motion pictures as an aid to the study of these special aspects of motor behavior, we can here but record the fact of its prominence in the culture, and its pervasiveness in the life of the people.

Great competence in a variety of media characterizes the graphic and plastic arts. Wood carving is the best known of African arts, though among the Ashanti other techniques take prior rank. These people are supremely competent weavers (as is to be seen in the discussion of their silk and cotton cloth designs by Rattray[28]), and they are also famous for the metal gold weights they cast from bronze, accurate to the fraction of an ounce and fashioned in a wide range of representative and geometric figures. In Dahomey, the high degree of economic specialization permits art to find expression in numerous forms. Wood carving, not as well known as it deserves among devotees of what, in art circles, is called "African sculpture," reaches a high degree of perfection. Stylistically, these carvings are especially interesting because of strength of line and balanced proportions which characterize the statuettes found in shrines of the gods or otherwise employed in the cult life of these people. Brass castings are made by a family guild which is differentiated from other metalworkers. These figurines, resembling our own art objects in that they have nonutilitarian value after a fashion not often encountered among primitive peoples, are prized essentially for the aesthetic pleasure they give and as a mark of leisure-class status, since only the wealthy can today afford them and since, in the days of Dahomean autonomy, to own them was a prerogative of royalty. Clothworkers make distinctive appliquéd hangings which, in the manner of the brass figures, are valued for their beauty alone.

The wood carvings of the Yoruba are known much more widely, and the Yoruban area has long been recognized as one of the principal centers of this form. Not only does one encounter single three-dimensional figures of considerable size, but also "masks," as the representations of human and other heads worn atop the heads of dancers are termed, bas-relief carving on doors, houseposts with human and animal forms superimposed one on the other, objects used in the Fate cult, and the like. In addition, however, these people, like the Dahomeans, do ironwork of distinction, weave cloth of cotton and raffia, and produce minor art forms in basketry, pottery, and other media.

4

A final point on the African background must be considered at this time. This concerns the extent to which the cultures that have been described—those, that is, of this focal area of slaving, which research has empirically demonstrated to have set the pattern followed by survivals of African custom in New World Negro life—differ from or are similar to the cultures of those other portions of West Africa which also exported large numbers of Negroes to the United States, the Caribbean, and South America.

That our information on the peoples of the slaving belt outside the area with which we have thus far been dealing is not as extensive as that which we have to draw on from within the area is quite true, though this does not mean that we are by any means exclusively left with gleanings from the writings of those who, as missionaries or government officials or travelers or traders, had other than scientific concerns. The scientific periodical literature makes numerous contributions of high competence available.[29] Only here as yet can we find reports of the recent field work among the Nupe of northern Nigeria by Nadel,[30] among the Tallensi of the Gold Coast by Fortes,[31] among the Dogon of the French Sudan by the various field parties of the Musée de l'Homme,[32] among the Niger Delta folk by Forde and Harris,[33] or, similarly, the materials gathered by those who have studied various Congo tribes.[34] Some monographic literature is available. The work of the French, especially of Labouret, on the tribes of Senegal and the interior of French West Africa,[35] of Thomas and Westermann,[36] and of earlier German missionaries[37] on the folk of the Liberia and Sierra Leone, and of Tauxier on the Ivory Coast tribes (of whom the Agni are of especial importance)[38] all fall in this category. On the eastern side of the "focal" region, the works of Meek on the Ibo as well as on various folk of the Nigerian hinterland[39] are to be remarked, together with the volumes by Talbot, both those derived from the Nigerian Census of 1920 on the peoples of the forested coastal belt and the descriptions more specifically directed toward the cultures of the Niger Delta region.[40] The earlier reports of Thomas, dealing with the same area,[41] and Mansfeld's account of the tribes of the Cross River region[42] give further resources. Such German works as those of Tessmann[43] and other German writers on the Cameroons folk can be consulted with profit in order to fill out the picture.

Deficiencies are greatest for Congo ethnography. In a general way, the outlines of Congo custom are known, but the poor quality of the reporting, especially the fact that except perhaps for the studies published by Torday and Joyce[44] on the Kasai river tribes or by Hambly on the Ovimbundu[45] there are no field data gathered purely for scientific ends, places great difficulties in our way when we search for detail. A long series of volumes was published some years ago in Brussels as a part of an ambitious plan, devised by Cyr. van Overbergh,[46] whereby political officers made returns on the basis of a rigid outline, thus allowing possible direct comparisons between the peoples reported on. Questionnaire ethnography of this type is, however, unacceptable in terms of modern ethnographic method. Not only are the facts gathered by untrained observers, but the procedure rules out any consideration of the place of the individual in his culture, and reduces civilizations to systems of institutions that give no sense of the variation about these cultural norms inevitably encountered in the life of any group. The writings of Weeks, an English missionary, can be used, but all caution must be allowed for obvious bias.[47] Moreover, like works by administrators interested in the natives,[48] they handicap the student because the data are not presented in terms of the rubrics generally accepted as representing the aspects of culture to be treated in any systematic description.

Nonetheless, the available resources are quite sufficient to establish the two major hypotheses on which the position taken in our discussion is based. In the first place. the data demonstrate the validity of our reasoning as to the relatively greater effectiveness of the "focal" cultures as against these "outlying" ones in establishing the patterns of New World Negro behavior. And they also demonstrate a sufficient degree of similarity in the cultures of the entire area so that a slave from any part of it would find little difficulty in adapting himself to whatever specific forms of African behavior he might encounter in the New World.

Language offers an excellent opportunity to document this latter point, which, as will become apparent in succeeding chapters, is the crux of the matter. It will be remembered that general opinion holds the destruction of aboriginal linguistic tradition to have begun as soon as the slaves arrived in the New World. Separated from fellow tribesmen on the plantations as a matter of policy, the slave, it is argued, had no means of communication with his fellows except in the language of the masters, and hence no linguistic vehicle was at hand to establish in the New World customs known in the Old. With this hypothesis, which emphasizes the linguistic diversity of Africa, in mind, we may consider the linguistic situation as it actually exists. The best summary is a short work published in 1930, the text of a series of lectures delivered to British Colonial Office probationers at Oxford, Cambridge, and London universities.[49] A simple introduction to the outlines of the languages with which these officials must cope with in Africa, it is nontechnical and succinct. The standing of its author, a distinguished scholar in the field, guarantees its authority, while the fact that it was written with no thought of the New World Negro makes its testimony as regards findings in the Western Hemisphere the more impressive.

The discussion opens with a statement concerning the principal types of African languages:

There are, we may say, three families of languages indigenous to Africa. . . . These are, the *Sudanic, Bantu,* and *Hamitic.* . . . The *Sudanic* Languages . . . constitute an organic family, extending in an irregular zone across Africa, from Cape Verde to the Highlands of Abyssinia. . . . Some typical *Sudanic* languages are: Twi, Ewe, Yoruba, in West Africa; . . . The name *Bantu* was adopted . . . to denote those languages of South and Central Africa which had been discovered to resemble one another so closely in structure as to constitute a singularly homogeneous family. . . . Among the most important Bantu languages are . . . Kongo.[50]

From this it would seem that the apparent linguistic differences found between the tribes of the slaving area are in reality but local variations of a deeper-lying structural similarity. Such mutual unintelligibility as existed is thus to be regarded as irrelevant to the basic patterns which, under contact, afforded a grammatical matrix to facilitate communication.

It is to be noted that the "typical" Sudanic forms of West Africa mentioned in this citation—Twi, the language of the Ashanti-Fanti people of the Gold Coast, Ewe, the name given the Togoland languages closely related in form and even in vocabulary to Fon, the speech of Dahomey, and Yoruba—are the principal linguistic stocks of our "focal" area. This means that the slaves who came from out-side this focus spoke tongues related to those found at the center of slaving operations. Among the more important of these found in regions to the west of the "core" are the languages of the Gambia and Senegal (Wolof or Jolof), Sierra Leone (Temne and Mende), and the middle Sudan (Mandingo) ; to the east are Ibo, Nupe, and Efik. To the north of the forested coastal belt Sudanic dialects also are spoken—Mossi, Jukun, and Kanuri among others.[51]

The Congo tribes are all Bantu speaking, and though there are considerable differences between the Sudanic and Bantu stocks, re-semblances also exist which, under mutual contact with Indo-European tongues, would loom large. The system of classifying forms which is the primary mark of the Bantu languages could not, in any case, be carried over into Indo-European speech, but other traits, such as the absence of sex gender, and those "vocal images," "ono-matopoetic words," and "descriptive adverbs," noted as of equal importance in the Sudanic and Bantu languages[52] could readily be employed by English-, French-, Spanish-, and Portuguese-speaking New World Negroes, whatever their African linguistic background.

We need not here document reservations to the conclusions reached by students of Negro speech as concerns African survivals in the United States.[53] Let us but point out how the problem of African survivals is affected by the existence of similarities and differences in underlying pattern that characterize the tribes in all the region where important slaving operations were carried on. Naturally, if each tribelet was linguistically quite independent, this would have made communication in the New World a matter of the utmost difficulty for the slaves, who would have been far more dependent than otherwise on the entirely new language that had to be learned. But if mutually unintelligible dialects were not under-standable because of differences in vocabulary rather than in con-struction, mutual understanding after a relatively short period of contact would be a simple matter. Analyses of various New World Negro forms of speech, as well as of West African and Congo "pidgin" dialects, show how importantly this common structural base functioned. For whether Negro speech employs English or French or Spanish or Portuguese vocabulary, the identical construc-tions found over all the New World can only be regarded as a re-flection of the underlying similarities in grammar and idiom, which, in turn, are common to the West African Sudanese tongues. And this, again, made it possible for men and women of different tribes to communicate with one another as soon as they had learned a modicum of the master's language, with a facility never recognized in the many discussions of the loss of African background that have been cited in the preceding pages.

If one may compare similarities in the grammar of language over the entire West African region with what may be termed the grammar of culture, one finds a similar situation. One indication of this is the tendency of all students to consider the west coast a unit, and of some to group it with the Congo in comparing its cultures with those of the north or east or southwestern parts of the African continent.[54] As in all culture-area analyses, a classification of this type entails an evaluation of the degree of differences to be recognized as significant. Thus, as has been said, certain very general aspects of the cultures of Europe, Asia and Africa are held in common in an Old World cultural province. In like manner, certain traits, such as the counting of descent in a unilateral line with strong emotional attachment to the families of both parents or the fact that ancestors function importantly among the supernatural forces of the universe, are found over most of the continent. Beyond this, however, are those characteristics which mark off the cultures of one part of the continent from those of other areas, while in each area local variations on the central themes are found, the local cultures becoming more and more specialized until one reaches the ultimate fact of individual variation in behavior.

What, then, are the characteristics of this West African-Congo area? To what extent do they agree with those given in the outline of the cultures found in the focal area of slaving? In all West Africa south of the Sahara, and in the Congo, agriculture is the mainstay of the productive economy, though in the northern savanna country herding is also important. In all the area gardening is done with the hoe, the heavy agricultural work being performed by the men, the crops being tended by the women. Cooperative labor is everywhere used to break the soil. Ownership of land is regulated by the larger relationship groups, but tenure during use is assured the occupant of a given plot of ground. In addition to the basic agricultural organizations are various craft groupings, which reflect a division of labor that makes for specialization in various callings—ironworkers, cloth weavers, wood carvers, traders, dealers in objects of supernatural moment, potters, basketmakers. These specialists commonly acknowledge affiliation to family guilds, which are ever present, though in some of the communities they are less closely organized than in others.

Social organization is unilineal and patrilocal. Polygyny exists everywhere in the area, and though the line of descent varies, the closeness of personal ties with the parent to whom one is "unrelated" is in accord with African custom elsewhere. It is becoming increasingly evident that division of social units based on kinship in terms of immediate family, extended family, and sib is widespread over the African continent; it is similarly found in those parts with which we are now concerned. In all West Africa, also, the rule of the elders within the larger family is paramount. Their power is based on the closeness of their relationship to the ancestors, who give them their authority. The rule of discipline enforced within the family as previously described likewise holds, which accounts for the efficiency with which these groupings exercise economic and political controls. Whether or not sibs are totemic in the entire area is for future research to decide. The question is open as concerns the western part of the slaving belt; in the central portion the incidence is varied; in eastern Nigeria it is present in some tribes and absent in others, and there is evidence that this is also the case in the Cameroons and the Congo.

Variation in nonrelationship groupings is considerable. "Secret societies" among the Yoruba and Dahomeans have proved on closer investigation to be either religious cult organizations or family aggregates. Secret societies seem to be lacking in any form among the Ashanti-Fanti peoples; in all this central portion of the slaving region "associations" thus take the form of work groups, insurance societies, mutual-aid organizations, and the like. Secret societies do flourish at both ends of the belt, however. The Leopard Societies of the Congo, various Ibo and Ibibio secret organizations, the Poro and Sande of Liberia and Sierra Leone suggest that similar societies, exercising political power as well as enforcing conformity to the mores by extra-legal methods, may have existed in all the western part of the continent before the dynastic controls of the more closely organized political entities were established. Such organizations may today be regarded as merely specialized manifestations of the underlying pattern of directed activity, which makes for the presence of many kinds of associations, having secret or known membership, in all parts of the area.

Though everywhere in the region of slaving operations the local unit is dominant and loyalties are toward such units, large variation is found in political organization. The power of the various kingdoms which existed rested always in their ability to mobilize the support of the local chieftains who, by negotiation or conquest, had been brought under control of the central power. In the northwestern portion of the slaving belt, among the Bambara and inland among the Wolof, kingdoms of some size had long been established when the period of slaving began, while farther to the east the Fulani kingdom was likewise of impressive dimensions. In Sierra Leone, Mandingo control has long been known, but in the rest of this territory and in Liberia and the Ivory Coast small autonomous units were the rule. Between the kingdoms of Ashanti and Dahomey numerous minute independent entities existed, while the Yoruba, who constitute a cultural unit, were divided into at least ten political subdivisions.

As we move eastward, the size of these units becomes smaller, so that as the Niger is reached a cluster of villages becomes the characteristic self-governing form. In this region Benin alone, noted for its priest-kings, constitutes the exception to this rule. Large kingdoms were not numerous in the Congo, though tightly knit political structures existed everywhere. The kingdom of Kongo, which was functioning when the Portuguese made their appearance in the fifteenth century and of which we know much through the writings of early travelers, was never of impressive size. Inland, the Bushongo and Lunda dynasties are to be cited; but again, the pattern of the local unit as the one on which all larger political structures were reared is apparent in their organization.

Yet whatever the size of the unit, in all this vast area the people looked to their "king" for direction, and everywhere his rule, and the counsel of his elders, assured the reign of law. The "legal genius of the African," so often mentioned in works dealing with the continent—and almost entirely disregarded by students in the United States who have attempted to describe African societies—is nowhere more manifest than in the universality of the institution of courts and the manner in which native courts functioned. Indeed, it is difficult to find such a congeries of societies anywhere in the world, literate or not, who are farther removed from the fang-and-claw concept of savage justice than those of the slaving area of Africa.

The general outline of religious life that has been given for the core of our area in the main applies to the other cultures of the total region. Everywhere some conception of the universe as ruled by Great Gods, customarily associated with the forces of nature, is found. The pervasiveness of divination would indicate a world view that implies beings whose decisions can be ascertained, thus making it possible to carry on activities in harmony with their desires by proper manipulation of the accepted tribal techniques of foretelling the future. Everywhere the ancestors are sacred. They may, in some regions, be regarded as the real owners of the land; they may be looked to as the possessors of such peculiar powers or abilities as families and sibs may be endowed with. But they are always the stabilizing force in the organization of society and are unfailingly consulted before important decisions are reached. They are, in short, respected and worshiped as those who, constituting the interested intermediaries between this world and the next, can most affect the fortunes of their descendants. Magic is likewise universal in the area. Charms themselves are as different as the varied situations of life, but the use of certain materials in their manufacture, such as pointed objects, or colored cloth, or white clay, or spines and strong hairs, is encountered over all the region.

Cult practices differ greatly. They vary from organized groups of worshipers with well-executed rituals by disciplined corps of singers and dancers in the larger population aggregates of the Gold Coast, Dahomey, Benin, and among the Yoruba to simple family rituals for gods and ancestors which, in the smaller communities, comprise almost the only type of ceremonials. The names of deities are as numerous as the localities with which they are identified; it is from this fact that the most reliable testimony of the origin of New World Negro groups derives. For despite the multitude of designations for the great numbers of gods that must have been worshiped by the varied tribes from which came the slave population, few deities except those from the central region have present-day devotees on this side of the Atlantic. Zambi, Simbi, Bumba, Lemba, who are worshiped in the Congo, are exceptions to this rule, but there are few others. It is possible that greater knowledge of the deities of other portions of the slaving belt will reveal survivals hitherto unrecognized; yet we know enough about the gods of peoples outside this "core" to be struck by the paucity of correspondences to them found in the New World, especially when this is compared to the wealth of carry-overs of Ashanti, Dahomean, and Yoruban supernatural beings.

The aesthetic aspects of life in the slaving region present an underlying unity, whatever the variations of local styles. Song and dance are everywhere found to play significant and similar roles in the daily round. The rattle, the drums, and the gong are always found in the battery of instruments employed, though in the Congo the sanza, the xylophone, and elsewhere certain string devices supplement the percussion units. Rhythm is invariably complex, and the convention of alternation of leader and chorus in singing likewise the rule. The more technical musicological problems in the study of similarities and differences over the area cannot be discussed for lack of data. Yet, again, enough is known to justify the conclusion that in musical style and rhythmic treatment—to say nothing of the sociological problem of the cultural setting of the music—fundamental structure is everywhere similar.

Over the entire area the graphic and plastic arts are of great importance. Indeed, the region of slaving includes most of those parts of the continent that have become famous for their art. The importance and quality of wood carving in the Congo, the Cameroons, eastern and western Nigeria, among the tribes living north of the coastal forested belt, the Ivory Coast, and Liberia are too well known to require more than mention here. The development of art forms in other media is similarly important, though less well known; along the vast stretch from the Gambia to the Congo one finds the techniques of ironworking, cloth weaving, basketry, beadwork, silver- and goldsmithing, and calabash decorating employed in the production of beautiful objects.

The aesthetic drive is equally manifested in the literary field, though in folklore, as in religion, the closeness of these interests to the concerns of everyday life makes their aesthetic aspects appreciable only on close acquaintance. Tales, proverbs, and riddles are the three major forms of this art and they function constantly and variously as educational devices, as a means of amusement, to make a case in court, to point a conversation, as sanctions for social institutions and world view, and as integral elements in funeral rites. Widespread are both animal stories and tales involving human and supernatural characters, and the basic unity in this as in other art forms is amply apparent even though only a relatively small sampling of the artistic resources of this type are available for the entire region.

These points, which suggest the underlying similarities between the cultures of the area where slaving was carried on, could be documented almost indefinitely. From the point of view of our present interest, the greater store of data they represent proves that emphasis on tribal differences in culture has been placed by those who have written of the cultural background of Negroes living in the United States with as little justification as the presumed linguistic dissimilarities have been emphasized.

Chapter IV

ENSLAVEMENT AND THE REACTION TO SLAVE STATUS

The story of slavery is in need of much revision, for there is great variation of fact about the patterned concept of the trade and the fate of those brought to the plantations. Historical truth, as evidenced in contemporary accounts, demands a realization that all types of individuals—slaves, captains of slaving vessels, overseers, and plantation owners—were concerned in this chapter of our past, and that these individual differences bulked large in determining the total situation.

To reevaluate the evidence will require the work of specialists for some time to come; here we are only concerned with general outlines in so far as the picture has significance for the past of the New World Negro. As has been indicated, the current point of view, which emphasizes the acquiescence of the Negro to slavery, is an integral part of the "mythology" sketched in our opening pages. As such, it reinforces certain attitudes toward the Negro and is thus of practical as well as scientific importance, the latter deriving principally from the fact that this phase of the Negro past aids in understanding the rate and the nature of the acculturative process prior to the abolition of slavery. Slaves who acquiesced in their status would be more prone to accept the culture of their masters than those who rebelled; hence, if the slaves were restless, as recent studies have indicated, and if this restlessness caused revolt to be endemic in the New World, then the reluctance to accept slave status might also have encouraged the slaves to retain what they could of African custom to a greater extent than would otherwise have been the case.

Other aspects of this historical problem also call for study. In the analysis of a given acculturative process, it is important to know as much as possible of the actual precontact status of the individuals party to it. For though acculturation is essentially an attempt to understand the mechanisms and results of contact between the carriers of differing cultures—which is to say, between manifestations of two different configurations of institutionalized modes of behavior—it must always be borne in mind that the carriers themselves are the crucial elements. In the "Memorandum for the Study of Acculturation" already referred to, one of the entries under the heading "Psychological mechanisms of selection and integration of traits under acculturation," is "differential selection and acceptance of traits in accordance with sex lines, differing social strata, differing types of belief, and occupation."[1] This means that the individual backgrounds of those party to the contact must be understood as completely as possible in terms of their particular group mores and interests, social status, class affiliations, and the like. In the case of those who were party to the contacts between Negroes and whites in the New World, this task is beset with enormous difficulties, yet a determined and systematic attack on the problem has already yielded some results that are of use in its analysis.

2

It needs no great probing of the literature of slaving to become aware that, from the beginning, vast numbers of Negroes refused to accept the slave status without a struggle. Contemporary accounts are so filled with stories of uprisings and other modes of revolt, cases of voluntary starvation and more direct forms of suicide, that it is surprising that the conception of the compliant African ever developed. A committee of the House of Commons investigated the slave trade in 1790 and 1791, and its report is replete with testimony concerning the difficulties caused the traders by the Negroes. Ships had to be "fitted up with a view to prevent slaves jumping overboard"; slaves on occasion would refuse "sustenance, with a design to starve themselves"; at times they also refused "to take medicines when sick, because they wished to die." The persistent attempts of certain slaves at suicide are in themselves eloquent of their grim determination. Thus one man, sold with his family on a false accusation of witchcraft, attempted to cut his throat. The wound was sewed by the ship's surgeon, whereupon the man tore out the sutures during the night, using his fingernails since nothing else was at hand. Ten days later he finally died of starvation, after what would today be termed a hunger strike. Again, the report tells of a woman who, rescued after an attempt to drown herself, was chained to the mast for four days; she jumped into the water as soon as she was released, "was again taken up, and expired under the floggings given her in consequence."[2] A passage written at about the same time by Falconbridge, out of his firsthand experience with the trade, may also be quoted here:

As very few of the negroes can so far brook the loss of their liberty, and the hardships they endure, as to bear them with any degree of patience, they are ever upon the watch to take advantage of the least negligence in their oppressors. Insurrections are frequently the consequence; which are seldom suppressed without much bloodshed. Sometimes these are successful, and the whole ship's company is cut off. They are likewise always ready to seize every opportunity for committing some act of desperation to free themselves from their miserable state; and notwithstanding the restraints under which they are laid, they often succeed.[3]

It may be argued that the use of such sources must allow for abolitionist bias. Yet other materials, written with no political purpose in mind or even presented by supporters of the slave regime, make such an argument less impressive than it would otherwise be. The work by Captain Snelgrave, who was a believer in slavery, tells tales of slave revolt experienced by himself or witnessed at first hand that carry conviction even beyond the dramatic quality of the narrative.[4] Phillips, who was but little concerned with Negro reactions, so completely accepts the danger of revolt during the voyage as a fact that he merely remarks in passing, when describing the trade, "the negro men were usually kept shackled for the first part of the passage until the chances of mutiny and return to Africa dwindled and the captain's fears gave place to confidence."[5] A recent systematic analysis of the materials by Wish illuminates the refusal of the slaves, from the very inception of their captivity in Africa, to accept their status as bondsmen. This is demonstrated, for one thing, in the slave revolts on shipboard enumerated by him:[6]

Year	Number of revolts	Year	Number of revolts
1699	1	1733	1
1700	1	1735	1
1703	1	1737	1
1704	1	1747	1
1717	1	1750	3
1721	2	1754	1
1722	1	1759	1
1730	2	1761	2
1731	3	1764	4
1732	1	1765	2

Year	Number of revolts	Year	Number of revolts
1773	1	1797	3
1776	3	1799	2
1785	2	1804	1
1787	1	1807	1
1789	1	1808	2
1793	2	1829	1
1795	1	1839	1
1796	1	1845	1

On the basis of these findings, obtained from published materials, not derived from an analysis of archival data, and dealing almost entirely with revolts on British ships, there is little reason to doubt that more extensive research would greatly expand the list. This is also indicated by the little-known fact, brought out by Wish, that advance precautions of a pecuniary nature were taken by owners of slaving vessels against revolt:

There is evidence of a special form of insurance to cover losses arising specifically from insurrections. An insurance statement of 1776 from Rhode Island, for example, has this item: "Wresk of Mortality and Insurrection of 220 slaves, Value £9000 Ste'g at 5 per cent is Pr Month = £37,10s." A Captain's statement of August 11, 1774, contains a request for insurrection insurance. In a Negro mutiny case of May 3, 1785, the court awarded payment in conformance with a policy provision for insurrection insurance. Sometimes the captain of a slaver would throw sick Negroes overboard to profit by the insurance payments given in such contingencies.[7]

3

Slave protest on the west coast of Africa and on shipboard is thus seen to have been regarded as a commonplace. In analyzing the hypothesis of Negro subservience to slavery, however, the possibility must next be considered that this characteristic developed later when, in the New World, having to cope with the stern controls and continued vigilance of the masters, and in enforced submission to the powers of European culture, Negroes became resigned to their fate and made the best of whatever life might hold for them. The available data do not make this assumption any more persuasive than that point of view which holds for Negro acquiescence to slavery on first contact with Europeans in Africa. In the face of materials from all over the New World showing what determined resistance was offered by the slaves to their status when even the slightest opportunity afforded, it is difficult to understand how the Negro obtained any reputation for docility. This may, of course, be due in part to the outer aspects of accommodation whereby, following the patterned flexibility of African tradition, the slave told his master what he believed his master desired, and for the rest kept his counsel and bided his time until he could make good an effective protest, or escape.

It is possible, also, that the stereotype of the pliant Negro has derived from the oft-repeated story which contrasts him to the Indian, who is held to have died rather than suffer enslavement. But this assertion also needs reinvestigation, since at the present state of our knowledge there seems some reason to believe that it was more than his wounded pride or a broken heart that carried off the enslaved Indian. In Haiti, for example, the Negroes were imported because work in the mines had almost exterminated the Indians, as it likewise did the Negroes who were imported for this purpose; and it was the discovery that the cultivation of sugar was more profitable than gold mining that allowed the Negroes to do the agricultural labor that was far more conducive to survival. In the United States, again, it would seem that Indians had a lower resistance to bacterial diseases borne by Europeans than did Negroes, which permitted the sturdier Negroes to survive where the Indians died off. It is also entirely possible that the Indians were regarded as unsatisfactory slaves because their simpler aboriginal economic system gave less preparation for the disciplined regime of the plantation than did the African background.[8]

Whatever may have been the case as concerns the Indians, there was no lack of protest in the New World by Negro slaves. It began at the very earliest period of enslavement:

. . . it is not generally known how early in the history of Negro slaving revolts did occur. The Negro slave-trade began with shipments of slaves to Haiti in 1510; the first slave uprising in Haiti, in 1522, thus took place only twelve years after the commencement of the traffic. In the New World possessions of Spain eleven other rebellions are recorded between the years 1522 and 1553, of which those of 1533, 1537, and 1548 occurred in Santo Domingo. During the following century two revolts took place at Haiti, one at Port-de-Paix in 1679 and another in 1691.[9]

These sixteenth century uprisings occurred before the introduction of slavery into North America, and the others took place before the slave trade to the colonies was in significant operation. It will be seen how faithfully the example set by the slaves who, in 1522, revolted against their masters was followed, however undesignedly, by many thousands of those coming after them.

Only the outstanding slave revolts outside the United States can be indicated here, since systematic research into the problem of the "Maroon," as the runaway slave can generically be termed, is for the future. Enough is known from study of the available facts, however, to indicate the richness of the field and its potentialities in giving us perspective on the reaction of the Negro to slavery. Over the entire New World, in so far as is known, the earliest prolonged protests were in the southernmost parts of the slaving area. One of the most famous of these is the Palmares "republic." In 1650 some Brazilian slaves in the province of Pernambuco, all native Africans, fled to the near-by forest. As news spread of their escape, other Negroes joined them; as a measure of prudence this larger group moved farther into the bush. From their settlement, named Palmares, they raided the plantations for women, eventually setting up an ordered society. Slaves who escaped to them were recognized as free citizens, but those who were captured in raids continued as slaves, since they had lacked the courage to achieve their own freedom. As the population grew, subsidiary villages were established. At its height, the town of Palmares is said to have had a population of about 20,000, with a hinterland which gave a total fighting force of some 10,000 men. Because of the increasing danger to the white settlements, the Portuguese in 1696 assembled an army of nearly 7,000 men for the attack. Palmares was surrounded by a stockade, but lacked the artillery necessary for defense, and was finally taken. Most of the warriors committed suicide, and those who were captured, being deemed too dangerous to be reenslaved, were killed.[10]

To the north, the Bush Negroes of Dutch Guiana offer another example of what determined men could accomplish when faced with the prospect of a life of servitude. Slave revolts, beginning about the middle of the seventeenth century, here as in Brazil resulted in escape to the surrounding jungle where, far in the interior of the country, the refugees set up their villages. The Bush Negroes, as these escaped slaves are called, thereafter descended periodically to the coastal cultivations, raiding slave barracoons and masters' houses. Only those born in Africa—"salt-water" Negroes, as they were termed—were taken away, since it was feared that to take Creoles, or those born in the colony, would vitiate the singleness of Bush Negro purpose and dilute the African character of their customs. So serious were these depredations that at about the time of the American Revolution the Dutch engaged a considerable force of mercenaries to subdue the revolted slaves. The account of the warfare against the Negroes given by Captain Stedman pays tribute to the military ability of his opponents, who were so successful in keeping off the attacking force that few members of the expedition survived. In 1825 the Dutch government concluded a treaty of peace with the Bush Negroes whereby, in consideration of their agreement to refrain from pillaging the coastal region and to return to their masters such runaway slaves as came to them, their own freedom was guaranteed in perpetuity. Today the tribes of Bush Negroes carry on a civilization they realize to be African and insist will remain so. They are determined that the whites shall not again enslave them—they refuse to believe that this will not be attempted at some future time— and in 1930, when the Netherlands government proposed to license their guns, they were fully prepared to resist, and their chiefs and village heads returned Dutch uniforms and other insignia to Paramaribo, the capital of the colony.[11]

Many revolts occurred in the Caribbean, where the planters were so apprehensive of Negro uprisings that this feeling persists to the present among the whites resident there. The Virgin Islands, now possessions of the United States, have had their share of slave unrest. In 1733 an insurrection occurred on the island of St. John that almost achieved the death of the governor. The planters fled, but not before the revolt cost the lives of a considerable number of whites. The only fort on the island had but a small garrison. On the day planned, the Negroes who customarily brought wood concealed knives and cutlasses in their bundles and, at a prearranged signal, fell on the soldiers. As soon as they were in possession of the fort they fired a gun, whereupon the plantation slaves arose. The survivors of the initial attack embarked for Tortola and St. Thomas, and though the Danish troops in the other islands recovered the fort, the slaves were so well organized that the recapture of St. John was impossible without reinforcement. The Royal Council obtained the help of 60 men from a vessel lying in the harbor of St. Thomas, but they also were repulsed. Finally, 400 French soldiers were sent from Martinique, and these were able to isolate the revolters on the northeast side of the island, where, however, the slaves held off this superior force for six months. When at last they were defeated, 300 Negroes threw themselves from a precipice, while the seven leaders shot each other.[12] Another large revolt occurred on the island of St. Croix, in this group, in 1759.[13] This tradition of rebellion was not forgotten by the whites when, almost a century later, the slaves became so threatening in their protest against the proposal to enforce a twelve-year period of probation prior to emancipation that the governor, on July 3, 1848, decreed an immediate end to slavery.[14]

The slave revolts in Haiti that culminated in the establishment of the present "Black Republic," whose independence has now been maintained for almost a century and a half, have been recounted so often that it is not necessary to tell the tale here. Less well understood is the role in the history of that country of the constant smaller uprisings that laid the groundwork for the final thrusts which drove the whites from the island. The initial revolts have already been enumerated;[15] that the potential power of the slaves was early recognized is to be seen from a report made to the Minister of Colonies in Paris which, dated 1685, states, "In the Negroes we possess a formidable domestic enemy." A hundred years later, an army officer, also a plantation owner, wrote: "A colony of slaves is a city under constant threat of assault; there one walks on barrels of powder." *Marronage,* as running away was termed, was successful enough so that these escaped slaves had their freedom formally recognized in 1784. In 1720 alone a thousand Negroes made off, and in 1751 at least three times this number. The name of Macandal, a "Guinea" Negro, who was the leader of one of these bands, has gone down into Haitian lore. Moreau de St. Méry recounts the strength of runaway groups living in the mountains behind the great central plain of the island; he also tells of a group in the south who, when finally subdued, was found to number among its members men of fifty years and more who had been born in the freedom of their retreat.[16]

Slave uprisings occurred everywhere in the British West Indies. On the island of St. Vincent, Negroes joined the aboriginal Carib Indians in action against the masters. They were eventually defeated and transported to the mainland, where today in British Honduras their descendants still live. Known as Black Caribs, this unstudied people constitutes one of the strategic points for future attack on New World Negro acculturation, since they represent an Indian-African amalgam that should establish a further control in the historical laboratory where this problem is to be studied.[17] In Trinidad a Dahomean named Daaga, enslaved by the Portuguese but released by the British contraband control, joined the West India Regiment with the purpose of eventual revolt. He was aided by the other Dahomeans he met in the island, and members of African tribes bordering on his own. His planned uprising was only put down with a considerable loss of life to the blacks.[18]

The treatment of the slaves in Barbados was notably harsh, and it is not surprising to read that there were "frequent slave revolts and projected (alas!—one feels inclined to exclaim—seldom accomplished) massacres of the whites." An initial rising in 1649 "was abortive, for as usual one tender-hearted negro could not bear to think of his white master (a judge) being murdered, so revealed the plot in time for measures of repression to be taken." Twenty-five years later another revolt was planned, this time under the leadership of "the warlike Kromanti slaves"—members of the Ashanti-Fanti tribes of the Gold Coast—but this uprising was likewise betrayed and suppressed. The same story was repeated in 1692, and again in 1702. During the latter part of the seventeenth century slaves are said to have taken canoes to escape to the French islands or to find refuge among the Caribs, perhaps of Trinidad. Finally, in 1815, a free mulatto named Washington Franklin began to circulate among the slaves the abolitionist speeches made in Britain, also telling them of the success achieved by the Negroes in Haiti. The uprising that came the following year took a severe toll before it, too, was put down "with great loss of life to the negroes," the rebels being deported to British Honduras.[19]

The history of slave revolt in Jamaica is a long one. The story involves successful rebellion, and a mass return to Sierra Leone in Africa by way of Nova Scotia, where descendants of the revolters live to the present time. Some of the Maroons, as these revolted Jamaican Negroes are termed, elected to remain in their Cockpit Hill country, and here their descendants are to be found to the present day.[20] Their separate corporate entity is recognized by the British government under terms of a treaty of peace signed at the conclusion of fighting between these escaped slaves and the government forces, whereby they also live untaxed by the central government, selecting their own headmen and having the right to hold their own courts and compel obedience to their own laws.

Other islands, French as well as British, provide further instances. The data concerning the revolts in most of these islands have never been published, so that such materials exist only in manuscript form; such hints of revolt and other forms of protest as are come upon in the literature indicate how rich this vein may prove to be.[21] Cuba, likewise, affords a fruitful field for future study. Mrs. Frederika Bremer, a Scandinavian traveler in the southern United States and Cuba shortly before the Civil War, gives us one of the few contemporary descriptions of slavery in that island. She tells of the many difficulties she encountered in obtaining permission to witness the religious rites carried on by the colonies of free Negroes, since "the government is very suspicious of strangers":

> The slave disturbances of 1846 are still fresh in the minds of people, and they originated in this part of the island. These disturbances, which gave rise to such cruel proceedings on the part of the Spanish government have also caused severe restrictions to be laid upon the occupations and amusements of the free negroes. Formerly, it is said, might be heard every evening and night, both afar and near, the joyous sound of the African drum, as it was beaten at the negro dances. When, however, it was discovered that these dancing assemblies had been made use of for the organization of the disturbances which afterward took place, their liberty became very much circumscribed.[22]

According to this same observer, resistance to slavery in Cuba by means of suicide and the use of magic was common:

> When the negroes become accustomed to the labor and life of the plantation, it seems to agree with them; but during the first years, when they are brought here free and wild from Africa, it is very hard to them, and many seek to free themselves from slavery by suicide. This is frequently the case among the Luccomees, who appear to be among the noblest tribes of Africa, and it is not long since eleven Luccomees were found hanging from the branches of a guasima tree. . . . They had each one bound his breakfast in a girdle around him; for the African believes that such as die here immediately rise again to new life in their native land. Many female slaves, therefore, will lay upon the corpse of the self-murdered the kerchief, or the head-gear, which she most admires, in the belief that it will thus be conveyed to those who are dear to her in the mother-country, and will bear them a salutation from her. The corpse of a suicide-slave has been seen covered with hundreds of such tokens.[23]

4

The reaction of the slaves to slavery in the United States has been given serious attention only in recent years. Most earlier historians took it for granted that the slaves were merely passive elements in the historical scene. The political history of the period of slavery could understandably be written without considering the Negroes, though the influence of potential revolts on policy might perhaps have been profitably taken into account. Disregard of the Negro in the field of social and economic history, and where the history of ideas is under consideration, is more serious, since here the influence of the Negroes was an immediate factor. Conventionally, however, ideas concerning slavery that are treated are those of pro- and anti-slavery whites; the social institutions of the pre-Civil War South analyzed are the institutions of the whites; the economics of slavery consists largely of prices of slaves and the productivity of the Negro in various employments. Even such outstanding social historians as Charles and Mary Beard make no mention of slave revolts in their principal work; the forces in shaping the trend of events in the United States before 1860, at least, are for them to be understood in terms of white thought and white action.[24]

This approach, however, seems to be slowly giving way to a different tradition in historical scholarship. The following passage from a recent economic history is suggestive of the wider perspectives of more recent research:

Slave Conspiracies. The constant fear of slave rebellion made life in the South a nightmare, especially in regions where conspiracies were of frequent occurrence. . . . In Colonial days there had been several uprisings where white people lost their lives.[25]

This passage is followed by accounts of the New York uprising of 1712, of the South Carolina rebellions of 1720 and 1739, of Gabriel's insurrection of 1800, and of three nineteenth century revolts. Or, as another instance, the reaction of the South toward the slave during the thirties and forties of the past century as summarized by Fish may be taken:

Nor was the fear of property loss the only or the greatest of Southern apprehensions. One of the strongest points in Southern culture was its acquaintance with the elements of classical literature. To them the history of the servile wars in Rome was a familiar topic. Nor was it ancient history alone which alarmed them. Fresh in their memory were the horrors of the Negro revolution in Haiti. Toussaint L'Ouverture, who to Wendell Phillips was an apostle of liberty, was to them a demon of cruelty. How far the Negroes who surrounded them, who cooked their food and nursed their children, had been affected by civilization, and how far they retained the primitive savagery they were presumed to have brought from Africa, they did not learn until the Civil War.[26]

In an even more recent textbook of American history, consideration of the slave regime is oriented to include the "numerous insurrections" that "bear witness to maladjustments among the slaves," specific mention being made of nine of these revolts; while a recent history of North Carolina before the Civil War devotes an entire section to this aspect of the past of that state.[27]

The most systematic study of slave uprisings in the United States, however, is to be found in certain papers published during the past five years, wherein the source materials have been reinvestigated with the problem of the Negro's reaction to slavery as the primary objective. Wish, in the paper already referred to, and Aptheker, in a study published at about the same time, demonstrated how often slave dissatisfaction was translated into active revolt, and how accurate such a description of ever-present fear on the part of the whites just given in the passage from the work of Fish may be regarded.[28] The mere number of these attempts as given in these contributions is impressive, especially when it is remembered that news of a slave uprising was usually not published unless it was of some magnitude. Aptheker, in his most recent publication, tells of the first of these revolts in continental North America in the following passage:

The first settlement within the present borders of the United States to contain Negro slaves was the victim of the first slave revolt. A Spanish colonizer, Lucas Vasquez de Ayllon, in the summer of 1526, founded a town near the mouth of the Pedee river in what is now South Carolina. The community consisted of five hundred Spaniards and one hundred Negro slaves. Trouble soon beset the colony. Illness caused numerous deaths, carrying off, in October, Ayllon himself. The Indians grew more hostile and dangerous. Finally, probably in November, the slaves rebelled, killed several of their masters, and escaped to the Indians. This was a fatal blow and the remaining colonists—but one hundred and fifty souls—returned to Haiti in December, 1526.[29]

This was but an eddy in the main current of American history, but it was a portent of things to come. Six uprisings in continental United States are listed by this author for the period between 1663 and 1700, fifty during the eighteenth century, and fifty-three between 1800 and 1864.[30]

Wish has described, with rich documentation, the panic that swept over the South in 1856:

In the fall of 1856 a series of startling allegations regarding numerous slave insurrections broke the habitual reserve maintained on the topic by the Southern press. Wild rumors of an all-embracing slave plot extending from Delaware to Texas, with execution set for Christmas day, spread through the South. Tales were yet unforgotten of Gabriel's "army" attempting to march on Richmond in 1800, of Denmark Vesey's elaborate designs upon Charleston in 1822, of Nat Turner's bloody insurrection at Southampton, Virginia, in 1831, and of the various other plots and outbreaks that characterized American slavery since the days of the early ship mutinies. Silence in the press could not stem the recurrent fears of insurrection transmitted by the effective "grapevine" intelligence of the South.[31]

Fear was translated into action in various sections—as, for example, in the organization of special vigilante bands in Texas—as one discovery followed another. In Tennessee and Kentucky actual plots were exposed, Missouri and Arkansas reported projected uprisings, Mississippi, Florida, Georgia, Maryland, and Virginia experienced an increase in slave repression, while the demand grew for the enslavement of the free Negroes to wipe out a possible focus of infection. The situation is thus summarized:

Although the thesis of an all-embracing slave plot in the South shows remarkable cohesion on the whole as far as geographic and chronological circumstances are concerned, much can be explained away by a counter-thesis of a panic contagion originating in the unusual political setting of the year. It seems probable, however, that a large number of slave plots did exist in 1856. The situation in Kentucky and Tennessee particularly seemed to involve authenticated stories of proposed insurrections. It also seems apparent from the news items and editorials of the contemporary press that the year 1856 was exceptional for the large crop of individual slave crimes reported, especially those directed against the life of the master. This fact would suggest a fair amount of reality behind the accounts of slave discontent and plotting. The deep-seated feeling of insecurity characterizing the slaveholder's society evoked such mob reactions as those noted in the accounts of insurrections, imaginary and otherwise, upon any suspicion of Negro insubordination. The South, attributing the slave plots to the inspiration of Northern abolitionists, found an additional reason for the desirability of secession; while the abolitionist element of the North, crediting in full the reports of slave outbreaks, was more convinced than ever that the institution of slavery represented a moral leprosy.[32]

It is not necessary here to detail the revolts which became most famous in the South—those of Gabriel in 1800[33] and of Nat Turner in 1831, in Virginia, and the South Carolina uprising led by Denmark Vesey in 1822. The tendency to revolt was unremitting, covering all the southern states, and those northern ones as well during the period they sanctioned slavery. It was more than the sporadic and insignificant phenomenon it is sometimes dismissed as having been in passages like the following:

Insurrection was more of an anticipated danger than an actual one. As soon as the negro population became at all formidable, energetic measures were taken to prevent the possibility of revolt, and they were largely successful. Though a number of attempted or supposed conspiracies were discovered during the seventeenth and eighteenth centuries, no actual insurrection worthy of the name occurred until the nineteenth, when the rigor of slavery and slave legislation was past. Absconding and outlying servants and slaves or assemblies, incited and aided by Indians, whites—especially convicts and foreigners—and free negroes were a convenient nucleus for combined action, and for this reason restrictive and punitive legislation was especially directed toward them. In this connection developed a system of police patrol known and feared among the negroes as the "Paterollers."[34]

Small most of the revolts were, yet in their aggregate and persistence over the entire period of slaving[35] they give point to the comments made by a recent Netherlands observer of the interracial situation in America, that today one of the keys to an understanding of the South is the fear of the Negro, a legacy of slavery.[36]

5

The Negro registered his protest against slavery in other ways than by open revolt, for, where organization was not feasible, individuals could only protest as best they might. Outstanding were the methods of slowing down work, and what seems to have been calculated misuse of implements furnished the slave by his master. These latter methods, though often commented on, have almost never been recognized as modes of slave protest. They are adduced as evidence of the laziness and irresponsibility of the African, when they are not merely cited without comment as an element in the added economic cost of slavery as against a system of free employment. Once the interpretation of such behavior as sabotage is employed, however, instance after instance comes to mind where contemporary writers tell how the slaves did no more work than they were compelled to do and had to be watched incessantly even at the simplest tasks; but how, when competent in skilled trades and permitted to attain worth-while goals, such as the purchase of their own freedom, they worked well without supervision. Similarly, where slaves cultivated their own plots of ground after hours, the energy they put to such tasks is remarked on again and again. The comment of one slaveowner to Olmsted, "If I could get such hired men as you can in New York, I'd never have another nigger on my place,"[37] indicates sufficiently how great was the problem of forcing an unwilling worker to perform his stint.

Olmsted's works may be quoted further for other examples of

slave protest. In the following passage, he shows the attitudes a slave-owner had to combat:

The treatment of the mass must be reduced to a system, the ruling idea of which will be, to enable one man to force into the same channel of labor the muscles of a large number of men of various, and often conflicting wills. The chief difficulty is to overcome their great aversion to labor. They have no objection to eating, drinking, and resting, when necessary, and no general disinclination to receive instruction.[38]

"The constant misapplication and waste of labor on many of the rice plantations," he tells us in another work, "is inconceivably great." He expands his initial statement as follows:

Owing to the proverbial stupidity and dogged prejudice of the negro (but peculiar to him only as he is more carefully poisoned with ignorance than the laborer of other countries), it is exceedingly difficult to intro-duce new and improved methods of applying his labor. He always strongly objects to all new-fashioned implements; and if they are forced into his hands, he will do his best to break them, or to make them only do such work as shall compare unfavorably with what he has been accus-tomed to do without them. It is a common thing, I am told, to see a large gang of negroes, each carrying about four shovelsful of earth upon a board balanced on his head, walking slowly along on the embank-ment, so as to travel around two sides of a large field, perhaps for a mile, to fill a breach—a job which an equal number of Irishmen would accomplish, by laying planks across the field and running wheelbarrows upon them, in a tenth of the time. The clumsy iron hoe is, almost every-where, made to do the work of pick, spade, shovel, and plow. I have seen it used to dig a grave. On many plantations, a plow has never been used; the land being entirely prepared for the crop by *chopping* with the hoe, as I have described.[39]

In the case of one experience, he further documents his assertions:

On the rice plantation I have particularly described, the slaves were, I judge, treated with at least as much discretion and judicious considera-tion of economy, consistent with humane regard for their health, com-fort, and morals, as on any other in all the Slave States; yet I could not avoid observing—and I certainly took no pains to do so, nor were any special facilities afforded me to do it—repeated instances of waste and misapplication of labor which it can never be possible to guard against, when the agents of industry are slaves. Many such evidences of waste it would not be easy to specify; and others, which remain in my memory after some weeks, do not adequately account for the general impression that all I saw gave me; but there were, for instance, under my observa-tion, gates left open and bars left down, against standing orders; rails removed from fences by the negroes, as was conjectured, to kindle their fires with; mules lamed, and implements broken, by careless usage; a flat-boat carelessly secured, going adrift on the river; men ordered to cart rails for a new fence, depositing them so that a double expense of labor would be required to lay them, more than would have been needed if they had been placed, as they might almost as easily have been, by a slight exercise of forethought; men, ordered to fill up holes made by alligators or crawfish in an important embankment, discovered to have merely patched over the outside, having taken pains only to make it *appear* that they had executed their task—not having been overlooked while doing it, by a driver; men, not having performed duties that were entrusted to them, making statements which their owner was obliged to receive as sufficient excuse, though, he told us, he felt assured they were false—all going to show habitual carelessness, indolence, and mere eye-service.[40]

These passages must not be regarded as indicating any organized system of sabotage such as might conceivably be read into this and other specific evidences of willful waste found in the writings of the times. Refusal to accept the plow in place of the hoe may, indeed, have been a direct result of labor patterns brought from Africa itself, since there the hoe is the primary agricultural tool, the plow nowhere being known. Certainly the slave's ineptness cannot be laid to inherited incompetence. Free Negroes, or slaves, released for work in the towns on condition that they return to their masters a percentage of their wages, had no difficulty in successfully employing implements far newer and more complicated than the plow, or a pick or a shovel. A possible survival of an aboriginal work habit, plus what would seem to be an unformulated drive to see to it that the master profit no more than a minimum from the slave's labor, would seem to account for the wastefulness so clearly revealed. In any event, to show that under slavery the Negro was a reluctant, willfully inefficient worker is to adduce further proof that he did not exhibit that docility held to be so deep-rooted a part of his nature.

Malingering and temporary escape also were common methods of avoiding labor; Olmsted can again be called on for testimony concerning these methods:

The slave, if he is indisposed to work, and especially if he is not treated well, or does not like the master who has hired him, will sham sickness—even make himself sick or lame—that he need not work. But a more serious loss frequently arises, when the slave, thinking he is worked too hard, or being angered by punishment or unkind treatment, "getting the sulks," takes to "the swamp," and comes back when he has a mind to. Often this will not be till the year is up for which he is engaged, when he will return to his owner who, glad to find his property safe, and that it has not died in the swamp, or gone to Canada, forgets to punish him, and immediately sends him for another year to a new master.[41]

Other ways of accomplishing this same end are also indicated:

He afterwards said that his negroes never worked so hard as to tire themselves—always were lively, and ready to go off on a frolic at night. He did not think they ever did half a fair day's work. They could not be made to work hard: they never would lay out their strength freely, and it was impossible to make them do it. This is just what I have thought when I have seen slaves at work—they seem to go through the motions of labor without putting strength into them. They keep their powers in reserve for their own use at night, perhaps.[42]

How carefully and constantly watch had to be kept over slaves is suggested in this passage:

The overseer rode among them, on a horse, carrying in his hand a raw-hide whip, constantly directing and encouraging them; but, as my companion and myself, both, several times noticed, as often as he visited one end of the line of operations, the hands at the other end would discontinue their labor, until he turned to ride towards them again.[43]

It is by methods such as these that the defenseless everywhere protect themselves; and, as always, the master is helpless against passive resistance of this sort, even more than against more active forms of protest. The situation described in the citations from Olmsted could be matched with others, from both the United States and the West Indies, and helps us understand the exasperation that drove slaveowners to inexpressible cruelties. It is but the repetition of a well-worn truth to indicate that systems based on force must resort to force to make them work at all; yet it must be indicated again that to approach the matter solely from the viewpoint of the slaveowner, after the manner of most discussions of slave life under slavery, has obscured the importance of these forms of protest and has done much to establish the stereotype of the innately irresponsible and innately lazy Negro.

Suicide, infanticide, and poisoning were often resorted to as a means of avoiding slave status. Bruce speaks of "suicide among adults" as "not unknown."[44] Ball, speaking from experience, says:

Self-destruction is much more frequent among the slaves in the cotton region than is generally supposed. . . . Suicide amongst the slaves is regarded as a matter of dangerous example, and as one which it is the business and the interest of all proprietors to discountenance and prevent. All the arguments which can be derived against it are used to deter negroes from the perpetuation of it and such as take this dreadful means of freeing themselves from their miseries, are always branded in reputation after death, as the worst of criminals, and their bodies are not allowed the small portion of Christian rites which are awarded to the corpses of other slaves.[45]

It is not without interest to learn that Nat Turner, the leader of one of the most important slave revolts, was almost a victim of his mother's frantic refusal to bring another slave into the world. Herself African born, she is said "to have been so wild that at Nat's birth she had to be tied to prevent her from murdering him."[46] Bassett speaks of the task of the overseer "to see that the women were taken care of that childbirth might be attended by no serious mishap." He continues:

The ignorance of the women made it necessary to take many precautions. A large number of children died soon after being born. In many cases it was reported that the mothers lay on them in the night. How much this was due to sheer ignorance, how much to the alleged indifference of the slave women for their offspring, and how much to a desire to bring no children into the world to live under slavery it is impossible to say. Perhaps each cause contributed to the result.[47]

Modern historical research has made evident the affection of the slave mother for her children; ignorance of child care is a difficult explanation why many mothers killed their children by lying on them at night. That the third cause listed in the preceding quotation was by far the most valid would seem to be the likeliest conclusion. Poisoning is not often mentioned in the literature, but Brackett[48] writes of a number of cases of Maryland slaves who were brought before the courts charged with attempts on the lives of their masters.

Running away, another form of protest, is well recognized. To cite Olmsted once more, it was "so common that southern writers gravely describe it as a disease—a monomania, to which the negro race is peculiarly subject"; to which this writer adds the aside— "making the common mistake of attributing to blood that which is much more rationally to be traced to condition."[49] It is but necessary to read any of the numerous biographies of escaped slaves or the tales of escape contained in reports of the "underground railroad"[50] to realize the strength of the compulsions toward freedom. Risks were great and punishment in the event of capture of the severest, yet men and women by the thousands took the risks.

One commentary is found in the reports of the operations of the underground railroad regarding the type of Negro who refused to accept the status of slave where he could possibly escape. For it is apparent that a large proportion of those who attempted and achieved flight were just those most-favored Negroes who might be expected to be the least moved to resentment—the educated men who could read and write, the skilled craftsmen, the house servants. The case of two slaves, hired out by their owner, a widow named Mrs. Louisa White of Richmond, Virginia, is typical. Both were skilled; William, a baker, was worth $1,200 to his owner, while James was equally valuable. Editorial comment in the Richmond *Despatch* read:

. . . These negroes belong to a widow lady and constitute all the property she has on earth. They have both been raised with the greatest indulgence. Had it been otherwise, they would never have had an opportunity to escape, as they have done. Their flight has left her penniless. Either of them would have readily sold for $1200; and Mr. Toler advised their owner to sell them at the commencement of the year, probably anticipating the very thing that has happened. She refused to do so, because she felt too much attachment to them. They have made a fine return, truly.[51]

Knowing how to read and write, James was able from Canada to answer a letter from his late owner asking that he return to save her from the distress and financial embarrassment his escape had caused her:

Instead of weeping over the sad situation of his "penniless" mistress and showing any signs of contrition for having wronged the man who held the mortgage of seven hundred and fifty dollars on him, James actually "feels rejoiced in the Lord for his liberty," and is "very much pleased with Toronto"; but is not satisfied yet, he is even concocting a plan by which his wife might be run off from Richmond, which would be the cause of her owner (Henry W. Quarles, Esq.) losing at least one thousand dollars.[52]

It will probably never be known how many slaves did make good their escape; that protest continued to the very end of the period is shown by recent historical studies into the behavior of the slaves during the Civil War.[53] It is widely held that most slaves refused to desert their masters and mistresses, preferring to remain with their "white folks" rather than risk the life of free men. Not only in the Sea Islands, where the Negroes were first liberated, did they refuse to accompany their masters fleeing from the northern troops,[54] but elsewhere slaves helped the Union troops wherever possible, as either soldiers or scouts, or by carrying information from behind the Confederate lines, or by performing manual labor on fortifications and other military works. Naturally, many slaves did remain with their masters during these times of stress, and this suggests that the range of variation in human temperament to be found everywhere existed in the large Negro population of the time.

6

The widespread and often successful character of organized revolts by Negro slaves indicates that among the Africans brought to the New World there must have been leaders able to take command when opportunity offered, and whose traditions of leadership were passed on to those who came after them. Africa had military specialists and, scarcely less important, those whose duty it was to see that the supernatural forces were rendered favorable before a given campaign was undertaken. The problem which we must attempt to answer, then, is the extent to which a process of selectivity was operative during enslavement that favored or tended to eliminate such specialists; whether, as eventually constituted, the slave population represented a cross section of the West African communities from which it was derived or was weighted toward either end of the social scale.

Opinion most generally has it that there was a strong weighting caused by the fact that Africans of least worth in their own countries were sold as slaves; or, because of lack of ability, fell most ready prey to the slaver. There is evidence to prove that some representatives of the upper socio-economic strata of African societies, at least, were sold into slavery; and there is some reason to believe that certain of these men or their descendants, such as Christophe, became leaders in the organized slave revolts of the New World. Mrs. Bremer gives an instance of a slave of noble African blood:

Many of the slaves, also, who are brought to Cuba have been princes and chiefs of their tribes, and such of their race as have accompanied them into slavery on the plantations always show them respect and obedience. A very young man, a prince of the Luccomees, with several of his nation, was taken to a plantation on which, from some cause or other, he was condemned to be flogged, and the others, as is customary in such cases, to witness the punishment. When the young prince laid himself down on the ground to receive the lashes, his attendants did the same likewise, requesting to be allowed to share his punishment.[55]

Again, Moreau de St. Méry tells of the respect shown in Haiti to members of African royal families:

The Mina Negroes have even been seen to recognize princes of their country . . . prostrating themselves at their feet and rendering them that homage whose contrast to the state of servitude to which these princes have been reduced in the colony offers a striking enough instance of the instability of human greatness.[56]

Field work in West Africa has made available some further information regarding the inclusion of upper-class persons in the slave cargoes. These data come entirely from Dahomey—no comparable materials have been collected elsewhere—so that any generalizations drawn from them must be made with all reservations. Yet the fact that in this instance traditional history is so specific might indicate that the mechanisms involved were more widely operative in West Africa than the facts previously in hand suggest.

Tradition has it that, in Dahomey, considerable numbers of persons were enslaved as a result of dynastic quarrels. When a new king was to ascend a throne, his right to the kingship was sometimes disputed by a brother, the son of a different wife of their common father. If revolt broke out, whatever the result, the winner had at hand the slave market to dispose of his rival. This not only meant that the unsuccessful contender was safely and profitably disposed of, but that his family and supporting chiefs, his diviners and the priests who had advised and aided him were also enslaved. The extent to which this account represents an exaggeration of oral tradition must not be lost sight of; it is well known that one of the Dahomean kings, Glele, was for many years held prisoner under the regency of an uncle, who sold the queen mother into slavery, and probably others of the royal compound. It is an historical fact that Da Souza, the Portuguese mulatto confidant of Glele, after his friend had regained his throne, traveled to Brazil to find and bring back the deported mother of the king. Tradition says he was successful; history says the mission failed.

This tradition concerning dynastic disputes would account for members of the Dahomean nobility being slaved; a further tradition tells the circumstances under which priests of the native cults were sold away in considerable numbers. The explanation of this fact is likewise political, though intertribal rather than intratribal differences are involved. Local priests, especially those of the river cults, are held to have been the most intransigeant among the folk whom the Dahomeans conquered. While it seems to have been policy to spare compliant priests so that the gods of a conquered people would not be unduly irritated and thus be rendered dangerous, priests who refused to submit, or who were detected in intrigue against the conquerors, were disposed of through sale to the slavers. It is firmly believed in Dahomey today that one of the reasons why the French conquered their country was that, having sold away all those competent to placate the powerful river-gods, these beings finally took their vengeance in this manner.

It is apparent that here is a mechanism which may well account for the tenaciousness of African religious beliefs in the New World, which, as will be seen in later pages, bulk largest among the various elements of West African culture surviving. What could have more effectively aided in this than the presence of a considerable number of specialists who could interpret the universe in terms of aboriginal belief? What, indeed, could have more adequately sanctioned resistance to slavery than the presence of priests who, able to assure supernatural support to leaders and followers alike, helped them fight by giving the conviction that the powers of their ancestors were aiding them in their struggle for freedom?[57]

On the basis of such evidence from Africa and the New World as is available, then, a prima-facie case can be made that the slave population included a certain number of representatives from African governing and priestly classes. What of the vast majority of slaves? Were they criminals and malcontents or derived from those incapable of carrying on in their aboriginal cultures? Contemporary testimony seems to be unanimous that there was no selective process that would have taken such types rather than others into slavery. Père Labat, who lived in the French West Indies for many years at the height of slavery, listed four classes in the African population from which slaves were drawn. First, he says, were those whose punishment against native law had been commuted to perpetual banishment—that is, to slavery "for the private profit of the kings." The second group were prisoners of war; the third those who, already slaves, were sold to meet the need of their masters for money. The fourth group, which Labat says comprised by far the greatest number, were those captured by marauding bands of robbers who, with the connivance of native rulers, carried on these raids to satisfy the demands of the European dealers.[58]

Snelgrave, active himself in the trade, gives the following ways in which the Negroes were enslaved:

As for the *Manner* how those People become Slaves; it may be reduced under these several Heads.

1. It has been the Custom among the *Negroes*, time out of Mind, and is so to this day, for them to make Slaves of all the Captives they take in war. Now, before they had an Opportunity of selling them to the white People, they were often obliged to kill great Multitudes, when they had taken more than they could well employ on their own Plantations, for fear that they should rebel, and endanger their Masters safety.

2dly. Most Crimes amongst them are punished by Mulcts and Fines; and if the Offender has not the wherewithal to pay his Fine, he is sold for a Slave: This is the practice of the inland People, as well as of those on the Sea Side.

3dly. Debtors who refuse to pay their Debts, or are insolvent, are likewise liable to be made Slaves; but their Friends may redeem them: And if they are not able or willing to do it, then they are generally sold for the benefit of their Creditors. But few of these come into the hands of the *Europeans*, being kept by their Countrymen for their own use.

4thly. I have been told, That it is common for some inland People, to sell their Children for Slaves, tho' they are under no Necessity for so doing; which I am inclined to believe. But I never observed, that the People near the Sea Coast practise this, unless compelled thereto by extreme Want and Famine, as the People of *Whidaw* have lately been.[59]

Falconbridge repeatedly stressed enslavement by kidnaping, and also describes the method called *boating*, whereby sailors would put off a small boat from their ship, load it with supplies, and, sailing up the rivers, take on whatever natives came into their hands, whether by sale or by capture:

I have good reason to believe, that of one hundred and twenty negroes, which were purchased for the ship to which I then belonged, then lying at the river Ambris, by far the greater part, if not the whole, were kidnapped. This, with various other instances, confirms me in the belief that kidnapping is the fund which supplies the thousands of negroes annually sold off these extensive Windward, and other Coasts, where boating prevails.[60]

In West African wars, no persons were considered noncombatants; everyone encountered by a conquering army was captured, and either retained for use by his captors, sacrificed in their religious rites, or in the vast majority of instances, sold away to the New World. The kidnaper also was no respecter of persons; one of the reasons why this technique was so feared was that it took its heaviest toll from the most defenseless, the young folk. How vivid the remembrance of these depredations remains even at the present time in West Africa is evidenced by the ability of a Togoland native, living in

Dahomey, to give a clear account of their consequences for his family. The people of this man had fled before the Ashanti, after a war in which many of his ancestral relatives had been lost in battle, either killed or carried off into captivity. Later, when a new home had been established, they had once more to migrate eastwards, for their enemies still raided them. Finally there was nothing to do but to put up such resistance as well as they could:

People we call *Aguda* [Portuguese], they buy plenty. If they buy they put for ship. That time no steamer. If man go out, man who be strong catch him go sell. My grandfather he say *Aguda* buy we people in Popo, then take go 'way to place they now calls Freetown. *Aguda* make village there, then make we people born children. When children born, *Aguda* take away to sell.[61]

The hypothesis that the selective character of the slave trade operated to bring the least desirable elements of Africa to the New World is thus neither validated by the reports of those who wrote during the days of the trade nor by the traditions of slaving held in the area where it was most intensive. That debtors were enslaved means nothing in terms of selectivity for, as was the case with white debtors, deportation merely acted to rid a country of persons lacking financial responsibility. Criminals likewise constituted a class determined by arbitrary definition, set up here as in every culture; it is to be doubted whether in West Africa, any more than in Europe, an inherent tendency to depravity can be ascribed to those who were deported because of their crimes. All agree, moreover, that debtors and criminals were but a small proportion of the cargoes of the slaving vessels, which means that the dominant factors were nonselective—warfare and kidnaping. And though it may be maintained that those Africans who were most able escaped in warfare, this point would be highly difficult to establish; while it would be even more difficult to prove that those kidnaped were possessed of any particular incapacity.

Chapter V

THE ACCULTURATIVE PROCESS

In considering the accommodation of Africans to their New World cultural milieu, differentials in the degree of contact between bearers of European and African traditions have often been recognized, even though most students of the Negro, in the United States at least, tend to limit their researches to the borders of their own country. Yet such attention as they have paid to other parts of the New World has made it plain that Africanisms have not survived to the same degree everywhere in the area. An example of this may be taken from a passage in the most recent work of Frazier, who, as will be remembered, completely rejects the thesis that any elements of African culture are to be found in the United States:

> Recent students . . . have been able to trace many words in the language of Negroes in the West Indies, Suriname, and Brazil to their African sources. There is also impressive evidence of the fact that, in the West Indies and in parts of South America, African culture still survives in the religious practices, funeral festivals, folklore, and dances of the transplanted Negroes. . . . Even today it appears that the African pattern of family life is perpetuated in the patriarchal family organization of the West Indian Negroes.[1]

Quotations from the writings of Weatherly, Park and Reuter to similar effect, given in the first chapter of this work will also be recalled.[2]

A second factor influencing acculturation, the situations under which certain types of slaves had greater opportunity for contact with their masters than others, has received more study. This point of attack has been most sharply defined in studies of slavery in the United States. On analysis, the approach is seen to be based on the "plantation portrait" that has played so large a role in shaping concepts not only of the institution of slavery, but also of the antebellum South in general. Most often, this differentiation takes the form of contrasting the intimacies of the relation between house servants and their masters to the slight contact and lack of personal feeling between the mass of rude field hands and their owners or overseers. In considering this phase of the problem, we are therefore dealing with a stereotype which causes the dichotomy between the two groups customarily to be taken for granted.

The third point to be considered concerns the manner in which the slaves accommodated themselves differently to various aspects of the European culture they encountered. Here we are breaking new ground, since few if any studies have even envisaged this approach. Yet it is important if the situation is to be analyzed adequately. For as has been indicated previously, and as will later be demonstrated in detail, an outstanding fact of New World Negro culture is that nowhere do Africanisms manifest themselves to the same degree in the several parts into which any human culture can be divided. It soon becomes apparent that, while Africanisms in material aspects of life are almost lacking, and in political organization are so warped that resemblances are discernible only on close analysis, African religious practices and magical beliefs are everywhere to be found in some measure as recognizable survivals, and are in every region more numerous than survivals in the other realms of culture.

With these three phases of the problem in mind, then, we may turn to a consideration of the acculturative mechanisms which endowed New World Negro tradition with the forms it was later to take. This can best be done by analyzing the interracial situation as it existed during the period of slavery, leaving to succeeding chapters the documentation of Africanisms found in Negro culture today.

2

What caused the differences between the several parts of the New World in retention of African custom? Though the answer can only be sketched, especially since adequate historical analysis of the data concerning plantation life outside continental North America and Brazil is quite lacking, yet the effective factors are discernible. They were four in number: climate and topography; the organization and operation of the plantations; the numerical ratios of Negroes to whites; and the extent to which the contacts between Negroes and whites in a given area took place in a rural or urban setting.

It is patent that the natural environment influenced the life led by various communities of slaves in the New World. Negroes

brought to those parts lying between the tropics were not faced
with the same need for readjustment as those taken to more temperate climates; those whose destination was the smaller islands
found new conditions of life, especially if they came from inland
tribes. Yet aside from these obvious and immediate adjustments,
the environment also influenced the type of contact which the Negro
had with Europeans when he lived out his life as a slave. The very
nature of the controls under which the slaves lived were dependent
to a large extent on the kind of crop most important in a given
area; that this, in turn, was affected by such factors as temperature, rainfall, soil conditions, and all those other aspects of the
natural setting that determine what can best be raised in a region
goes without saying. In the United States, for example, a man's life
on the rice plantations of the Carolina coast or in the cotton country
of Alabama and Louisiana was quite different than if his lot made
him the property of a master interested in the production of turpen-
tine or one whose farm was in the uplands.

The point becomes even more important if we contrast the United
States as a whole with the West Indies and South America, where
the slave regime took as its primary economic base the production
of sugar. Here the plantation system reached its most elaborate development; on these sugar estates, where slaves were especially
numerous, Negro contacts were confined to Negroes far more than
elsewhere. The significance of this will be taken up later; it must
be mentioned here, however, to point out how, despite its importance in influencing the acculturative processes, it is itself to be considered a reflex of conditions set by the natural environment which
made this type of large-scale agricultural production imperative if
the system was to function efficiently. This is to be seen in considering the economic order in the West Indies even today. For wherever
any measure of efficiency in the production of these staples for the
world market has survived the slave regime, the estates system has
been perpetuated; where the production of these crops has been attempted on the basis of small-scale individual ownership of land,
failure has been so obvious that enlightened procedure today dictates the encouragement of subsistence farming as the only solution for the many social and economic problems that beset the
region. The encouragement given by such an environment to large-
scale agricultural production thus made for the concentration of
large numbers of Africans and of persons of African descent in
natural settings where the greatest possible use could be made by
them of their aboriginal cultural heritage. And this, in turn, per-

petuated this heritage to a far greater extent than where numbers were smaller and major adjustments to an almost completely strange environment were essential if the slave was to achieve even a personal survival.

The importance of the environment in furthering the success of slave revolts, or in aiding the masters to suppress uprisings and to capture runaway slaves, is in itself of some moment. Thus, as a simple example, it is clear that individual escapes were more likely to be made good where natural obstacles to pursuit were the most severe; in the United States, swamps always invited running away, permitting the slave a measure of protection from his pursuers that open country could never have afforded him. In the tropics, dense jungles aided revolters, both because of the similarity between the conditions of life the runaways had to meet in these forests and the setting of their lives in Africa and because of the difficulties they presented to Europeans who were tempted to track the fugitives through the high bush.

Mountainous country, and other natural aids to concealment where escaped slaves did not have adequate weapons, or regions which were of strategic importance when they were armed, must likewise not be undervalued. For everywhere in the New World where slave revolts were successful, the Negroes had the jungle or the mountains, or both, as a refuge, wherein they might establish and consolidate their autonomous communities. This was true in Guiana, in Haiti, in Brazil, in Jamaica; it is striking that in smaller, less heavily forested islands such as Barbados, St. Vincent, and the Virgin Islands, or in the United States (with the one exception of the Maroons of Florida, whose salvation lay in their joining the Indians rather than through their own unaided efforts), serious revolts were put down. And since escape in numbers invariably meant the preservation of Africanisms in greater quantity and purer form than was otherwise possible, particularly where the revolts came early in the history of slaving, as in Brazil, or where, as in Guiana and Haiti, continuous recruitment through raids on the plantations brought a constant supply of newcomers fresh from Africa into the communities of the revolters, the significance of this as influencing the acculturative process is clear.

The manner in which the plantations were organized and operated was, in the main, similar in all the New World. This is understandable, if only because the system of slavery was everywhere oriented toward producing for the world market; which means that unskilled labor had to be directed toward growing the principal crop in which the plantations of a given area specialized. Sugar in the West Indies and South America and cotton in the United States were most important, but some plantations and some regions were devoted to growing other major crops, such as coffee and tobacco and rice and indigo. The routine of work varied from region to region, but everywhere processes were simplified so as to restrict the number of operations. Fields had to be prepared before planting, either by breaking in virgin land or working over land cultivated the preceding season. Planting, hoeing, and reaping made up the second stage, while preparation of the product for the market was the final step. In the case of cotton, this last step required ginning and baling; on the sugar plantations, boiling and cooling of the sirup and refining the crystals was necessary. Some slaves had to be trained to perform such special tasks, but everywhere careful and constant supervision assured that all aspects of the work went forward, the white overseers being assisted by slaves who were charged with supervising the labor of a smaller group than the white superintendent could effectively control.

There were historical as well as economic reasons why the plantation system was so unified. Contacts between the various parts of the entire New World were continuous, and the manner in which work was done in one region influenced the mode of operations in another. Thus, it is pointed out:

Planters coming to South Carolina from Barbados, Jamaica, Antigua, and St. Kitts brought with them the ideas of the plantation regime with slavery as the basis of the labor system.[3]

Similarly, French colonists fleeing the slave revolts of Haiti brought their ways of working their slaves as well as the slaves themselves to Louisiana, while, in the opposite direction, the Tory sympathizers for whom the American Revolution was unpalatable moved their establishments to the Bahamas and elsewhere in the British West Indies. But these were only a few of the unifying movements. The Jews, expelled from Brazil in the seventeenth century, moved to Dutch Guiana and Curaçao with their slaves, setting up new estates in these Dutch colonies; the settling of Trinidad by planters from neighboring French, English, and Spanish possessions fixed the plantation economy firmly on that island. In consequence, an underlying unity, the result of historical and economic forces, developed in the New World as a background for Negro acculturation to European patterns.

The frequency with which slaves were trained as specialists on plantations in the various parts of the New World differed greatly, but it must be assumed that wherever large numbers of slaves were employed, certain individuals were assigned to what may be termed the service of supply. Carpenters, blacksmiths, and others who could repair broken implements were essential; some had to be entrusted with the care of infants where mothers with young children were put to work in the fields; while specialists were required to handle the apparatus used in preparing the plantation produce for the market, particularly on the sugar estates. House and personal servants of the masters must also be included, though their numbers, duties, and manner of selection varied.

It does not follow, however, that the training of those who operated and repaired sugar mills, or acted as house servants, or otherwise followed a routine different from that of the great body of field hands was in itself sufficient to cause them to give up their African modes of thought or behavior. Though greater personal contact with the masters was the direct route to greater taking over of European modes of behavior, the manner of life led by the whites, certainly in the West Indies, was not such as to inculcate either love or respect for it on the part of those who viewed it closest. Furthermore, in all the slaving area, the slaves closest to the masters were those most exposed to their caprice; and the situation of the house slave, if not involving as hard and continuous physical labor as that required of the ordinary field hand, had serious drawbacks in the constant exposure to the severe punishment that followed even unwitting conduct that displeased the ever-present masters.

It has already been remarked how, in the United States, many of the slaves aided by the underground railroad had had the greatest opportunities to learn the white man's way of life; and in the case of such as these, acculturation to European patterns had proceeded far. But in the West Indies and South America, it was in many instances just those experiencing the closest contact with the whites who steadfastly refused to continue in this way of life when freedom could be attained through manumission or self-purchase or escape. On the basis of studies made in the United States, it is generally held that European habits were most prevalent among the free Negroes. Yet though this may be true of the United States—and further research by students with adequate background of West African and West Indian Negro cultures must precede any final word on the point—in the rest of the New World it by no means follows. Thus, in Brazil, it was just among those Negroes who, either for the account of their owners or because they had pur-

chased their own freedom, followed specialized callings, that the
stream of African tradition was kept free-flowing to reach in relatively undiluted form to the present day.[4] In Haiti, again, it was
those who had had considerable opportunity to acquaint themselves
with European methods of life—and warfare—who were prominent
among the leaders of the most successful slave revolts.

The ratio between Negroes and whites must be kept in mind
when attempting to assess the mechanisms of accommodation. Acculturation, it must be remembered, occurs as a result of contact,
and it is the continuing nature of the contact and the opportunities
for exposure to new modes of life that determine the type and intensity of the syncretisms which constitute the eventual patternings
of the resulting cultural orientations. That racial ratios varied greatly
in various parts of the New World must be taken into account; even
in the United States the differences in the numbers of whites and
Negroes from one portion of the slave belt to another were so striking that they attracted the attention of contemporary observers no
less than of present-day students.

Bassett, for example, phrases this in the following terms:

The planters, that is the owners of large farms, were but a small part
of the white people of the old South. The great mass were small farmers,
owners of small groups of slaves or of none at all, men who had land
and lived independently without leisure, education, or more than simple
comforts. . . . It was from this class of small farmer that the overseer
came. He was often a man whose father had a few slaves, or some ambitious farmer youth who had set his eyes upon becoming a planter and
began to "manage," as the term was, a stepping-stone to proprietorship
in the end.[5]

The analysis made by Gaines of American literature dealing with
the South, as it reflects and distorts the social realities of the slaving period, leads him to the following conclusion:

One of the most common misrepresentations is in the matter of the
size of estates. Almost unfailingly the romancers assume a great realm
bounded only "by blue horizon walls." There were, as a matter of fact,
some large holdings . . . but colossal estates were the exception, not
the rule. Over certain zones, as most of North Carolina and Georgia,
there were few big places. Page justly affirms that the average Southern
estate was small, and few Southerners owned negroes, that most of these
possessed but a small number.[6]

Statistics of Negroes and whites in Maryland reflect opportunities
for contact, and the resultant possibilities for taking over European
modes of life by Negroes:

The Governor of Maryland wrote, in 1708, that the trade had been rising and was then a "high" one; that some six or seven hundred blacks had been imported in the ten months past. Two years later, came word that the negroes were increasing. The Public Record Office in London had a list of the "Christian" men, women and children and also of negro slaves, in Maryland, in 1712. The whites numbered nearly thirty-eight thousand, the negroes over eight thousand. In three of the Southern counties, the blacks far outnumbered the whites. In the years following, both races increased fast, but the blacks faster than the whites. By 1750, the whites may have been nearly a hundred thousand, the blacks nearly forty thousand. In 1790, there were over two hundred and eighty thousand whites, and nearly half as many slaves; the eight thousand and odd free blacks making the proportion of white to black as less than two to one.[7]

The Carolinas have likewise been subjected to careful study from this point of view, and are especially pertinent to our inquiry because of the variation from one district to another in the proportion between masters and slaves. The fact that some of the most extensive retention of Africanisms in all the United States is found in the Sea Islands may be coupled with the following statements in pointing a conclusion as to causation. In the first place, we are informed that:

On St. Helena Island, where there were some two thousand slaves to a little more than two hundred whites, the Negroes learned very slowly the ways of the whites. Their mastery of English was far less advanced than that of the Piedmont slaves. They spoke a garbled English, imperfect words and expressions which they and their parents and grandparents had learned from the few whites with whom they came in contact.[8]

The statement of a slaveowner of the region throws further light on the process involved:

A Charleston planter told his English guest, Captain Basil Hall, in 1827, that he made no attempt to regulate the habits and morals of his people except in matters of police, "We don't care what they do when their tasks are over—we lose sight of them till the next day," he said.[9]

The sense of these regional differences for North Carolina in general was expressed in language such as the following:

The *Carolina Cultivator* divided North Carolina farmers in 1855 into "two well-known classes." One class owned slaves who were "ragged, filthy, and thievish"; the other, slaves who were well clothed, fed, and housed, cheerful, industrious, and contented.[10]

Statistically, this is represented by numbers of slaves owned on the great estates and smaller farms:

While the average number of slaves per slaveholding family in North Carolina seemed high, it must be pointed out that more than half, 67 per cent, of these families held less than ten slaves. The large slave-holders, those few who owned from 50 to 200 slaves, give color to the picture of ante-bellum North Carolina, but the small slaveholders actually shaped the character of slavery in the state because they were in the majority. Equally important . . . were the families, 72 per cent of the total in 1860, who owned no slaves.[11]

The most extended discussion of slave ownership is that of Phillips. By 1671 the population of Virginia, he tells us, was estimated at 40,000, including 6,000 white servants and 2,000 Negro slaves.[12] These proportions changed rapidly, as shown by a census taken in "certain Virginia counties" during 1782-1783 which enumerated the numbers of slaves owned in the Tidewater and lower Piedmont regions:

For each of their citizens, fifteen altogether, who held upwards of one hundred slaves, there were approximately three who had from 50 to 99; seven with from 30 to 49; thirteen with from 20 to 29; forty with from 10 to 19; forty with from 5 to 9; seventy with from 1 to 4; and sixty who had none.[13]

At the end of the slaving period, the "greatest of the tobacco planters" was reported in 1854 to have slave populations of some 1,600 individuals in his "many plantations lying in the upper Piedmont on both sides of the Virginia-North Carolina boundary."[14] In Maryland,

. . . the ratios among the slaveholdings of the several scales, according to the United States census of 1790, were almost identical with those just noted in the selected Virginia counties. . . . In all these Virginia and Maryland counties the average holding ranged between 8.5 and 13 slaves. In the other districts in both commonwealths, where the plantation system was not so dominant, the average slaveholding was smaller, of course, and the non-slaveholders more abounding.[15]

In South Carolina, the concentration of slaves was somewhat higher, as might have been expected, though not so great as is ordinarily thought:

In the four South Carolina parishes of St. Andrew's, St. John's Colleton, St. Paul's and St. Stephen's the census-takers of 1790 found 393 slaveholders with an average of 33.7 slaves each, as compared with a total of 28 non-slaveholding families. In these and seven more parishes, comprising together the rural portion of the area known politically as the Charleston District, there were among the 1643 heads of families 1318 slaveholders owning 42,949 slaves. . . . Altogether there were 79 separate parcels of a hundred slaves or more, 156 of between fifty and ninety-nine, 318 of between twenty and forty-nine, 251 of between ten and nineteen, 206 of between five and nine, and 209 of from two to four, 96 of one slave each, and 3 whose returns in the slave column are illegible.[16]

The contemporary comment of Sir Charles Lyell, though couched in the style of the period, is germane:

When conversing with different planters here, in regard to the capabilities and future progress of the black population, I find them to agree very generally that in this part of Georgia they appear under a great disadvantage. In St. Simon's island, it is admitted, that the negroes on the smaller estates are more civilized than on the larger properties, because they associate with a greater proportion of whites. In Glynn County, where we are now residing, there are no less than 4,000 negroes to 700 whites; whereas in Georgia generally there are only 281,000 slaves in a population of 691,000, or more whites than colored people. Throughout the upper country there is a large preponderance of Anglo-Saxons, and a little reflection will satisfy the reader how much the education of a race which starts originally from so low a stage of intellectual, social, moral and spiritual development, as the African negro, must depend not on learning to read and write, but on the amount of familiar intercourse which they enjoy with individuals of a more advanced race. So long as they herd together in large gangs, and rarely come into contact with any whites save their owner and overseer, they can profit little by their imitative faculty, and can not even make much progress in mastering the English language. . . .[17]

For Alabama, the situation has been summarized as follows:

The cotton counties were the chief slave counties and are consequently the plantation counties. They are Dallas, Marengo, Greene, Sumter Lowndes, Macon and Montgomery. Sections of the state, poorer as well as less adapted to cotton cultivation, had an average of 1.4 slaves to the household. Madison County in the northern part of the state, actually had over two families to each slave. The average number of slaves per plantation for the state was 4.5. However, in the cotton counties the large plantations set the prevailing patterns. In 1805 there were 790 owners of from 30 to 70 slaves; 550 owners with from 70 to 100 slaves; 312 with from 100 to 200; 24 with from 200 to 300 and 10 with from 300 to 500. Thus, some 150,000 slaves were on plantations of 50 or more even though only a third of the white people were directly interested in slavery.[18]

Phillips, though not directly concerned with the mechanisms of acculturation, has given a useful summary of the matter as concerns the United States in general:

It is regrettable that data descriptive of small plantations and farms are very scant. Such documents as exist point unmistakably to informality of control and intimacy of black and white personnel on such units. This is highly important in its bearing upon race relations, for according to the census of 1860, for example, one-fourth of all the slaves in the United States were held in parcels of less than ten slaves each, and nearly another fourth in parcels of from ten to twenty slaves. This means that about one-half of the slaves had a distinct facilitation in obtaining an appreciable share in the social heritage of their masters . . . the very fact that the negroes were slaves linked them as a whole more closely to the whites than any scheme of wage-labor could well have done.[19]

The contact between Negroes and whites in continental United States as compared to the West Indies and South America goes far to explain the relatively greater incidence of Africanisms in the Caribbean. In the earliest days, the number of slaves in proportion to their masters was extremely small, and though as time went on thousands and tens of thousands of slaves were brought to satisfy the demands of the southern plantations, nonetheless the Negroes lived in constant association with whites to a degree not found anywhere else in the New World. That the Sea Islands of the Carolina and Georgia coast offer the most striking retention of Africanisms to be encountered in the United States is to be regarded as but a reflection of the isolation of these Negroes when compared to those on the mainland.

Certainly the opportunity of the slave who was the sole human possession of his master to carry on African traditions was of the slightest, no matter how convinced such a Negro might have been that this was a desirable end. Even where the slaves on a farm numbered ten or fifteen, it was difficult to achieve continuity of aboriginal behavior. Unless such a group was in the midst of a thickly settled area, which could afford them constant contacts with other slaves, it would be well-nigh impossible for them to live according to the dictates of a tradition based on large numbers of closely knit relationship groupings organized into complex economic and political structures. How might the specialists in technology, in magic, in manipulating the supernatural powers carry on their work among such small groups existing, as these had to exist, under the close scrutiny and constant supervision of the slaveowners?

Matters were quite different in the Caribbean islands and in South America. Here racial numbers were far more disproportionate; estates where a single family ruled dozens, if not hundreds, of slaves were commonplace and the "poor white" was found so seldom that he receives only cursory mention even in such meticulous historical treatments as those of the French scholars concerned with social conditions, economic status, and political developments of pre-revolutionary Haiti. The white man with but a few slaves was likewise seldom encountered. The requirement of conformity to the plantation system was far more stringent than in the United States, where a certain degree of economic self-sufficiency was sought after, no matter to what extent the raising of a single cash crop was the major objective. The territory itself liable for exploitation on the mainland was almost literally without limit, and the constant migration of the planters and their slaves to the west made for the growth of population centers and the development of means of communication that made the type of isolation experienced by West Indies and South American planters a rarity. A passage such as the following, taken from the letter of a Haitian planter, describes a situation quite foreign to the experience of the southern slaveholder:

Have pity for an existence which must be eked out far from the world of our own people. We here number five whites, my father, my mother, my two brothers, and myself, surrounded by more than two hundred slaves, the number of our negroes who are domestics alone coming almost to thirty. From morning to night, wherever we turn, their faces meet our eyes. No matter how early we awaken, they are at our bedsides, and the custom which obtains here not to make the least move without the help of one of these negro servants brings it about not only that we live in their society the greater portion of the day, but also that they are involved in the least important events of our daily life. Should we go outside our house to the workshops, we are still subject to this strange propinquity. Add to this the fact that our conversation has almost entirely to do with the health of our slaves, their needs which must be cared for, the manner in which they are to be distributed about the estate, and their attempts to revolt, and you will come to understand that our entire life is so closely identified with that of these unfortunates that, in the end, it is the same as theirs. And despite whatever pleasure may come from that almost absolute dominance which it is given us to exercise over them, what regrets do not assail us daily because of our inability to have contact and correspondence with others than these unfortunates, so far removed from us in point of view, customs, and education.[20]

Plantation owners in the Caribbean islands and in North America differed in attitude, primary affiliation, and adjustment. Where the setting was as strange to Europeans as was the tropics, thought persistently turned to the homeland; in the United States, the life of a planter, who regarded himself as permanently settled on his plantation, took on more normal routines. The influence of such orientations on the Negroes in terms of affording opportunities for acculturation is apparent. Closer contact with the whites in the United States made for greater familiarity with white customs and facilitated their incorporation into Negro behavior; nothing could contrast more to this than the situation as it affected the slaves where greater numbers of Negroes lived a life on the estates which removed them from their masters far more than on the North American continent. Consequently, one can set off the United States from the rest of the New World as a region where departure from African modes of life was greatest, and where such Africanisms as persisted were carried through in generalized form, almost never directly referable to a specific tribe or a definite area.

The matter of urban and rural residence of the slaves is closely related to the point just considered. Information is much less available on this than are data concerning ratios between Negroes and whites in general; yet the situation can be broadly summarized as it encouraged or hindered intensity and speed in acculturation, and thus contributed to the differences in Negro behavior both within and between various regions.

The urban centers of the United States were more numerous and of greater size than anywhere else in the New World. Only Brazil approaches the United States in this respect. In the slaving area of northern South America, one finds Paramaribo in Dutch Guiana and Georgetown in British Guiana, both but small towns when compared to the cities at the extremities of the New World slaving belt. The coastal towns of Venezuela had considerable Negro populations; for the rest of the Latin-American republics our information is so scanty that nothing can be said. In the Caribbean islands, such centers as existed were at best but settlements where planters might have establishments to spend their time when not in residence on their estates, or when business brought them to these seaports. Port-of-Spain, Trinidad, was during the entire period of slavery but a small town, located on an island where, in any event, slavery had a secondary place in the economic life. Bridgetown, Barbados, was somewhat larger, but has always been an administrative and shipping center. The islands of the Windward and Leeward chains, themselves minute, could support no settlements of any size; hence even for small cities in the Caribbean we must turn to Cuba, Jamaica, Haiti, and perhaps Puerto Rico and Martinique.

This meager list is to be contrasted to the centers of considerable size in the United States, in many of which Negroes had numerous opportunities for contact with whites. In the North were Boston, New York, and Philadelphia, all possessing appreciable Negro populations from relatively early days of the slave trade. Besides these northern centers were many smaller towns, such as Albany and Jersey City and Fall River, whose Negro populations in the aggregate were large enough to make them a real influence in bringing the customs of the white man to the Negro community as a whole. On the borderline between slave and free states were Washington, Baltimore, and farther west, Cincinnati and Louisville, whose large Negro groups were in constant contact with the much greater numbers of whites among whom they lived, and lacked both the motivation and opportunity of living in terms of a peculiar pattern such as did the Negroes of Brazilian or West Indian communities. Finally, there were many centers in the South itself where Negroes and whites had close association. Richmond, Charleston, and Atlanta, to the east, and New Orleans, to the west, are but the outstanding example of cities whose smaller counterparts existed elsewhere in the South.

Life in all these communities differed strikingly from that in centers outside the United States. The householders of these cities were permanently established; their dwellings thus being in no sense residences to be occupied only during certain portions of the year, or mere stopping places when business brought a planter to town. The southern plantation owner lived on his estate, and his contacts with the city, except for visits to friends, were brief and infrequent. The transfer of slaves from country to town and back again to the country with the travels of the master, as was common in the islands, was much rarer on the continent, if it occurred at all; which means that in the United States the town Negro was a permanent resident, something that in turn made for a sharper differentiation between him and the rural slave than elsewhere. In the North, of course, where no plantations existed, slave-owning was practiced on only a small scale. This in itself affected the acculturation of the northern Negro, for whether in town or country, the relatively slight number of slaves, or of freedmen after slavery was outlawed in the North, made it inevitable that Negroes everywhere experienced much more intensive exposure to European patterns of behavior than in any other part of the New World.

It is important, however, in considering how urbanism affected the acculturative processes, to refrain from making any a priori assumptions. In cities the Negroes led a life of their own, and just because their opportunities for contact with the whites were greater than on isolated plantations, this does not necessarily mean that they followed the lead to behavior placed before them. What is the more likely is that, where retention of Africanisms or adoption of European customs is concerned, either or both were accelerated. Popular belief holds it almost as an axiom that to find aboriginal custom in its purest form one must go to the most remote districts; though this may be to a certain extent true, it is by no means an acceptable generalization. The Bush Negroes of the Guiana forests manifest African culture in purer form than is to be encountered anywhere else outside Africa, yet in such centers as Bahia—especially in the Negro quarters on the outskirts of the city—or Paramaribo or Port-of-Spain one also finds retentions of a surprising degree of purity. Puckett has put the matter similarly for the United States:

Even today almost three-fourths of the Southern Negroes live in a rural environment, and in this relative isolation the more primitive type of superstitions are generally preserved more easily than in an environment where outside contact is greater. That this is not always true is shown by the apparently greater prevalence of superstition with the New Orleans city Negro as compared with many of his rural kinsmen—a fact probably accounted for by the voodoo traditions of that city and the more frequent interchange of such ideas through a multitude of people all clinging to the same old beliefs.[21]

Certainly the anonymity of city life is often conducive to carrying on outlawed customs and beliefs when they can be quietly pursued, or to furthering activities under a ban in disguised form. Even were this not the case, an important economic factor, almost entirely neglected, would come into play. Where the more dramatic African survivals, such as possession dances and other manifestations of religious belief and of magic, are concerned, it is essential that enough wealth be at hand to allow adequate support for the specialists who direct these rites and control the supernatural powers. This is strikingly exemplified by the situation in Trinidad, where the rites of the Yoruban Shango sect can only be found on the outskirts of the capital and principal seaport, Port-of-Spain. In the interior this cult is entirely absent. Folk living there are vague about its ritual or beliefs, because in these outlying districts there are not enough persons or enough wealth to support the extensive ceremonies. Analogous is the case of the "shouting" churches of the United States, where the forms of spirit possession represent one of the most direct African carry-overs to be encountered in this country. Though churches of this kind are numerous in the South, if one wishes to hear the "hottest" preaching and to witness the greatest outbursts of hysteria one must go to such great Negro centers as New York or Chicago or Detroit. Good preachers are in demand and, in accordance with the economic pattern of our culture, they accept calls where their services can be most adequately compensated. In the South, by and large, except the most populous Negro communities, congregations cannot meet the terms offered by the richer churches of the North.

It is thus unjustifiable to make the assumption that mere contact, such as was engendered by city life, brought about the suppression of Africanisms. That it discouraged their retention in pure form is undoubtedly true; yet this does not mean that white patterns were taken over without serious revision. Rather, it means that rural and urban Negro cultures took on somewhat different shadings; that the impact of European custom on African aboriginal traditional values and modes of behavior was directed along divergent courses. And hence it is that, while Africanisms are to be found both in cities and in the countryside all over the New World, they differ in intensity and in the specific forms they take.

3

The difference in the opportunities afforded various types of slaves to acquaint themselves with the behavior of their masters and model their own conduct in terms of these conventions is the second point to be considered in analyzing the acculturative process. As has been stated, the assumption that significant differences existed between the manner of life and treatment of the house servant and the field hand is widely held. One of the reasons frequently advanced for this is a presumed difference in color between these two groups. It is therefore necessary at the outset to enter a caveat against injecting any such biological explanation into the analysis of a situation that can be accounted for adequately on the historical and cultural level.

That color differences play a role of some importance within the Negro community is well known, and has been demonstrated as important in the selection of mates, where it follows a well-defined pattern whereby lighter-colored women tend to marry darker-colored men.[22] Two possible explanations of this can be offered that highlight the difficulty arising when the presumed differences between house servants and field hands during slavery are referred to, or correlated with, the biological fact of color. One explanation, essentially psychological, rests its argument on the tendency of any under-privileged minority to set up desired goals in accord with some outstanding characteristic of the dominant population among whom it lives; in the case of the Negroes, an absence of heavy pigmentation.[23] The other explanation is concerned only with the historical experiences of the Negro since his arrival in North America. It holds that the lighter-colored Negro, today socially and economically more favorably situated than his darker brother, is the offspring of the white master; that he is the descendant of the manumitted mulatto, who, in his capacity as house servant or personal retainer, came into closest contact with the whites and thus achieved an earlier and more effective acculturation to the patterns of the majority group.[24]

It would not seem unreasonable to hold that both these explanations are pertinent, while recognizing that neither can account entirely for the facts as established. What must be guarded against in either case is that any explanation of such color differences as mark off present socio-economic groupings, or did mark them off in earlier times, be not drawn in terms of differing innate capacity, resulting from different degrees of racial crossing. To be specific, in the case at hand the prestige accorded lighter color by Negroes must not be interpreted as something indicating a recognition on their part of the superiority derived from the presence of "white blood."

With this caution in mind, we may turn to the matter of differentials in opportunity for acquiring the culture of the whites accorded various types of slaves. The need for such an analysis becomes apparent as one reexamines the literature with the accepted point of view in mind, since the distinctions drawn between the mode of life of house servant and field hand are not as apparent in the contemporary accounts as the stress laid on them by present-day writers might lead one to believe. The remarks of Phillips concerning the acculturative process during the early days of slavery are pertinent:

. . . for two generations the negroes were few, they were employed alongside the white servants, and in many cases were members of their masters' households. They had by far the best opportunity which any of their race had been given in America to learn the white man's ways and to adjust the lines of their bondage in as pleasant places as possible.

Their importation was, for the time, on but an experimental scale, and even their legal status was during the early decades indefinite.[25]

According to his summary, the differentation between slaves was not much greater in later times:

The purposes and policies of the masters were fairly uniform, and in consequence the negroes, though with many variants, became largely standardized into the predominant plantation type. The traits which prevailed were an eagerness for society, music and merriment, a fondness for display whether of person, dress, vocabulary or emotion, a not flagrant sensuality, a receptiveness toward any religion whose exercises were exhilarating, a proneness to superstition, a courteous acceptance of subordination, an avidity for praise, a readiness for loyalty of a feudal sort, and last but not least, a healthy human repugnance toward overwork.[26]

Lyell did not find any great differences between the manner of life of slaves employed at various tasks on the larger plantations:

The out-door laborers have separate houses provided for them; even the domestic servants, except a few who are nurses to the white children, live apart from the great house—an arrangement not always convenient for the masters, as there is no one to answer the bell after a certain hour.[27]

Mrs. Smedes, whose book is almost a caricature of the "standard" idyl of slavery, tells a story which indicates that house servants did not always appreciate the opportunities of a presumably favored position:

It may not be out of place to give an illustration of how one of the Burleigh servants carried her point over the heads of the white family. After the mistress had passed away, Alcey resolved that she would not cook any more, and she took her own way of getting assigned to field work. She systematically disobeyed orders and stole or destroyed the greater part of the provisions given to her for the table. No special notice was taken, so she resolved to show more plainly that she was tired of the kitchen. Instead of getting the chickens for the dinner from the coop, as usual, she unearthed from some corner an old hen that had been setting for six weeks, and served her up as fricassee! We had company to dinner that day; that would have deterred most of the servants, but not Alcey. She achieved her object, for she was sent to the field the next day. . . .[28]

Nor were mulatto offspring of the masters always given the favored treatment envisaged for them as members of the household:

One might imagine, that the children of such connections, would fare better, in the hands of their masters, than other slaves. The rule is quite the other way; and a very little reflection will satisfy the reader that such is the case. A man who will enslave his own blood, may not safely be relied on for magnanimity. Men do not love those who remind them of their sins—unless they have a mind to repent—and the mulatto child's face is a standing accusation against him who is master and father to the child. What is still worse, perhaps, such a child is a constant offense to the wife. She hates its very presence, and when a slave-holding woman hates, she wants not means to give that hate telling effect. . . . Masters are frequently compelled to sell this class of their slaves, out of deference to the feelings of their white wives; . . . it is often an act of humanity toward the slave-child to be thus removed from his merciless tormentors.[29]

A modern summary of the situation in South Carolina indicates how the common conception of distinctions made between the two classes of slaves tends to overlook the manner in which the task assigned a given slave, except in the largest establishments, was subject to change:

On every plantation where there were more than twenty slaves at least one was set aside as a house servant. The very young and the old were usually engaged in the house, while the full "taskables" were more profitably employed in the field. For instance, the house servants on Henry C. Middleton's Weehaw plantation near Georgetown, South Carolina, were "a cook that is not a full task, a girl of twelve and a boy of fourteen." An old man was "stable boy" and coachman for the family and an old woman was gardener. Stephen A. Norfleet of Wood-bourne in Bertie County often put his house servants at other work during the rush season; and when his wife became ill in 1858, he employed a white housekeeper. In some families, however, the household retinue was large: a cook and assistant, a butler in uniform, a parlor maid, a personal maid, a "boy" to serve the master, a nurse if there were children, a liveried coachman, a gardener, and a stable boy.[30]

Another reason why the difference in opportunities for acculturation between house servants and field hands was not as great as is supposed is found in the accounts of the life of Negro children during their early, most formative years. How similar were the early conditionings of all slave children may be inferred from this comment of a contemporary observer:

It is a universal custom on the plantations of the South that while the slaves, men and women, are out at labor the children should all be collected at one place, under the care of one or two old women. I have sometimes seen as many as sixty or seventy, or even more together, and their guardians were a couple of old negro witches, who with a rod of reeds kept rule over these poor little black lambs, who with an unmistakable expression of fear and horror shrunk back in crowds whenever the threatening witches came forth flourishing their rods. On smaller plantations, where the number of children is smaller, and the female guardians gentle, the scene, of course, is not so repulsive; nevertheless it always reminded me of a flock of sheep or swine, which were fed merely to make them ready for eating.[31]

It may, of course, be objected that descriptions of this sort concern precisely the children of field hands, while the offspring of house servants were cared for differently, being permitted to play about the great house and thus from their infancy imbibing the master's ways of life. In the light of the variations in handling the slaves that obtained over the entire slaving area of the United States, this undoubtedly was true in many cases; yet the passages cited in preceding pages would seem to indicate that such differentiation was far from universal.

The number of estates that could support a trained retinue of house servants was relatively small, and it is thus likely, as Johnson indicates, that on the majority of those plantations where the Negroes were employed at housework, those too old or too young to work in the fields, or incapacitated for some other reason, were assigned to the task. And this being the case, it would not be surprising if the following description of table service in Jamaica was not closer in many cases to the North American reality than the traditional liveried retainers and turbaned maids and cooks of the fabled "great house":

. . . it is very common to see black boys and girls, twelve or thirteen years of age, almost men and women, in nothing but a long shirt or shift, waiting at table; so little are the decencies of life observed toward them.[32]

Certainly on the smaller farms there was no dissimilarity in whatever conditioning to white norms the young slaves might have experienced from intimate association with their master or, what is more to the point, with their master's children. Lyell gives some indication of the reciprocal nature of this relationship, which, it must be pointed out, constitutes an element in the total acculturative situation that has received far less attention than it deserves:

In one family, I found that there were six white children and six blacks, of about the same age, and the negroes had been taught to read by their companions, the owner winking at this illegal proceeding, and seeming to think that such an acquisition would rather enhance the value of his slaves than otherwise. Unfortunately, the whites, in return, often learn from the negroes to speak broken English, and, in spite of losing much time in unlearning ungrammatical phrases, well-educated persons retain some of them all their lives.[33]

Another instance, where this same writer observed intimacy of contact between the young of both races, may be given:

We were passing some cottages on the way-side, when a group of children rushed out, half of them white and half negro, shouting at the full stretch of their lungs, and making the driver fear that his horses would be scared. They were not only like children in other parts of the world, in their love of noise and mischief, but were evidently all associating on terms of equality, and had not yet found out that they belonged to a different caste in society.[34]

It is also sometimes overlooked that in the early days of slavery, and in certain parts of the South at a relatively late date, field hands had opportunities to learn the ways of life of whites with whom they were associated in their jobs. Those whose conception of "culture" is not that of the scientific definition which comprises all aspects of behavior, but who use the term in an evaluative sense of signifying behavior approved by those of "gentler" origin, would perhaps feel that this type of association was of little advantage to the slave in exposing him to the "best" of European behavior. The following passage, which refers to the earlier days of slavery, implies that European influence was present when the initial patterns of Negro behavior in this country were laid down:

Side by side in the field, the white servant and the slave were engaged in planting, weeding, suckering, or cutting tobacco, or sat side by side in the barn manipulating the leaf in the course of preparing it for market, or plied their axes to the same trees in clearing away the forests to extend the new grounds. The same holidays were allowed to both, and doubtless, too, the same privilege of cultivating small patches of ground for their own private benefit.[35]

An instance can be given of how in later times, also, contacts on this level, though more restricted, took place more frequently than is customarily recognized:

Yesterday I visited a coal-pit; the majority of the mining laborers are slaves . . . but a considerable number of white hands are also employed, and they occupy all the responsible posts. . . . The white hands are mostly Englishmen or Welchmen. One of them, with whom I conversed, told me he had been here several years . . . he was not contented, and did not intend to remain. On pressing him for the reason of his discontent, he said, after some hesitation, that he had rather live where he could be more free; a man had to be too "discreet" here. . . . Not long since, a young English fellow came to the pit, and was put to work along with a gang of negroes. One morning, about a week afterwards, twenty or thirty men called on him, and told him that they would allow him fifteen minutes to get out of sight, and if they ever saw him in those parts again, they would "give him hell." . . . "But what had he done?" "Why, I believe they thought he had been too free with the niggers; he wasn't used to them, you see, sir, and he talked to 'em free like, and they thought he'd make 'em think too much of themselves."[36]

In another passage, Olmsted again indicates closeness of Negro-white contact on the humbler levels of white society:

The more common sort of habitations of the white people are either of logs or loosely-boarded frame, a brick chimney running up outside, at one end; everything very slovenly and dirty about them. Swine, fox-hounds, and black and white children, are commonly lying very promiscuously together, on the ground about the doors. I am struck with the close co-habitation and association of black and white—negro women are carrying black and white babies together in their arms; black and white children are playing together (not going to school together); black and white faces constantly thrust together out of doors to see the train go by.[37]

In the preceding excerpts, the situations in the United States where contact between whites and Negroes occurred have been indicated, especially as regards the way in which differences in opportunity to cope with European custom made for differences in rapidity of acculturation and the absorption of these new habits in customary behavior. We have also been concerned to discover to what extent the opportunities to learn and imitate white behavior were different for differing categories of slaves, or were spread evenly over the Negro population as a whole. This leaves us with a third point to be considered—those mechanisms which, in the case of Negroes most exposed to white contact, permitted them to retain Africanisms.

As with the other points raised, the data are scanty and scattered, and intensive research will be needed before such a question can be answered in any adequate measure. Nonetheless, hints in the literature amply justify asking the question here. It is customarily assumed that during the "seasoning" process, whereby newly arrived Africans were taught the manner of life on the plantations, the scorn of the teachers for savage ways prevented any interchange that could have reinforced Africanisms present in the behavior of those in charge of the newcomers. Yet the relationship between Africans and their teachers, except as concerns plantation routine, has never been systematically studied. Such a quotation as the following indicates the accepted treatment:

Planters learned early in the use of slave labor that it was necessary to give certain trusted Negroes limited authority over the others so that with a change of overseers the plantation routine might be disturbed as little as possible. On the large plantations the seasoned Negroes trained the new ones and were responsible for their behavior. In the early days of the plantation regime, when a gang of fresh Africans were purchased, they were assigned in groups to certain reliable slaves who initiated them into the ways of the plantation. These drivers, as they were called, had the right of issuing or withholding rations to the raw recruits and of inflicting minor punishments. They taught the new slaves to speak the broken English which they knew and to do the plantation work which required little skill. . . . At the end of a year, the master or overseer for the first time directed the work of the new Negro who now had become "tamed," assigning him to a special task of plantation work along with the other seasoned hands who had long since learned to obey orders, to arise when the conch blew at "day clean," to handle a hoe in listing and banking, to stand still when a white man spoke.[38]

What magic formulae might not have been transmitted by these newly arrived Africans to a receptive ear? What discussions of world view might not have taken place in the long hours when teacher and pupil were together, reversing their roles when matters only dimly sensed by the American-born slave were explained in terms of African conventions he had never analyzed? Certainly during much of the slave period the masters did little to care for whatever needs their slaves might have had for instruction as to the nature of the world and the forces that actuate it; the numerous complaints which the lack of adequate religious teaching for the slaves inspired in the earlier period of slavery, and the lax manner in which religion was later taught them, give ample justification for asking whether African belief and African methods of coping with supernatural forces might not have been taught and thus perpetuated on this more humble level. And the same method of retransmitting and reinforcing aboriginal customs may well have been in force as regards such other African culture elements as dancing and singing and the telling of folk stories, while African patterns envisaged in the terms "morality," "etiquette," and "discretion" may also have been discussed enough to act as a brake on too rapid or too complete an adoption of white values.

This would mean, then, that not only field hands, but all slaves were exposed to forces making for the retention of Africanisms. House servants had contact with newly arrived Africans when such persons were employed as helpers in the kitchens of the great house, or, as an actual instance recounted in the Sea Islands had it, when such persons were themselves in an emergency put to cooking.[39] That to assume a continuous process of mutual influence between Negroes born in this country and those freshly arrived from Africa, in all aspects of belief and behavior, is not unreasonable is further indicated by such a citation as the following:

The "swonga" people were the drivers who took their orders directly from the overseer, the house servants who were intimately associated with the master's and overseer's families, the mechanics who were permitted to hire their own time from their masters and work in Beaufort or Charleston. To this group also belonged any among them who from superior rank or intelligence acted as their official or self-appointed leaders. The religious leaders and the plantation watchmen were usually "swonga" Negroes, as were also the witch doctors and those who could boast of physical prowess.[40]

Other occasions when Negroes in close contact with whites could reabsorb Africanisms were when slaves were released from immediate supervision, as when they worked by themselves to supplement what was provided them by their owners or when they were released for holiday celebrations. Under the former heading would come those numerous instances when slaves were permitted to take produce or chickens and eggs to market; or where they were allowed to go into swamps or uncleared forests to gather wood or trap possum and other game. In the latter category fall such festivities as those at Christmas, a holiday whose celebration on some plantations and in some regions extended into the New Year. These occasions were marked by songs and dances and games and tales, many of which, being African in character, were thus passed on from one generation to the next. These gatherings also afforded unusually good opportunities for other African cultural elements, such as world view and magical practices, to be learned and thus kept living.

One aspect of Negro experience in the United States that is significant in the total acculturative situation concerns the results of Negro-Indian contact. That in most of the New World, as well as in the United States, this contact was continuous from the earliest days of Negro slavery has not been as well recognized as might be. Students of the American Indian have speculated on the amount of

Negro influence to be discerned in the present-day tribal customs of certain Indian groups who either possessed Negro slaves they later absorbed, or offered haven to escaped Negroes, giving them places as members of the tribe. But the possibility of Indian influence on Negro behavior, either directly or at second hand through the taking over of Indian customs already accepted by the whites, has not figured in terms of the possibilities that have been envisaged in analyzing the forces which impinged on the Negro in his new habitat.[41]

It is apparent that in merely raising these questions, the need for reexamination of the problem is indicated, especially in view of the essentially simplistic character thus revealed of most statements concerning the nature, rate, and intensity of the acculturative process undergone by Negroes in the United States. As has been shown, in the West Indies the mechanisms are to be more clearly seen because the situation was such that Africanisms could be retained in great enough measure to permit the student more surely to assess the means whereby this end was achieved. In the United States, where the acculturative process went much farther than in the islands, the intellectual provincialism that has held students of slavery and of the Negro to preoccupation with the problems of the mainland alone has caused them to develop hypotheses that reflect their lack of acquaintance with the comparative data which allow the problem to be phrased in more realistic terms.

Two or three instances may be offered from the rich documentation available, to illustrate again the nature of the customary approach. The following summary statement is typical:

As individuals, the mulattoes always have enjoyed opportunities somewhat greater than those enjoyed by the rank and file of the black Negroes. In slavery days, they were most frequently the trained servants and had the advantages of daily contact with cultured men and women. Many of them were free and so enjoyed whatever advantages went with that superior status. They were considered by the white people to be superior in intelligence to the black Negroes, and came to take great pride in the fact of their white blood. They developed a tradition of superiority. This idea was accepted by the black Negroes and consequently the mulattoes enjoyed a prestige in the Negro group. Where possible, they formed a sort of mixed-blood caste and held themselves aloof from the black Negroes and the slaves of lower station.[42]

Or, again, this passage, offered as background for a consideration of the Negro family as at present constituted, states the familiar point of the special opportunities of the house servant as against the field hand to acquire, in this case, the religion of the master:

Although the house-servants because of their favored position in relation to the master class were early admitted to the churches, it was only with the coming of the Methodists and the Baptists that the masses of slaves "found a form of Christianity that they could make their own."[43]

This same author is more emphatic when he discusses the differentials in acculturation that pertain to the family:

The examination of printed documents as well as those collected from ex-slaves gives evidence of a wide range of differences in the status of the Negro family under the institution of slavery. These differences are related to the character of slavery as it developed as an industrial and social system. Where slavery assumed a patriarchal character the favored position of the house servants, many of whom were mulattoes, facilitated the process by which the family mores of the whites were taken over. Thus close association of master class and the slaves often entailed such moral instruction and supervision of the behavior of the slave children that they early acquired high standards of conduct which seemed natural to them. Sexual relations between the white masters and the slave women did not mean simply a demoralization of African sex mores but tended to produce a class of mulattoes, who acquired a conception of themselves that raised them above the black field hands. In many cases these mulattoes either through emancipation or the purchase of their freedom became a part of the free class where an institutional form of the Negro family first took root.[44]

In the light of the data cited in earlier pages, however, it is apparent that the "favored position of the house servant" is taken for granted to a degree not justified by the facts. But beyond this, the assumption that in the slave cabins no "moral instruction" took place will strike the critical reader as a highly questionable assumption. Granting lack of contact by field hands with masters, is it to be inferred that the Negroes had no values of any sort to transmit to their children? Furthermore, would not a realistic appraisal of the morals of the masters, with whom the mulatto house servants are held to have been in such close contact, also be desirable in assessing the nature of the conventions which their personal retainers absorbed? It is likewise somewhat difficult to follow the assertion that sexual relations of slave women with masters did not demoralize African sex mores, but "tended to produce a class of mulattoes." Does this mean that the African sex mores were so loose that casual sexual relations were of no importance, when compared to the fact that offspring were born of matings?

Criticism along these lines indicates the manner in which adherence to the stereotyped view, that differentials in acculturation on the basis of differing opportunities for contact are essentially to be explained in terms of assignment to house or field, tends to dull perceptions of the possible variations in the historic situation, and to induce contradictions and irrelevances in subsequent analyses of data. Because of this, quite as much as because of the doubts raised by the consultation of contemporary accounts and reevaluation of points, often made in passing, by historians of slavery, it becomes apparent that the entire problem must be reexamined if we are adequately to understand how the Negroes took over the behavior of their masters.

4

It has been stated that everywhere in the New World Africanisms are manifested to different degrees in the several aspects of Negro culture. In discussing this phase of the acculturative process, we will here concern ourselves with the manner in which these various differentials were established, leaving to succeeding chapters the task of illustrating actual survivals to be found in each of these aspects. For this, we must again briefly turn to the sociological and economic matrix of plantation life, since this was not only the dominating factor in the experience of most Negroes during the entire period of slavery, but has also played an important part in the life of large numbers of Negroes over all the New World since that time.

In outlining this approach to the problem, the mistake must not be made of regarding the Negroes merely as passive elements in the situation. For study in Africa and in the New World has shown with great clarity that, as in all societies, certain aspects of culture are of greater concern to a people than others, which means that in every culture interests tend to center on certain activities more than on others. These conscious drives, directed toward a certain segment of the entire body of tradition, determine that area of the culture wherein the greatest elaboration of basic traditions is to be found at a given period in the history of a people; and under acculturation these interests come to be those held to with the greatest tenacity possible.

If, then, the acculturative situation be analyzed in terms of differing opportunities for retention of Africanisms in the various aspects of culture, it is apparent that African forms of technology, economic life, and political organization had but a relatively slight chance of survival. Utensils, clothing, and food were supplied the slaves by their masters, and it is but natural that these should have been what was most convenient to procure, least expensive to provide, and, other things being equal, most like the types to which the slave-owners were accustomed. Thus African draped clothes were replaced by tailored clothing, however ragged; the short-handled, broad-bladed hoe gave way to the longer-handled, slimmer-bladed implement of Europe; and such techniques as weaving and iron-working and wood carving were almost entirely lost. Except for such poor barter as the slaves could contrive among themselves, or in so far as they were permitted to sell in the markets, no remnant of the economic complexities of Africa remained on the plantation. Such widespread institutions as pawning had no opportunity to function in the New World, nor could more than a few of the most rudimentary economic devices be carried on outside the all-encompassing dictates of the master.

The extinction of African political institutions also resulted from the situation of the slaves. Only in the most informal, even secret ways could African legal talent find expression or African political genius express itself. These traditions, it is true, persisted during the earlier periods of slavery even in the face of suppression, as is attested by the prompt organization of groups on African lines wherever Negro revolt was on a large enough scale to permit any stability in social structure to be realized. But such cases were exceptional, and with the continuation of the slave status down the generations, aboriginal traditions in these aspects of culture tended to become more and more dilute, until today, as will be seen, they exist only rarely in immediate African form.

In religion and magic, on the other hand, and as concerns certain nonmaterial aspects of the aesthetic life, there was greater reason, in terms of both aboriginal interest under slavery and the masters' requirements, for retentions. For, as it has been put:

. . . The Africans were brought over to be industrially exploited, and the white master was careful to see that American farming practice was followed by the slaves. He cared less about the amusements and religion of the Negro so long as they did not affect his working ability.[45]

Slaves were proselytized by both Protestants and Catholics—if compulsory baptism of Negroes in the Catholic countries of Hispaniola, Cuba, Brazil, and elsewhere can be described by this term—but whatever the attention given to "religious instruction" of the slaves in various areas and at various periods of slavery, the freedom of the slaves to conduct their own services without white supervision was always greater than their freedom to work or organize politically in the African manner. Magic was almost by its very nature adapted to "going underground," and was the natural prop of revolt, as the following passage shows:

> Gullah Jack (one of the leaders in Denmark Vesey's Insurrection in South Carolina in 1822) was regarded as a sorcerer. . . . He was not only considered invulnerable, but that he could make others so by his charms (consisting chiefly of a crab's claw to be placed in the mouth); and that he could and certainly would provide all his followers with arms.[46]

Secret in its manipulation, it came to be feared almost as widely among the masters as among the slaves.

The attitudes of the masters toward song and dance and folk tale varied throughout the New World from hostility and suspicion through indifference to actual encouragement. Such a trivial matter as the extent to which recreational forms tended to interfere with the master's personal convenience was important; such apparently irrelevant features as the amount of noise made in dancing or while singing tended to influence white attitudes. The quiet with which tales are told, plus their appeal to the whites as stories for children, made the retention of this element of African culture as ubiquitous as it is in the New World. African types of dancing and singing were allowed when they did not interfere with work or were performed on holidays; at such times, according to numerous accounts, they were enjoyed by the masters who watched them as much as by the slave dancers and singers. The rhythmic accompaniment to song which, as we have seen, is fundamental in African musical expression, was forced into various channels. The slave-owners found to their cost that drums which beat for dances could also call to revolt, and thus it came about that in many parts of the New World the African types of hollow-log drums were suppressed, being supplanted by other percussion devices less susceptible of carrying messages and could thus be restricted to beating dance rhythms.

The disappearance of another outstanding form of African aesthetic expression, wood carving, is to be attributed to a number of causes, not the least of which was economic. Slaves were bought to be worked, and the leisure essential to the production of plastic art forms was entirely denied them. Furthermore, there was little if any demand for what might have been carved, since European patterns of art appreciation were at that time hardly such as to encourage the production of such exotic art forms. Why one special style of African carving should have survived in one part of the

New World, and there alone, we do not know, but the special circumstances under which Yoruban carving survived in Brazil, should these ever be discovered, will throw light on why similar survivals are not found elsewhere.

Institutions in the field of social organization stand intermediate between technology and religion in respect to retention in the face of slavery. It goes without saying that the plantation system rendered the survival of African family types impossible, as it did their underlying moral and supernatural sanctions, except in dilute form. Only where Negroes escaped soon enough after the beginning of their enslavement, and retained their freedom for sufficiently long periods, could institutions of larger scope such as the extended family or the clan persist at all; and even in these situations the mere breakup in personnel made it unlikely that some manifestation of European influence should not be felt. In Dutch Guiana alone has the clan persisted; what forms the social structures of present-day Negro communities of Brazil take is unknown, but in Haiti and Jamaica larger groupings go no further than a kind of loosely knit extended family. Yet, on the other hand, slavery by no means completely suppressed rough approximations of certain forms of African family life. Even in the United States, where Africanisms persisted with greatest difficulty, such family organization as existed during slave times in terms of the relationship between parents and children, and between parents themselves, did not lack African sanctions.

Though slavery gave a certain instability to the marriage tie, in the New World as a whole the many persons who lived out their lives on the same plantation were able to establish and maintain families; even in the United States it is far from certain that undisturbed matings have not been lost sight of in the appeal of the more dramatic separations that actually did occur in large numbers. As will be indicated in the next chapter, certain obligations of parents to children operative in Africa no less than in the European scene were carried over with all the drives of their emotional content intact. The special kind of relationship between husband and wife characterizing the Negro family in all parts of the New World presents a problem whose historical solution can by no means be satisfactorily reached without reference to comparable patterns of life in preslavery Africa. The vivid sense of the power of the dead and the related feeling that the ancestors are always near by to be called on by their living descendants give a kind of strength to family ties among Negroes that can be traced in lessening degrees of intensity as one moves from West Africa itself to the New World areas where contact with European patterns was closest. It was but natural that all these elements of attitude, belief, and point of view concerning the ties of kinship should have been passed on as children were taught by their parents; to have been inculcated, moreover, without undue interference by the masters as long as they led to no action that would impede the smooth functioning of the estates.

The traditions underlying nonrelationship groupings of various kinds likewise survived the slave regime. The degree to which this is true varied with the function of a given organization, approaching the impossible in the case of those secret societies so widely spread in the parts of Africa from which the slaves were drawn. In the latter instance, this sort of organization could only go underground or disappear, but for other kinds of associations such drastic action was not necessary. The spirit behind the numerous types of cooperative societies of Africa tended to be kept alive by the very form of group labor employed on the plantations. The feeling for mutual helpfulness inherent in this tradition contributed directly toward the adjustment of the African to his new situation, for without some formula of mutual self-help he could scarcely have supported the oppression he suffered. And how strongly this formula did persist is indicated by the manner in which, on emancipation, cooperative organizations sprang up immediately in the Sea Islands, or how in the West Indies insurance organizations, of the kind common in Africa, at once came into being. The great number of Negro lodges in the United States, though outwardly following conventional white patterns, are by no means the same as their white counterparts. One factor in preserving African sanctions in institutions of this sort has been the sense of the importance of leadership that characterizes all kinds of African social institutions. The principle of order and regularity, induced by discipline exerted through responsible headship, permeates African life, and this, reinforced by the very submission to authority demanded of the slave, has in many ways flowered under freedom.

In seeking to understand the mechanisms making for the differing degrees of African elements found today in the several aspects of New World Negro culture, the forces which drove these folk toward the acceptance of European culture must also be evaluated. That is, we must consider those positive measures which made for acceptance of the master's way of life as well as the negative forces that, without conscious direction, operated to discourage the retention of aboriginal patterns. The difference between these two drives may be illustrated by an example. As has been stated, the economic system in the New World tended of itself to inhibit African material culture and technological capabilities. Ironworking, wood carving, basketry, and the like simply had no place in the new scene, and hence such techniques almost everywhere died out. On the other hand, proselyting among the slaves by Christian missionaries of various denominations constituted a positive drive. In the realm of belief, there is no logical reason why African world view might not have been continued in the same way as motor habits of African dancing were retained. Changes would undoubtedly have appeared of themselves, as they have in the dance, since some measure of innovation is the inevitable result of the contact stimulus. Yet in the case of the African world view, the efforts directed toward effecting change caused a premium to be placed by the whites on the overt acceptance of Christian religious beliefs and practices, and thus accelerated the disappearance of African forms.

A further factor in inducing acculturation was an unconscious identification by the Negroes of the better way of life with the customs of those who possessed the power to get for themselves the good things of existence. With passing generations, prestige values among the slaves, certainly of the United States, came more and more to be based on white values. Where contact was less immediate and less constant, as was the case in the Caribbean, the identification of these values with the traditions of the masters was not so strong, while in this area there were those who could function effectively in terms of African ways of life and could thus retain prestige in terms of these capacities in a manner not possible on the continent. The worker of magic, the wise old woman, the man whose personality made him a leader in cooperative effort or in successful revolt thus retained a hold on the people such as was impossible where the impact of European custom was such as to inhibit these individuals from employing African methods of coping with their problems.

This brings us to a final point in considering the forces that caused the differentials in Africanisms existing today in the various aspects of New World Negro culture in any given region—the effect of that resilience toward new experience which in itself is a deep-seated tradition of Africa. It has already been indicated how, in West Africa, it was common for both conquerors and conquered to take over one another's gods and how, in the course of a man's everyday experience, it was deemed more advantageous for him to give way to a point of view against which he could not prevail than to persist in his attitude, however firmly he might hold an opinion. This tradition underlies the soft-spoken politeness for which the Negro is famous, which, in the form of a code far more elaborated in Africa and among New World Negroes than it ever was in European life or in the behavior of whites in the Western Hemisphere, characterizes the relationships between Negroes quite as much as it does the behavior of Negroes toward whites. This tradition likewise gives historic validity to the circumspectness that has so often been interpreted by students of Negroes in the United States as a mere reflection of an accommodation to slavery that persisted because of the disadvantaged social and economic position of the Negro since emancipation. Yet we have seen how deeply rooted is this circumspection in Africa itself, and how it is found among Negroes in every part of the New World. Though undoubtedly reinforced by the exigencies of slavery, it is thus nonetheless to be considered the carry-over of an older pattern, rather than merely something which afforded a means to adjust to the difficulties of life where freedom of personal decision was not permitted.

Certain striking instances that document this tradition of pliability can be traced in the religious life of Negroes in those parts of the New World where, Catholicism being the official religion, numerous Negroes are members of the church while at the same time they continue African modes of worship. What seem to be far-reaching contradictions are reconciled without apparent difficulty, for the pagan spirit believed to control a given manifestation of the universe is merely identified with a given saint, and unless missionary pressure places the African spirit under a ban and removes the prestige it would normally receive as a functioning entity, no demoralization results. The fate of African percussion instruments offers another example of this process of accommodation on a less dramatic level. African drums have entirely disappeared in the United States, yet one who is familiar with African music in its original forms cannot hear "boogie-woogie" piano rhythms without realizing that there is little difference between the two except in the medium of expression. Dutch Guiana provides a hint of how such adaptations came about. It is forbidden to use African types of drums in the city of Paramaribo, except at certain specified times. Yet the proper rhythms must be beaten whenever a diviner determines that an illness has been caused by a spirit, since drumming the rhythms of that spirit is essential to effecting a cure. Adaptation to the legal ban is simple, employing objects of European manufacture never intended for such use. A metal washbasin is filled with water, and another caused to float in it upside down; the rhythms beaten on the bottom of this smaller basin give the sound of a hollow-log drum without the same carrying power. The curing rite is thus carried out quietly, and African medical practices continue despite the troublesome rule.

Chapter VI

THE CONTEMPORARY SCENE: AFRICANISMS
IN SECULAR LIFE

In the preceding pages the position taken by students of the Negro concerning the existence or nonexistence of African survivals among Negroes of this country has been considered, and reservations to this position advanced on the basis of an horizon broader than that encompassed by the boundaries of the United States alone. The problem of Negro origins has been outlined, and the state of our knowledge about the tribal ancestry of New World Negroes assessed. The cultural heritage brought by them to the New World has been sketched and compared to statements commonly made regarding African societies; and this has demonstrated how far the reality is removed from assertions found in the literature as written by those who, lacking firsthand experience in Africa, have in addition but slight control of the source materials. How the slaves were obtained, and the extent to which they constituted selected portions or represented a cross section of the West African population, has been analyzed, and the refusal of the Negroes to accept slavery, in the manner customarily attributed to them, indicated. Finally, the mechanisms whereby the adjustment of the Negroes was achieved in the New World have been detailed. And this brings us to the subject of our ultimate concern—the survivals of African traditions, attitudes, and institutionalized forms of behavior actually to be observed in present-day Negro life in the New World, particularly in the United States.

In the discussion which follows, certain points are to be kept in mind. In the first place, the attack is always on the broad front implied in the methodological principle that Africanisms in the United States can be understood only in terms of the increased intensity of their counterparts elsewhere in the New World. This method permits recognition of vestigial occurrences of African tradition in this country that might otherwise be overlooked; in the case of more obvious survivals, it affords the best clues to the processes which brought about the changes they have undergone. In short, though this analysis is primarily oriented toward the problem of Africanisms in the United States, this country will be treated as but a portion of the larger New World area which, from the point of view of the derivation of its Negro inhabitants, is largely an ethnic and cultural unit.

A second point to be remembered is that documentation rests almost entirely on the available literature. This means that the materials are weighted on the side of those aspects of Negro life that have attracted students of Negro custom. Because in this literature the same citations of fact are repeated from one work to another, it has not been considered necessary to employ any but the more comprehensive sources. Puckett's standard work on Negro folk customs, which, despite the emphasis on European provenience laid in its opening and concluding chapters, is filled with materials that point directly to Africa, is one such work. C. S. Johnson's study of the Negro rural population of Macon County, Georgia, and the reports on a small-town Negro group by Powdermaker and Dollard also have been found useful. For the kinship units in Negro social organization, Frazier's volume has been utilized, as have the works of Parsons and G. B. Johnson. Finally, the files of such periodicals as the *Journal of American Folk-Lore* and certain literary studies of southern Negro life which have been found to contain significant sociological and ethnographic materials complete this list of the kinds of sources primarily called on to supply our data.

A most serious handicap is the absence of adequate field reports based on a combined historical, demographic, and comparative ethnographic attack. Field work by Powdermaker, conceived in the broadest manner of any single study of a Negro group in the United States, suffers from lack of acquaintance with data from Negro societies in other parts of the New World and Africa. Interpretations in this work are therefore often speculative where they might be subject to historical control; more regrettable is the fact that some of the critical materials which would have given greater insight into the life of this group have been overlooked. The psychological preoccupations of Dollard in studying the same community have conditioned the frame of reference employed in gathering his data, and also frequently tend to becloud his interpretations. Moreover, the same criticism, on a somewhat different level, may be lodged against the results of most sociological research in this field, since, as in the case of Johnson, for example, control of requisite comparative materials is likewise lacking.

Finally, a caution is in order concerning the degree of purity assumed to exist in the African traits to be reviewed. Because of the emotional "loading" of attitudes toward the problem under discussion, the attempt to trace Africanisms is too frequently met with the counterassertion that the Negroes of the United States are not Africans, regardless of the fact that no implication of this kind is involved. In this discussion the point of view is held that, as in all scientific inquiry, the data must be followed wherever they lead; and that an open mind on all phases of the problem must be retained until all possibilities of analysis have been exhausted. Negroes in the United States are not Africans, but they are the descendants of Africans. There is no more theoretical support for an hypothesis that they have retained nothing of the culture of their African forebears, than for supposing that they have remained completely African in their behavior. The realistic appraisal of the problem attempted here follows the hypothesis that this group, like all other folk who have maintained a group identity in this country, have retained something of their cultural heritage, while at the same time accommodating themselves, in whatever measure the exigencies of the historical situation have permitted, to the customs of the country as a whole.[1]

2

Our analysis of African survivals may begin with a consideration of how certain isolated African traits have held over in American Negro behavior, most often in uninstitutionalized form. That more examples of carry-overs falling in this category are not found in the literature is probably due to the lack of acquaintance of observers with related New World Negro cultures and the African background, so that such points of significance are not reported. From a larger point of view, however, these instances, in the aggregate, contribute to no slight degree to the total pattern of Negro behavior and, because of their African character, help to distinguish it from the behavior of other elements in the population.

The retention of Africanisms in motor habits presents a vast field for study. Methodological difficulties in the way of such research are appreciable, since results having scientific validity can be obtained only by analyzing motion pictures of such routine activities as walking, speaking, laughing, sitting postures, or of dancing, singing, burden carrying, hoeing, and movements made in various industrial techniques. Yet on the basis of uncontrolled observation, it is a commonplace that many American Negro forms of dancing are essentially African; and this is confirmed by motion pictures taken of the Kwaside rites for the ancestors of the chief of the Ashanti village of Asokore, which include a perfect example of the Charleston, or by the resemblance to other styles of Negro dancing well known in this country included in films taken in Dahomey and among the Yoruba.

In another less well-known field, it may be indicated that the precise method of planting photographed in Dahomey and in Haiti was observed by Bascom in the Gulla Islands in the summer of 1939.[2] This method, already described and illustrated for these other two regions,[3] is to work down and back each pair of rows in a field. A container of seed is held under the left arm, and the right or left heel, as the case may be, is used to make a shallow depression in the soil. The seeds are dropped in this hole, and dirt to cover them pushed over it with the toes; this foot is then placed ahead of the sower, and the same movements performed in the opposite row with the other foot. Whether this method is used elsewhere in the United States cannot be stated, but where it does occur it constitutes a direct survival of a West African motor habit.

The description given of a Sea Island woman of the Civil War period may be cited as another instance of the survival of motor behavior:

It was not an unusual thing to meet a woman coming from the field, where she had been hoeing cotton, with a small bucket or cup on her head, and a hoe over her shoulder, contentedly smoking a pipe and briskly knitting as she strode along. I have seen, added to all these, a baby strapped to her back.[4]

The habit of carrying burdens on the head, so widespread in tropical countries, is favored in West Africa and the West Indies. To what extent it has survived in the United States cannot be said, but that the practice has had an important influence on walking style is apparent. Whether or not it is the factor that has given the Negro his distinctive walk is for future research to determine, but the point must be kept in mind as at least a somewhat more tenable hypothesis than that advanced by one Freudian disciple, who held the Negro "slouch" to be the manifestation of a castration complex! The ways in which southern rural Negro women habitually carry their infants do not today ordinarily include the method depicted in the quotation above; as described, however, it corresponds exactly to one manner in which infants are transported in West Africa. The other method, still commonly to be seen among persons of the lower socio-economic strata of Negro rural society in this country, is to use one arm to hold the child as it straddles the hip of its carrier.

The retention of certain industrial habits is hinted in the following passage:

Broughton was a rice island, and the garnered rice was carried from the fields to the flats, then towed to the mill, where it was threshed and loaded on ships to be carried to the city. The rice was not husked at the plantation mills. This was done in the city, as rice was not considered good in those days unless freshly beaten. On the plantation it was beaten fresh for dinner every day. For this purpose pestles and mortars, hewn from the trunks of trees, were used, these becoming smooth and shining like metal from constant use. Two boys or two women would seize the pestles together in the middle, raising and letting them fall so quickly and evenly that the beating of rice was not considered a difficult task. The children often tried it, but never succeeded, as the motion required a knack they did not possess. After the rice was loosened from the husks, it was placed in flat-bottomed baskets called fanners, held high, and allowed to fall into baskets placed on the ground, the wind blowing the chaff away. This process, which was called "fanning the rice," was repeated until the rice was perfectly clean.[5]

This technique is still employed in the islands, but the use of mortar and pestle elsewhere in the South has not been reported. Mortars and pestles of the type included in the collection from the Sea Islands at Northwestern University and woven trays used in winnowing the rice[6] are entirely African. Their use to shell cereals of one kind and another is ubiquitous throughout Africa, though of course not confined to that continent. The way in which these are used, however, shows a further retention of motor habit, especially in the tendency to work as rhythmically as possible; in the West Indies, Guiana, and West Africa, it takes some experience for the visitor to learn to distinguish the alternate strokes of two pestles in the mortar from the beat of a drum. The woven trays used in the Sea Islands are made with the sewing technique called coiling, which is paramount in West Africa; more interesting is the fact that, as in Africa, the coils, in all instances examined, are laid on in a clockwise direction. This is an excellent example of the way in which the determinants of behavior lying beneath the conscious level may be continued where the manipulation of materials is involved. And this point is of the utmost importance in assessing carry-overs in personal habits which, lying beyond the attention of those to whose advantage it was during slavery to mold Negro behavior, continued undisturbed until today, to be numbered among those intangibles which give to the expression of Negro motor habits their distinctive form.

The ways in which the hair of some Negro women, but particularly of small Negro girls, is dressed is so distinctive that only mention of this convention is necessary to bring it to mind. As far as is known, the one attempt to give a derivation for these styles of hairdressing has been made by Puckett:

> In Africa, decoration of the hair reaches a high point, often consisting in mixing some plastic material with the hair and shaping the whole into a highly fantastic coiffure. With the Negro woman of the South the hair is still a prime object of decoration as evidenced by the many elaborate coiffures and by the "Hair Dresser" signs on many a lowly Negro cabin; although there is a decided tendency to remove the kink, by odoriferous unguents of all kinds in imitation of the straight hair of the whites.[7]

Yet a statement of this kind is of but little help, for in many parts of the world men and women take pains in dressing their hair. That Negroes have many methods of hair straightening is well known; but this is decidedly not African, for nothing of the sort has been recorded from there. The most popular system of hair treatment used by Negroes is named "Poro," presumably after a Sierra Leone secret society of that name; but this merely indicates how this unit of American business conforms to the procedures of American business enterprise in general.

The correspondences to be found in hairdressing are, however, far more specific and definite than those mentioned in the preceding passage. Unfortunately, we do not know whether definite names are given the many patterns into which the hair of Negro women and children is braided, nor have the actual braid designs been systematically described. Yet the multiplicity of these is the outstanding feature of the hair-braiding pattern; unbroken parts running the length of the head, lengthwise parts broken by lateral lines, and many other combinations emphasize the contrasts between the whiteness of the scalp and the blackness of the hair, while the units into which the hair is gathered for braiding are frequently so small that one wonders how the braids can be achieved. These modes of hairdressing are ubiquitous in West Africa, while everywhere in the West Indies Negro girls and women dress their hair in a similar manner, with similar designs based on the whiteness of the lines when the scalp shows between the numerous parts. In Dutch Guiana, these designs are frequently given names; among the Bush Negroes of that country men as well as women part and braid their hair in this fashion. This, however, is a local elaboration, since in West Africa and the rest of the New World men customarily cut the hair close and wear it unparted, in the manner to be seen among the rural Negroes in this country.

"Wrapping" the hair is part of the head-dressing complex as is the wearing of kerchiefs. This wrapping has been recorded both in the Sea Islands and farther west. Parsons has remarked the custom in the former locality:

Women, old and young, quite commonly wear kerchiefs around the head and tied at the back. Underneath, the hair is likely to be "wrapped." You "wrap um" (i.e., wrap strings around wisps of hair), beginning at the roots of the hair, and winding to the ends, "to make um grow."[8]

Again, from Missouri, the existence of this custom during a somewhat earlier period is vouchsafed:

There was nothing Aunt Mymee desired less than a "head-handkerchief," as she wore her hair (except on Sundays, when it was carded out in a great black fleece), in little wads the length and thickness of her finger, each wad being tightly wrapped with white cord.[9]

Concerning the wearing of headkerchiefs in the United States, another Africanism, we have but little knowledge. The headkerchief was common enough so that it came to be accepted as an integral part of the conventional portrait of the Negro "mammy," and a pre-emancipation passage hints at reasons for this earlier importance:

Precedence and rank were respected among the slaves. In Charleston Ferguson noted that the married women were distinguished by a peculiarly-tied kerchief they wore upon their heads.[10]

In recent decades the wearing of headkerchiefs has greatly decreased in the United States, but they are to be seen everywhere in the West Indies, while, as we move southward to Guiana, they are found to function importantly in the everyday life of women through the varying significance of the names given kerchief designs and styles of tying.[11] A West African distribution cannot be given on the basis of our present knowledge, but that a considerable number— over fifty—proverb-names for styles of tying kerchiefs could be recorded among the Ashanti of the Gold Coast is to be regarded as of some significance. Informants there maintained that the custom was one of long standing. In Haiti a white headkerchief marks the *mambo*, or woman officiant in the *vodun* cult, and elderly people in general, men as well as women, often wear kerchiefs bound about their heads.

3

Outstanding among the intangible values of Negro life in the United States is strict adherence to codes of polite behavior. Comments on the etiquette of Negro slaves are numerous, and some of these may be cited as illustrating the point. Botume, who worked among the Negroes released by Union troops on the Sea Islands, gives one aspect of the code:

> Before I had gone far I discovered that as I had begun to make calls, I must not omit one house, nor fail to speak to a single person, from the oldest grandparent to the youngest child. Their social rights were inexorable. My guide said, "All them people waits to say how d'ye to you," so I went on.[12]

Doyle, whose discussion is primarily concerned with the canons of interracial behavior, quotes a contemporary statement which shows how readily the pattern of politeness among whites was taken over by the slaves who accompanied their masters to the health resort at White Sulphur Springs:

> If you would take your stand near the spring where they come down after pitchers of water you would witness practical politeness. The courtesy of Samuel, coachman of Dr. W—— to Mary, the maid of Mrs. Colonel —— . . . The polite salaams of Jacob to Rachel, the dressing woman, and of Isaac, the footman, to Rebecca, the nursery maid, would charm you.[13]

That this behavior did not merely imitate that of the whites, but had a solid foundation in the mores of the slaves themselves is to be seen from the following, wherein Douglass tells how politeness was exacted in the cabins:

> . . . These mechanics were called "uncles" by all the younger slaves, not because they really sustained that relationship to any, but according to plantation *etiquette*, as a mark of respect, due from the younger to the older slaves. Strange, and even ridiculous as it may seem, among a people so uncultivated, and with so many stern trials to look in the face, there is not to be found, among any people, a more rigid enforcement of the law of respect to elders, than they maintain. I set this down as partly constitutional with my race, and partly conventional. There is no better material in the world for making a gentleman, than is furnished in the African. He shows to others, and exacts for himself, all the

tokens of respect which he is compelled to manifest toward his master. A young slave must approach the company of the older with hat in hand, and woe betide him, if he fails to acknowledge a favor, of any sort, with the accustomed "tank'ee," etc. So uniformly are good manners enforced among slaves, that I can easily detect a "bogus" fugitive by his manners.[14]

This strict ordering of conduct is by no means a matter that began with slavery nor has ended with it. Puckett in several passages comments on Negro etiquette, attempting to account for it in a number of ways, among which is the importance of taking adequate precautions against magic:

Many of these taboos have to do with matters of etiquette and seem to be in reality a linking of unpleasant results with uncouth manners in an attempt to frighten the young into a quicker acquisition of American good-breeding.[15]

As will be seen shortly, however, the elements in the Negro code differ somewhat from patterns followed by the American majority, so that an explanation in terms of drives to acquire these new modes of conduct is not entirely satisfactory. Puckett's analysis of the respect accorded elderly folk is more to the point. Here he notes "the practice of calling all old people 'Uncle' and 'Aunty' whether they are relatives or not,"[16] and in the following passage, which affords testimony of how viable has been the custom noted for a preceding generation by Douglass, he says:

. . . it is considered bad luck to . . . "sass" the old folks. This latter idea may have at one time had a real meaning, since the old folks were "almost ghosts," and hence worthy of good treatment lest their spirits avenge the disrespect and actually cause bad luck to the offender.[17]

The validity of this explanation is best indicated by referring the assertion that "old folks" are "almost ghosts" to the tenets of the ancestral cult which, as one of the most tenacious Africanisms, has left many traces in New World Negro customs. For, as has been shown, the belief in the power of the ancestors to help or harm their descendants is a fundamental sanction of African relationship groupings, and this has influenced the retention of Africanisms in many aspects of Negro life in the New World.

Another specific survival of African etiquette is the matter of turning the head when laughing (sometimes with the hand over the mouth), or in speaking to elders or other respected persons of averting the eyes and perhaps the face. The clue to this correspondence came when working with a native of the Kru tribe of Liberia, who,

while demonstrating his language to a university class, performed
what was thought to be this characteristically American Negro gesture as the group before him laughed at a joke he had made. On
inquiry the nature of this as a form of politeness was made clear,
the theory behind it being that it is rude to laugh in the face of
another. This convention was later found general in other portions
of West Africa; unfortunately, the literature does not deal with
minor matters of personal conduct such as this, and other comparative data are therefore lacking.

In Guiana, not only does one not laugh in the face of another,
but a young man does not even look at the elder to whom he is
speaking. Moreover, he speaks in a low voice, and introduces a conventionalized stammer into his speech. How this pattern has carried
over into Negro behavior in this country is to be seen from the experience of a colored principal in a northern school, where many
children, recent migrants from the South, had to be dealt with. It
was only when this officer learned that to turn the head is a mark of
respect and not a sign of inattention, that the injustice that had been
done to a number of these southern Negro children sent to the school
office for discipline was realized.

The manner in which, in many Negro churches, the sermon forms
a kind of litany between preacher and congregation represents the
reworking of still another form of African polite behavior. In these
discourses, it will be recalled, the words of the preacher are constantly interspersed with such expressions as "Yes, Lord," or "Oh,
Jesus," and those other numerous phrases that have come to be
standard in such rituals. Insight into the African nature of this
convention came during field work in the interior of Dutch Guiana,
where a running series of assents to what is being said by a man of
rank or age punctuates his speech, the responses being the more
frequent and fervent the more important the person speaking, and
the greater the respect to which he is entitled. "Yes, friend," "So it
is," "Ya-hai," "True, true," are some of the expressions which are
as standardized as are the interpolations of Negro worshipers during
the sermons of their ministers.[18] The same trait marks West Indian
Negro churches, while in the Caribbean there is also a tendency to
interject stylized assents into what is often no more than give-and-
take between two acquaintances. And, completing the sequence, it
may be noted that the same rule of polite conduct characterizes the
African scene, both as regards the responses made by common persons to those of rank and between persons of equal position, it being

explicitly stated there that to listen passively to the words of another is to be guilty of rudeness.

A different kind of carry-over from an earlier tradition is found in such an intangible as the concept of time held by Negroes. By this is not meant the disregard for punctuality so often made the occasion for joking when reference is made to such an hour as "eight o'clock C. P. T." signifying that, since this is "Colored Peoples' Time," an hour or two later than the one named is actually meant. Disregard for punctuality is to be expected wherever timekeeping devices are lacking; which is to say that approximations of time rather than punctuality mark the life of most human beings. What is meant here is the way in which the day is divided, and the special significance for Negroes of terms such as "evening" and "morning." The point can be made by a quotation from Dollard which illustrates "the initial appearance of strangeness" he experienced in the manners of the Negro community he was to study:

I took my laundry one day to a Negro laundress . . . and asked her when it would be ready. She said, "Oh, tomorrow evening." After supper the next day I went back. She reproached me on the ground that it had been done for five or six hours and I could have had it earlier: "I expected you to come around about two o'clock this evening." Morning is from when you get up until around two, and evening is from then on. At first I thought only Negroes used the word in this way, but later found that white people do too.[19]

The same linguistic conception of time divisions is to be encountered throughout the West Indies, while the prevalence of a similar usage throughout West Africa traces them to their source.

At this point we may consider in some detail manifestations in the United States of the ability of the Negro to adjust to his situation by adapting himself to the requirements of the moment, a point that has been referred to several times before. Ordinarily, this is held to indicate the quickness of members of this underprivileged group to comprehend and acquiesce in the wishes of those over them, especially manifest in their circumspectness in handling whites. Some of the comments that have been made on Negro reticence and pliability may be indicated. Doyle analyzes the common reaction:

The Negro . . . "gets along" because, when in doubt as to what is expected of him, he will ask what is customary—not what is the law. He seems subconsciously to feel that custom is more powerful than law. And yet there are instances where no one can tell him just what is the custom or what will be accepted. In this case he falls back on old habits.

If these habits are not accepted, the Negro merely "turns on his personality" and, by apology, ingratiation, or laughter, will be able to turn even this hard corner.[20]

How one element in this technique was employed during slavery is recounted by this same writer in the following passage:

A slave, on occasion, might be impudent if he supported his impudence with a quotation from the Scriptures. A slave trader was unloading a carload of Negroes at a station in Georgia. As he stepped on the platform he asked if all the Negroes were there. Thereupon one slave replied: "Yes, massa, we's all heah." "Do dyself no harm, foh we's all heah," added another, quoting Saint Peter. . . . On other occasions slaves would improvise songs which were positively impudent, but which, clothed in the right forms, would pass unnoticed, or even provoke a smile or laughter.

> We raise de wheat, dey gib us de corn;
> We bake de bread, dey gib us de cruss;
> We sif' de meal, dey gib us de huss;
> We peal de meat, dey gib us de skin;
> And dat's de way dey takes us in.
> We skims de pot, dey gib us de liquor,
> An' say, "Dat's good enough fer a nigger."[21]

Puckett feels that perhaps "the opportunity of poking fun at the white race in an indirect way is the basis of the many Irishman jokes, so widespread among the Southern Negro," and indicates the form which such satisfactions take in the following passage:

. . . the Negro does love to laugh at the mishaps of his white master, as evidenced by such stories as that of the new field hand who did not understand the meaning of the dinner bell. His master found him in the field still working after the bell had rung, and angrily commanded him to "drop whatever he had in his hands" and run for the table whenever he heard it ring. Next day at noon he was carrying his master, taken sick in the fields, across a foot-log over the creek when the bell rang. He "dropped" the white man in the water and nothing was done to him for he had only done what the master had commanded.[22]

This complex of indirection, of compensation by ridicule, of evasion, and of feigned stupidity has obviously been important in permitting the Negro to get on in the different situations of everyday life he has constantly encountered. How this operated during the days of slavery has been summarized as follows:

. . . the Negroes are scrupulous on one point; they make common cause, as servants, in concealing their faults from their owners. Inquiry elicits no information; no one feels at liberty to disclose the transgressor;

all are profoundly ignorant; the matter assumes the sacredness of a "professional secret": for they remember that they may hereafter require the same concealment of their own transgressions from their fellow servants, and if they tell upon them now, they may have the like favor returned them; besides, in the meantime, having their names cast out as evil from among their brethren, and being subjected to scorn, and perhaps personal violence or pecuniary injury.[23]

Mutual aid on this level has continued to characterize Negro behavior. This passage explains reactions described in such a literary work as *Porgy*. In this play it will be remembered how the white man asks for Porgy at the cluster of dwellings where he lives; how, though Porgy himself is present, no one reveals that he has so much as heard of such a person until the good intentions of the inquirer have been established.

That reactions of this type are common among New World Negroes is shown by an incident which occurred a few years ago in the island of St. Vincent. During field work in Dutch Guiana, it had been discovered that a certain African game, named variously *adji boto* and *wari*, was played by the Negroes of the bush and coastal region there.[24] Because it was an important item in the list of New World African survivals, attempts were made to discover its further distribution in the Caribbean islands, since certain almost involuntary reactions of Trinidad Negroes who had seen a board collected in Guiana indicated that they were not entirely unacquainted with the game, despite the fact that none of them would admit knowing it. At St. Vincent, therefore, no time was lost in making inquiry of one of the boatmen who rowed passengers ashore. "Wari, wari?" He repeated the term. "Never heard of it." It was then explained that the game was played with "horse-nickel" seeds on a board having twelve holes, whereupon he replied, "Oh, you mean wari! I've heard of it, but we doesn't play it here. They plays it strong in Trinidad." It was then indicated that at Trinidad it had been said that the game was played in St. Vincent, and disappointment was expressed that it would not be possible to have a game before the ship left port. Since few whites know this game, the man looked up sharply. "You play wari? French or English?" On learning that either was acceptable, he pondered further. "I think I know a man who has a board like that. I'll see if I can find him." Eventually, the game was played —with this same boatman, in his own house, on his own board! The incident becomes the more significant when it is borne in mind that the game has no moral or political importance that would lead to its suppression; that it is not a gambling game, but is played for the prestige that goes with winning; that there is no reason why anyone should deny knowing it. Yet in the islands, unless one comes on men actually playing it, knowledge even of its existence is almost invariably denied; and not until it becomes clear that it is already known to an inquirer is there any relaxation of vigilance concerning this game, which is so closely identified with Negro life.

Yet why must reference be made to the African past in accounting for this pattern of indirection? Is it not true that all underprivileged peoples take recourse to subterfuge and concealment as their only weapon? Is it not true, in any event, that such lack of frankness merely represents the customary caution of the peasant mind? In short, cannot such a tradition of reserve be attributed to experience under slavery more immediately than any other part of Negro behavior? Whatever the African basis for this attitude, it must be clear that slavery did nothing to diminish the force of its sanctions. Nor have the disabilities under which the Negro has lived since slavery tended to decrease its appeal as an effective measure of protection. Nonetheless, certain characteristic reactions to life in Africa itself—on the part of upper class as well as ordinary folk, which even take certain institutionalized forms in the political system of at least one well-integrated African culture—make it essential that this tradition of indirection be regarded as a carry-over of aboriginal culture.

One instance where this view was clearly expressed was in the course of a discussion of nonesoteric aspects of Bush Negro burial rites—such matters as the disposal of the house of the dead, kinds of goods placed in coffin, and the like. The conversation dealt with no new points, but was merely incident to checking certain overt details of death rituals which had been jointly observed and participated in by questioner and informant. Suddenly, however, the conversation ended with a flat refusal on the part of the Bush Negro to discuss the matter further. Argument availed nothing, except to elicit a reply that was more enlightening than the information sought could possibly have been: "White man, long ago our ancestors taught us that a man must not tell anyone more than half of what he knows about anything. I have told you half of what I know." No better exposition of the point of view under discussion could be desired; that it could be expressed so succinctly is an indication of how consciously it is accepted by these people as a guiding principle in everyday relations, while the recognition of its applicability by American Negro groups to whom this incident has been recounted shows that it is not limited by any means to Dutch Guiana.

Numerous examples of the operation of the rule of indirection in West Africa could be given, but its institutionalized forms best demonstrate how congenial is the principle to the thought of the people. An outstanding instance is the role it played in the taxation systems of the various monarchies. These systems have been described in detail, both for the Ashanti of the Gold Coast and for the kingdom of Dahomey, and hence need only be outlined here.[25] Among the former people, the throne did not exact inheritance taxes except at several times removed from the original levy. When an ordinary man died, the local chief took control of the government's share, and retained this during his lifetime. The duties on the estate of such a local chief went to his superior, the district head; and it was only on the death of such a high official that the inheritance taxes of those under him who had died during his lifetime finally reached the central power. In Dahomey, the entire system of census enumeration and the taking of vital statistics, on which taxation was based, postulated the acquisition of the requisite information without the knowledge of those being counted. The identical principle operated in levying the taxes themselves, for necessary counting of resources and goods was similarly achieved by such devious ways that one can well believe the statement of members of the native royal family that the people rarely realized when or by whom the count was taken.

This was but a part of an entire system of control. Each official, through whose hands flowed the stream of wealth directed toward the royal palace, was "controlled" by a "wife" of the king who, as a member of the inner bureaucracy, was charged with seeing to it that not even the word of the highest officials was taken without independent validation. The attitude of the natives toward the straightforwardness of the European is revealed by current comment on the methods of the French colonial officials in administering taxation. The French have imposed a head tax which, like the taxes of the native kingdom, is based on census enumeration. Unlike native practice, however, the French query each compound head directly as to the number of people in his compound. It is well understood that the more truthful a man is, the more he will have to pay; the comment on the technique was: "Our ancestors may have had no guns and had to fight with hoe handles, but they were wiser than to ask directly that a man tell them something to his disadvantage!"

As has been stated, many other instances of the principle of indirection as this operates among Negroes might be given from the

West Indies and West Africa no less than from the United States, such as the oblique references in the "songs of allusion" that play an appreciable role in regulating social life. Certainly this principle is everywhere given clear expression as a guide to overt behavior. That as life is lived, it is a worth-while principle to speak with reserve, to hold back something of what one knows, to reveal no more than one must, can be immediately recognized; in the most ordinary dealings, the principle that one keep one's counsel and, as a minimum, offer only such information as may be requested, has been found to be not unprofitable. To ask a question such as Puckett poses, "May not the organized hypocrisy of the Southern Negro also be an adaptation forced upon the Negro by conditions of life?" shows how misinterpretation can easily arise where the force of traditional sanctions has gone unrecognized. For diplomacy, tact, and mature reserve are not necessarily hypocrisy; and while the situation of the Negro in all the New World, past and present, has been such as to force discretion upon him as a survival technique, it is also true that he came on to the scene equipped with the technique rather than with other procedures that had to be unlearned before this one could be worked out.

The principle of indirection, then, must be looked on as immediately descended from the African scene. The implications of this fact in giving form to Negro behavior, like other intangibles such as canons of etiquette and concepts of time also considered in this section, cannot be overlooked if a true picture of Negro life is to be had, either for scientific analysis or to help understand the present-day interracial situation.

4

We now move from less overt aspects of culture to more institutionalized forms, and consider first those elements in the organization of Negro society that are not dependent on relationship ties. The question at this point reduces itself essentially to what vestiges of African "associations," if any, are to be discerned—the extent to which such nonpolitical organizations as cooperative groupings of various kinds and secret or nonsecret societies have survived the experience of slavery.

It would be strange if African political forms had continued in any degree of purity except where successful escape from slavery rendered necessary some administrative arrangement to care for the affairs of the runaway group. In such cases as those of Brazil, Dutch

Guiana, and Jamaica, African political organizations were set up. But for other parts of the New World, little information of the controls that operate within Negro communities is to be had. Colonial administration, or the organization of national governments on the republican model, as in Haiti, effectively mask any extralegal institutions which may exist among the Negroes. In Trinidad, among such a group as the followers of the Shango cult, or in the "shouting" Baptist churches, little recourse is had to governmental instruments for the settling of disputes or for administering other measures of control. The leader and elders are entirely capable of handling such situations as arise within the community, and their decrees are followed by common consent. This is probably similar to what is found in more tenuous form among the Negroes of the southern part of the United States, where similar extralegal devices operate. These may perhaps represent a response to the conviction that justice is not to be found in the white man's courts and that it is therefore the part of wisdom to submit disputes to the arbitration of an impartial member of the group.[26]

Yet the question remains whether any survivals of African legal institutions are to be found beyond these informal methods of caring for situations that might otherwise fall into the hands of the law. Aimes has given the matter the most careful study of any student to date, but has found few clues except in the early history of the period of slavery.[27] The Negroes of New England, particularly of Connecticut, appear to have elected a headman or "governor." A record exists of a gravestone in the burial ground of Norwich, Connecticut, inscribed "In memory of Boston Trowtrow, Governor of the African tribe in this Town, who died 1772, aged 66."[28] Steiner, who takes it for granted that the election of such an official by the Negroes "showed the usual imitation of . . . white masters"— which Aimes disputes—quotes a description of this officer:

The negroes, "of course, made their election to a large extent deputatively, as all could not be present, but uniformly yielded to it their assent. . . . The person they selected for the office was usually one of much note among themselves, of imposing presence, strength, firmness, and volubility, who was quick to decide, ready to command, and able to flog. If he was inclined to be arbitrary, belonged to a master of distinction, and was ready to pay freely for diversions—these were circumstances in his favor. Still it was necessary he should be an honest negro, and be, or appear to be, wise above his fellows." What his powers were was probably not well defined, but he most likely "settled all grave disputes in the last resort, questioned conduct, and imposed penalties and punishments sometimes for vice and misconduct."[29]

It is understandable how the institutions Negroes set up to control their own affairs eventually came into conflict with the need for centralization of authority in the North; in the South, any toleration of such types of organization was unlikely. Aimes' findings confirm such an a priori judgment:

Considerable research has failed to reveal any very satisfactory material relating to these institutions in the South. The laws repressing meetings of negroes appear to have been severe. The following account of an African "wizard" is interesting and important, but the fact that he is said to have operated "many years ago" may detract somewhat from its value. An old Guinea negro, a horse-trainer and hanger-on of sporting contests, "claimed to be a conjurer, professing to have derived the art from the Indians after his arrival from Africa." The only use he made of this valuable accomplishment was "in controlling riotous gatherings" of negroes, and "in causing runaway slaves to return, foretelling the time they would appear and give themselves up." He would get the masters and overseers to pardon their erring slaves. This shows a powerful control in this man over his fellows, and one that could be put to good use if properly directed. The basis of his power undoubtedly lay in some combination of the mores of the negroes themselves. Traces of this individual power seem to be present in the Gabriel revolt in Virginia in 1800, and in the Nat Turner revolt at a later date. It is not to be supposed that the negroes would have submitted to a form of conjuration derived from Indians.[30]

It is thus understandable why few institutionalized survivals of the political systems of West Africa are to be encountered in this country. It is rather a tradition of discipline and organization that is found, a "feel" for the political maneuver apparent in operations marking the attainment of control within Negro organizations, or the shrewdness with which participation of Negro groups in the larger political scene is directed by Negro politicians so as to get the most out of the truncated situation.[31] Yet because in the main we find African sanctions rather than African political institutions does not mean that within the Negro group more specific manifestations of the African pattern of organized directed effort are lacking. In West Africa, these nonrelationship groupings have their most important manifestation in cooperative endeavor. It is therefore to various kinds of cooperative and mutual-aid effort among Negroes of this country that we must look for the survivals of the African tradition of discipline and control based on acquiescence and directed toward the furtherance of community needs.

The tradition of cooperation in the field of economic endeavor is outstanding in Negro cultures everywhere. It will be recalled that this cooperation is fundamental in West African agriculture, and in other industries where group labor is required, and has been reported from several parts of the slaving area.[32] This tradition, carried over into the New World, is manifest in the tree-felling parties of the Suriname Bush Negroes, the *combites* of the Haitian peasant, and in various forms of group labor in agriculture, fishing, house-raising, and the like encountered in Jamaica, Trinidad, the French West Indies, and elsewhere. This African tradition found a congenial counterpart in the plantation system; and when freedom came, its original form of voluntary cooperation was reestablished. It is said to have reappeared in the Sea Islands immediately after the Civil War,[33] but its outstanding present form is gang labor. It is the essence of this system that work is carried on cooperatively under responsible direction; by use of the precise formula under which cooperative work is carried on in all those other parts of the New World, and in Africa, where it has been reported.

Such instances of cooperative labor among Negroes of the United States as have been noticed have been dismissed as something borrowed from such forms in European tradition as the "bee." That these types of cooperation were important in frontier life is self-evident; it does not follow, however, that cooperation among Negroes is merely a reflection of these white manifestations of organized aid. The "bee," characteristic of white America, was, as a matter of fact, not current to any considerable degree in those parts of the country where Negroes were most to be found. The phenomenon characterized the northern and northwestern states rather than the southern; in a plantation slave economy, the necessity of calling in neighbors to help in doing work slaves could perform was obviated. This is especially true since the neighbors, themselves presumably slave-owners, had no great competence in the manual arts. It is thus much simpler to assume that resemblances existed between European and African patterns which tended to reinforce each other.

Cooperation among the Negroes of this country is principally found in such institutions as lodges and other benevolent societies, which in themselves are directly in line with the tradition underlying similar African organizations. The role of the secret societies in the parts of Africa from which the slaves were derived is well known, but has been stressed in favor of the large number of less sensational, but no less important, nonsecret associations. It is these more prosaic organizations, however, that in time of need assure their members access to resources greater than those of any individual, which give this type of society an especially significant part in assuring stability to African social structure. That in this country Negro assurance societies, especially burial societies, take on the form of lodges in so many cases, and that Negro lodges of various types represent such an exuberant development of the common American lodge, is to be explained in two ways. In the first case, the coalescence of the cooperative assurance and secret society traditions may be considered as developing out of a tendency, under acculturation, to blur distinctions which prior to contact were quite clear. Secondly, the psychological device of compensation through overdevelopment, so often encountered among underprivileged groups forced to adhere to majority patterns, and the failure of white lodges to accept their Negro counterparts brought it about that the initial stimulus was diverted from the channels it followed among the donor group and emphasized for the Negro lodge its distinctive traits.

Whatever the derivation of such organizations, their importance has long been recognized. Citations such as the following are typical of earlier studies:

Perhaps no phase of negro life is so characteristic of the race and has developed so rapidly as that which centers around secret societies and fraternal orders. . . . Scores of different orders are represented in Southern towns, with hundreds of local chapters. A special feature of the colored organizations is found in the local character of their orders. The majority have their home offices in the state in which they do business. Few extend over much greater territory.[34]

Continuing, this account becomes somewhat more specific:

Investigations show that other societies are in operation in Mississippi besides those chartered and recorded on the official lists. Some of these operate under secret rules and assess members according to their own agreement. The total number of such organizations, including the many little ephemeral societies operated wherever groups of negroes are found, would run into the hundreds. Sometimes they continue for a year, sometimes only for one or two meetings. . . . A study of the names of the societies . . . will reveal much of their nature. . . . They pay burial expenses, sick benefits, and small amounts to beneficiaries of deceased members. Such amounts are in many cases determined entirely by the number of members, the assessment plan being the most common and most practical one. Members are admitted variously according to a flexible constitution made to meet the demands of the largest number of people. There are non-paying members who receive only the advantages coming from the fraternal society; there are those who take insurance for sick benefits only, while others wish burial expenses also. Still others take life insurance, while some combine all benefits, thus paying the larger assessments and dues.[35]

Though couched in language not commonly employed at the present time by students of the Negro, the following further observations of this same student are to the point:

Some evidences of the higher forms of sympathy may be seen in the working of the fraternal societies in ministering to the sick, the widows and the orphans, and in paying off benefits. While the obligation of the society upon its members seems in every case to be the direct cause of a service, sympathy often grows out of the deed, and the members of such societies grow enthusiastic in their advocacy of the cause, giving these deeds of service as evidence. So it happens that the leaders of the various societies have come to feel, in addition to the personal gratification of succeeding in rivalry, an eager interest in their work.[36]

This explanation of how sympathy is aroused in this people may be dismissed as aside from the point; what is important for our purpose is the variety of ends which these societies fulfill in exercising their cooperative function.

Today this type of organization and its place in such a community as that studied recently by Powdermaker, corresponds closely to the traits mentioned in the earlier statement:

Three large insurance companies compete for the patronage of the Negroes in the community: The Afro-American, the Knights and Daughters of Tabor, and the Universal Life Insurance Co. The first two are also fraternal orders, with appropriate rituals and a pronounced social flavor. . . . Most of the local Negroes belong to at least one of the societies, and some belong to more than one. Twenty to thirty cents a week is a rough and conservative estimate of the average family contribution for insurance.[37]

Though in the district studied by Johnson, "a loss of confidence" in the insurance groups resulted from numerous failures in the early 1930's, and because of their "widespread exploitation by both whites and Negroes from the outside," it is still noted that:

There were 224 of the 612 families who now have, or have had, insurance, and 170 of these paid premiums of 25 cents a week or less. Twenty-one companies and lodges were represented in these numbers.[38]

The many functions of the various fraternal or insurance socie-

ties, secret and nonsecret, are suggested in the following passage
from a study made in 1906 of the various forms of economic cooperation among Negroes:

> No complete account of Negro beneficial societies is possible, so large
> is their number and so wide their ramification. Nor can any hard and
> fast line between them and industrial insurance societies be drawn save
> in membership and extent of business. These societies are also difficult
> to separate from secret societies; many have more or less ritual work,
> and the regular secret societies do much fraternal insurance business.[39]

That the incidence of these societies is not restricted to the South is
to be seen in the enumeration of organizations given by this author
for various towns and cities. Xenia, Ohio, which at that time had a
Negro population of 2,000, possessed eleven chapters of various
more or less national organizations. The following passage, concerning Philadelphia, is instructive:

> From general observation and the available figures, it seems fairly
> certain that at least 4,000 Negroes belong to secret orders, and that these
> orders annually collect at least $25,000, part of which is paid out in sick
> and death benefits and part invested. . . . The function of the secret
> society is partly social intercourse and partly insurance. They furnish
> pastime from the monotony of work, a field for ambition and intrigue,
> a chance for parade, and insurance against misfortune. Next to the
> church they are the most popular organizations among Negroes.[40]

It is impossible to read such an account of the development of
these cooperative groupings as is contained in the work cited, or in
Browning's analysis of their history,[41] without realizing that here
the student is face to face with one of the deep-seated drives in
Negro life; drives so strong, indeed, that it is difficult, if not impossible, to account for them satisfactorily except in terms of a
tradition which reaches further than merely to the period of slavery.
Allowing for the advantages of such organizations to any underprivileged group, facing the problem of existence in an economy such
as the one in which they live, this fact alone cannot explain why
cooperative institutions of the type found among Negroes flourish
to the extent they do, why they call forth such devotion, or why
they include so many noneconomic activities. Browning puts the
matter in these terms:

> The existence today of a Negro economy is the result of a long process
> of evolution caused by varied factors. On the one hand was pressure
> from the outside, and on the other a nationalism within the Negro
> group; but perhaps farthest removed in point of time was the cultural

heritage which was filled with the cooperative spirit. This spirit of co-operation was not crushed during the days before the Civil War but emerged in the form of a Negro economy.[42]

Some instances of insurance societies in aboriginal African groups, and in the New World outside the United States, may be cited to indicate why the institutionalization of this feature of Negro life in the United States must be referred to the stimuli of aboriginal custom. The cooperative work groups that are more or less *ad hoc*, such as the Dahomean *dokpwe* and the Haitian *combite*, have been mentioned; but these only begin the tale of cooperation. Almost all permanent groupings other than kinship units possess cooperative and even insurance features. Mutual self-help characterizes Dahomean iron-working guilds. Each member of a "forge" accumulates such scrap iron as he can, and the entire membership joins in turning this iron into hoes or other salable objects until the supply is exhausted, when they turn to the materials of the next member. What has been made from a man's iron is his to sell as he will, and from the proceeds he supports himself and gets the means to buy more iron to be worked when his turn is again reached. It makes no difference if he is ill when this comes, since all will work on his iron regardless of his presence or absence; in such a case his fellow members aid in disposing of his goods so that when he recovers he will be able to resume his normal place without any undue handicap.

The Yoruba of Nigeria have an organization called *esusu*, the exact counterpart of a Trinidad type of institution of the same name, *'susu*. Because the *gbe* and *so* types of Dahomean groupings have similar features, it is reasonable to expect that further research will reveal more arrangements of this kind elsewhere in West Africa and in the New World. In Trinidad, as among the Yoruba, it makes it possible for a person without the initiative to carry on a systematic program of saving to finance projects for which he does not possess the ready means. A stated number of persons agree to deposit a certain sum each week with one of their number who, taking nothing for his services unless the group is large and the amounts to be handled are considerable, undertakes to turn over the entire weekly collection as taken up to a different member of the group until all have realized on their "hands." Difficulties naturally enter, since there is an excellent opportunity for dishonesty, and some suspicion is roused when the collector takes the first "hand." Yet despite occasional mishaps, the system works well, and is recognized in Trinidad law.

Certain Dahomean forms of mutual-aid societies actually constitute permanent insurance societies, the *gbe* in particular being far removed from the type of organization customarily conceived as existing in nonliterate cultures. With elected membership and with ritual secrets in the manner of American lodges, such groups often have large followings and persist over long periods of time. Their primary purpose is to provide their members with adequate financial assistance so that at the funeral of a member's relative or, more importantly, of the parent of a member's spouse, he can make a showing in competitive giving that will bring prestige to himself and to his group. Each member must swear a blood oath on joining, and there are adequate controls over the treasurer. Each society has its banner, and indulges in public display of its power and resources in its processions, especially when it goes as a body to the funeral rituals. The prominence of assuring proper performance at funerals in this aboriginal insurance system is of special significance in the light of the important place held by burial insurance in Negro life in the United States, as testified by the presence of numerous of these "bury-leagues."

The lodge itself, aside from its insurance features, is another expression of the Africanlike flair for organization. Granting the elementary fact that Negroes in the United States, like all other persons here, tend to adapt their behavior to prevailing patterns, yet the divergences from the patterns that are found in the case of these lodges are especially cogent. For while it is true that many Negro fraternal organizations are the counterparts of white groups having similar names, rituals, and paraphernalia, yet the numbers of Negro lodges, including those which have no counterparts among the whites, and their role in everyday Negro life, which far transcends their importance for the vast majority of white lodge members, makes them distinctive in the American scene.

This is relevant to the fact that numerous other societies exist in Africa, taking forms and having objectives that resemble the aims of Negro lodges in the United States far more than is recognized. Not only do many of these societies have some religious basis, but many of them are essentially religious organizations. In one instance, groupings considered secret societies were found to be actually cult groups, whose secrets are religious secrets, whose initiatory rites are education in the ways of the gods, and whose public appearances in regalia are made on those occasions when the deities are worshiped. This recalls the structure and functioning of various New World Christian religious "orders" among Negroes, notably the Trinidad Baptist groups. While a direct relationship between this and the religious preoccupations of Negro societies of various sorts, either secret or economic, is difficult to envisage, it is yet entirely possible that something of the strong nonsecular bent of the Negro lodges in this country is a partial survival of this tradition. For again, it is the importance laid on this aspect of the "work" in the Negro lodges that in one respect differentiates them—in degree, it must be emphasized, not in kind—from societies having white membership.[43]

5

It is well recognized that Negro family structure in the United States is different from the family organization of the white majority. Outstanding are its higher illegitimacy rate and the particular role played by the mother. Certain other elements in Negro social organization also make it distinctive, and these will be considered later; but for the moment the more prominent characteristics must be treated in terms of the cognate African sanctions which make them normal, rather than abnormal, and go far in aiding us to comprehend what must otherwise, after the conventional manner, be regarded as aberrant aspects of the family institution.

At the outset, it is necessary to dismiss the legal implications of the term "illegitimate" and to recognize the sociological reality underlying an operational definition of the family as a socially sanctioned mating. In this case, illegitimacy is restricted to those births which are held outside the limits of accepted practice. The situation in the West Indies, projected against the African background of marriage rites and family structure, will here as elsewhere make for clarity. In West Africa, it will be remembered, preliminaries to marriage include negotiations between the families of the two contracting parties to assure all concerned that the young man and woman are ready for marriage, that they are competent to assume their obligations under it, and that no taboos in terms of closeness of actual or putative relationship stand in the way of the match. This done, the young man (and in some tribes the young woman) assumes certain obligations toward his prospective father- and mother-in-law, which in many instances continue after marriage. In all this area, it is further to be recalled, the family is marked by its polygynous character, and the manner of its extension into such larger kinship groupings as the extended family and the sib.

In the New World, these forms when brought into contact with European patterns of monogamy and the absence of wider social structures based on relationship have resulted in institutions which, however, though differing considerably from one region to another, have nonetheless become stabilized in their new manifestations. Thus the elaborateness of the betrothal mechanism has in several regions been translated into ceremonies which even when European in form are essentially African in feeling. The Haitian *lettre de demande*[44] and its counterpart in the British West Indian islands are, in their form and mode of presentation, entirely in the tradition of Africa. The survival of the polygynous marriage pattern is likewise found in Haiti in the distinction made between marriage and what is termed *plaçage*, a system whereby a woman is given a man by her father but without legal or church sanction. The similar means whereby a man and woman in the British West Indies may form regularly constituted unions without the approval of church or government is seen in the institution of the "keeper."

In Haiti, at least, actual polygyny is found, though as a practical matter it can be practiced only by men who are wealthy and powerful enough to manage their plural wives. For while it is a delicate task, at best, for a man to manage a polygynous household—even in Africa, "a man must be something of a diplomat," as one Dahomean put it—where invidious distinctions are set up between legal and free matings, the tensions become greatly heightened. Therefore, even in Haiti, actual polygyny is rare, while elsewhere in the New World it takes the form of what may be termed "progressive monogamy," not unlike that developed by the whites in recent years, though in this latter type formal divorce must precede socially sanctioned remating. Thus, while a Trinidad woman, once legally married, is always called "mistress," the fact that her union is legal does not mean that it will be any more enduring than if she were to take up housekeeping with a keeper. Nor does it often occur that she or her husband will go to the trouble of securing a legal divorce should the match be broken. They merely separate, and subsequent keepers are taken without regard for the legal niceties. The children of matings previous to or subsequent upon the "marriage" are under no social handicap, despite their legal illegitimacy as compared to those born of regularly married parents. For as elsewhere in the Negro New World, a child is rarely handicapped because of the nature of the relationship under which he was brought into the world; he stands on his own feet, and his parentage figures but slightly in establishing his social position.

Another aspect of West African social organization having important implications for the study of New World Negro kinship groupings concerns the place of women in the family. By its very

nature, a polygynous system brings about a different relation between mother and children than a monogamous type—a relationship
that goes far in bringing about an understanding of the so-called
"matriarchal" form of the Negro family in the United States, the
West Indies, and South America. The question most often raised in
accounting for any African derivation of this type of family,
wherein, unlike most white groups, the importance of the mother
transcends that of the father, is whether this may not reflect African
unilateral canons of sib descent. But while this fact may enter into
the traditional residue, it is not to be regarded as playing any considerable role. In West Africa, descent is counted more often on the
father's than on the mother's side and, as in other portions of the
continent, the parent socially unrelated to the child is as important
from a personal and sentimental point of view as is the one to whose
family the child legally belongs.

What is much more important for an understanding of the sanctions underlying this "matriarchal" Negro family type is the fact
that in a polygynous society a child shares his mother only with his
"true" brothers and sisters—everywhere recognized as those who
have the same father and the same mother—as against the fact that
in the day-to-day situations of home life he shares his father with
the children of other women. This means that the attachments between a mother and her child are in the main closer than those
between father and children; from the point of view of the parent,
it means that the responsibilities of upbringing, discipline, and supervision are much more the province of the mother than of the father.
In most parts of the African areas which furnished New World
slaves, the conventions of inheritance are such that a man may, and
often does make an arbitrary selection of his heir from among his
sons. Because of this, there is a constant jockeying for position
among his wives, who are concerned each with placing her children
in the most favorable light before the common husband. The psychological realities of life within such a polygynous household have yet
to be studied in detail; but that the purely human situation is such
as to make the relationship between a mother and her children more
intimate than that between the family head, and any but perhaps
one or two of the offspring of the various wives who share this
common husband and father, is a point which cannot be overesti-
mated.

Against this background the patterns of marriage and family organization prevalent in the Negro communities of the United States
may be projected, so as to indicate the points in the available litera-

ture at which the influence of African tradition can be discerned. The following summary statement as concerns mating and the family in the southern county studied by C. S. Johnson is to the point:

The postponement of marriage in the section . . . does not preclude courtship, but accentuates it, and gives rise to other social adjustments based on this obvious economic necessity. The active passions of youth and late adolescence are present but without the usual formal restraints. Social behavior rooted in this situation, even when its consequences are understood, is lightly censured or excused entirely. Conditions are favorable to a great amount of sex experimentation. It cannot always be determined whether this experimentation is a phase of courtship, or lovemaking without the immediate intention of marriage, or recreation and diversion. Whether or not sexual intercourse is accepted as a part of courtship it is certain no one is surprised when it occurs. When pregnancy follows pressure is not strong enough to compel the father either to marry the mother or to support the child. The girl does not lose status, nor are her chances for marrying seriously threatened. An incidental compensation for this lack of censuring public opinion is the freedom for the children thus born from warping social condemnation. There is, in a sense, no such thing as illegitimacy in this community.[45]

In studying a community such as this, we are therefore faced with a situation where acculturation has brought on disintegration—disintegration due to slavery, to the present economic background of life, and to those psychological reactions which are the concomitants of life without security. Reinterpretation of earlier, pre-American patterns has occurred, but readjustment to normal conditions of life has been inhibited. We thus must recognize that the elasticity of the marriage concept among Negroes derives in a measure, largely unrecognized, from the need to adjust a polygynous family form to patterns based on a convention of monogamy, in a situation where this has been made the more difficult by economic and psychological complications resulting from the nature of the historical situation.

A rich documentation exists in the way of indices which point the aspects of Negro social organization that differ strikingly from white patterns. It is only necessary to turn to the general study of the problem by Frazier[46] or such a specialized analysis as that of Reed[47] to realize to what an extent the incidence of productive matings without legal status is out of line with white practices. Yet when the emphasis laid on the proper type of marriage proposal in the Sea Islands, where there is some measure of stability in Negro society,[48] is compared with Frazier's statement that 30 per cent of the births on that island are illegitimate, it is apparent that here, at least, sanctions other than those of the European type are operative. Johnson's summary of the various forms of union found among the Negroes of Macon County, Georgia, provides further illustrative material:

Children of common-law relationships are not illegitimate, from the point of view of the community or of their stability, for many of these unions are as stable as legally sanctioned unions. They hold together for twenty or thirty years, in some cases, and lack only the sense of guilt. Again, there are competent, self-sufficient women who not only desire children but need them as later aids in the struggle for survival when their strength begins to wane, but who want neither the restriction of formal marriage nor the constant association with a husband. They get their children not so much through weakness as through their own deliberate selection of a father. Sexual unions for pleasure frequently result in children. There is a term for children born under the two latter circumstances. They are called "stolen children." "Stolen children," observed one mother, "is the best." A woman with children and who has been married though later separated from her husband may add other children to her family without benefit of formal sanctions. These are "children by the way." The youthful sex experimentation, which is in part related to the late marriages, often results in children. These are normally taken into the home of the girl's parents and treated without distinction as additions to the original family. Finally, there are the children who result from the deliberate philandering of the young men who "make foolments" on young girls. They are universally condemned. These children, as circumstances direct, may be placed with the parents of the mother or father of the child, an uncle, sister, or grandmother. They are accepted easily into the families on the simple basis of life and eventually are indistinguishable from any of the other children. Even if there were severe condemnation of true "illegitimates," confusion as to origin would tend both to mitigate some of the offenses and to obscure them all from specific condemnation.[49]

What is recognizably African in all this? The "common-law relationship" is merely a phrase for the recognition of the fact that matings not legally sanctioned may achieve enough stability to receive equal recognition with regularly performed marriages. In Africa, and in the West Indies where Africanisms persist, marriage is not a matter requiring approval of the state or of any religious body. Only consent of the families concerned is needed, while marriage rites depart from the secular only to the extent that they are directed toward obtaining the benevolent oversight of the ancestors. Therefore Negro common-law marriages in the United States conflict in no wise with earlier practices, while in so far as they require the approval of the families of the principals, they are, indeed, directly in line with African custom.

The "competent, self-sufficient women" who wish to have no husbands are of especial interest. The social and economic position of women in West Africa is such that on occasion a woman may refuse to relinquish the customary control of her children in favor of her husband, and this gives rise to special types of matings that are recognized in Dahomey and among the Yoruba, and may represent a pattern having a far wider distribution. The phenomenon of a woman "marrying" a woman,[50] which has been reported from various parts of the African continent and is a part of this same complex, testifies to the importance of a family type which might well have had the vitality necessary to make of it a basis for the kind of behavior outlined in the case of the "self-sufficient" woman who, in the United States, desires children but declines to share them with a husband. The same traditional basis exists for "children by the way," those offspring of women, once married, by men other than their husbands.

In the community studied by Powdermaker, types of mating and attitudes toward them have likewise been differentiated:

For this group, there are three ways in which a man and woman may live together: licensed marriage, solemnized by a ceremony, usually in a church; common-law marriage; and temporary association, not regarded as marriage. For the large majority of the households the form is common-law marriage, which is legally valid in Mississippi. Of the remainder, temporary matings are probably more numerous than licensed marriages. Most of the latter are in the upper and the upper middle class. Temporary mating is most easily countenanced in the lower class, though it is not uncommon in the middle class. A licensed marriage in the lower or lower middle class is extremely rare. A common-law marriage in the upper class is even more so; and in this class for two people to live together with no pretense of real marriage would be extremely shocking.[51]

The approach to this problem through the analysis of mores which differ according to classes within the Negro community is especially pertinent, for these classes represent differing degrees of acculturation to majority patterns. This being the case, then the variations in attitude and behavior concerning the family from one class to another reflect differentials in accommodation in so far as this institution is concerned.

This is made even clearer by the discussion of attitudes toward divorce:

Even the few members of the upper middle class who are regularly married do not as a rule consider it necessary to go through court procedure in order to be divorced from a former mate and free to marry another. It is not regarded as immoral to remarry without securing a divorce, since in this class the marriage license is not a matter of morals, and marriage itself is highly informal. Divorce proceedings are expensive, and involve dealing with a white court, which no Negro chooses if he can avoid them. Thus a legal divorce becomes something more than a luxury; it savors of pretension and extravagance.[52]

Here is evidence of lag under acculturation. Sanctioned divorce is a comparatively recent introduction into white mores, and has been superimposed upon a complex of quasi-puritanical religious and social prohibitions. This antecedent patterning being absent from aboriginal and early New World Negro conventions, the attitude toward legal divorce as a pretension and an extravagance is understandable. For under Negro conventions, operative in Africa and in the New World generally, there is little social disapprobation of divorce. Consequently, in terms of a carry-over of this point of view, legal divorce is needless, since separation and subsequent remating (if not remarriage) is taken more or less for granted.

The other major difference between Negro family organization and that of the white majority touches on the position of women within the family. So important is the role of the woman when compared to that of the man, in terms of common American convention, that the adjective "matriarchal" has come to be employed in recent years when describing this family type. Statistical reports bear out common observation concerning the phenomenon:

The 1930 census showed a larger proportion of families with women heads among Negroes than among whites in both rural and urban areas. Moreover, it also appeared that in the cities a larger proportion of Negro families were under the authority of the woman than in the rural areas. In the rural non-farm areas of the southern states from 15 to 25 per cent of the Negro families were without male heads; while in the rural-farm areas the proportion ranged from 3 to 15 per cent. In the rural-farm areas tenant families had a much smaller proportion with woman heads than owners, except in those states where a modified form of plantation regime is the dominant type of farming. For example, in the rural-farm area of Alabama between 13 and 14 per cent of both tenant and owner families were without male heads.[53]

Some further statistics are also relevant:

In southern cities the disparity between whites and Negroes in respect to the proportion of families with woman heads is much greater.

In the twenty-three southern cities with a population of 100,000 or more in 1930, from a fifth to a third of all Negro families had a female head. However, in most of these southern cities, the difference between owner and tenant Negro families in this regard was much greater than in northern cities.[54]

Of the several classifications of Negro family types which take the position of the woman into account, two may be cited. The first concerns the family as it exists at the present time among the Negro urban workers:

The status of husband and wife in the black worker's family assumes roughly three patterns. Naturally, among the relatively large percentage of families with women heads, the woman occupies a dominant position. But, because of the traditional role of the black wife as a contributor to the support of the family, she continues to occupy a position of authority and is not completely subordinate to masculine authority even in those families where the man is present. . . . The entrance of the black worker in industry where he has earned comparatively good wages has enabled the black worker's wife to remain at home. Therefore, the authority of the father in the family has been strengthened, and the wife has lost some of her authority in family matters. . . . Wives as well as children are completely subject to the will of the male head. However, especially in southern cities, the black worker's authority in his family may be challenged by his mother-in-law.[55]

Johnson has differentiated family types in the rural region studied by him into another set of categories. Noting the fact that in terms of the commonly accepted pattern wherein the father is head of the family, "the families of this area are, . . . considerably atypical," since, "in the first place, the role of the mother is of much greater importance than in the more familiar American group," he goes on to distinguish three kinds of families. First come those "which are fairly stable" and are "sensitive to certain patterns of respectability"; then there are those termed "artificial quasi-families" that "have the semblance of a normal and natural family, and function as one," except that "the members of the group are drawn into it by various circumstances rather than being a product of the original union"; and finally the form is found where "the male head remains constant while other types of relationship, including a succession of wives and their children by him, shift around him."[56] In addition to these, however, are the families headed by women:

The numbers of households with old women as heads and large numbers of children, although of irregular structure, is sufficiently important to be classed as a type. . . . The oldest generation is the least mobile, the children of these in the active ages move about freely and often find their own immediate offspring, while young, a burden, as they move between plantations. Marriages and remarriages bring increasing numbers of children who may be a burden to the new husband or a hindrance to the mother if she must become a wage-earner. The simplest expedient is to leave them with an older parent to rear. This is usually intended as a temporary measure, but it most often ends in the establishment of a permanent household as direct parental support dwindles down. The responsibility is accepted as a matter of course by the older woman and she thereafter employs her wits to keep the artificial family going.[57]

Powdermaker likewise notes the elasticity of families headed by women, and indicates how congenial this pattern is to Negroes living in various social and environmental settings:

The personnel of these matriarchal families is variable and even casual. Step-children, illegitimate children, adopted children, mingle with the children of the house. No matter how small or crowded the home is, there is always room for a stray child, an elderly grandmother, an indigent aunt, a homeless friend . . . The pattern of flexibility, however, expanding and contracting the household according to need is not restricted to the poorer and more crowded homes. A typical family of the upper middle class is headed by a prosperous widow, who in her early twenties married a man over sixty years old. He was considered very wealthy and had been married several times before. The household now includes his widow's eleven-year-old daughter (an illegitimate child born before she met her husband), the dead husband's granddaughter by one of his early marriages, and the granddaughter's two children, two and three years old. The granddaughter was married but is divorced from her husband. Everyone in the household carries the same family name.[58]

It is evident that this so-called "maternal" family of the Negro is a marked deviant from what is regarded as conventional by the white majority. Yet it must not be forgotten that the economic and social role of the man in Negro society is of the utmost significance in rounding out the picture of Negro social life. Though important from the point of view of the search for Africanisms, interest in the position of women in the family must not obscure perspective so as to preclude the incidence and role of those families wherein the common American pattern is followed. Despite the place of women in the West African family, the unit holds a prominent place for the husband and father who, as head of the polygynous group, is the final authority over its members, sharing fully in all those obligations which the family must meet if it is to survive and hold its place in the stable society of which it forms a part.

With this point in mind, certain further special characteristics of the Negro family may be considered before the causes which may best account for its place in Negro life are analyzed. Outstanding among these is the fact that an older woman frequently gives the group its unity and coherence. Frazier indicates the following sanctions in explaining the place of such elderly females in Negro families:

The Negro grandmother's importance is due to the fact not only that she has been the "oldest head" in a maternal family organization but also to her position as "granny" or midwife among a simple peasant folk. As the repository of folk wisdom concerning the inscrutable ways of nature, the grandmother has been depended upon by mothers to ease the pains of childbirth and ward off the dangers of ill luck. Children acknowledge their indebtedness to her for assuring them, during the crisis of birth, a safe entrance into the world. Even grown men and women refer to her as a second mother and sometimes show the same deference and respect for her that they accord their own mothers.[59]

The question whether or not an explanation of the importance of old women in these terms is valid may be deferred for the moment; that it is not only among the "simple peasant folk" of the country-side that she wields her power but in the city as well is to be seen from the following:

The Negro grandmother has not ceased to watch over the destiny of the Negro families as they have moved in ever increasing numbers to the cities during the present century. For example, she was present in 61 of the families of 342 junior high school students in Nashville. In 25 of these a grandfather was also present. But in 24 of the remaining 36 families, we find her in 8 families with only the mother of the children, in 7 with only the father, and in 9 she was the only adult member.[60]

How large these family groups headed by old women may be, and from how many sources their members may be drawn, is to be seen in the description of one such family given by Powdermaker:

A larger household is presided over by a woman of seventy-five. She has had two husbands, both dead now, and nine children, two of them born before she met her first husband. Her second husband had seven children by a previous marriage. She brought up three of them. Living with her now are the son and daughter of her second husband's daughter by a previous marriage. Each of these step-grandchildren is married. The two young couples pay no rent, but "board" themselves. In the house also is a nine-year-old boy, the illegitimate child of a granddaughter. After this child was born, his mother left his father and went north with another man. The grandmother paid the railroad fare for the child to be sent back to Mississippi.[61]

The fact, likewise noted by Powdermaker, that "among Negroes household and family are on the whole considered synonymous" indicates how far flexibility may go; only boarders were excluded from membership in the families studied by her.

What are the causes which, in the United States, have brought into being a type of family organization that is so distinctive when compared with the common family pattern? The preceding discussion makes it clear that no single reason will account for its establishment and persistence. Explanations based on assumptions of a theoretical nature concerning the origin of the human family may be dismissed out of hand, since the validity of such propositions has been successfully challenged many times both on methodological and on historical grounds. Thus when Puckett points out that,

It is also rather noticeable that in the Negro folk-songs, mother and child are frequently sung of, but seldom father—possibly pointing back to the African love for the mother and the uncertainty and slight consideration of fatherhood . . .[62]

the only possible comment is that his conception of African attitudes and the facts of African family life is false in the light of known facts. Similarly, when Frazier speaks of the "maternal family" as representing "in its purest and most primitive manifestation a natural family group similar to what Briffault has described as the original or earliest form of the human family,"[63] he is merely repeating poor anthropology.

One of the most popular explanations of the aberrant forms taken by the Negro family is by reference to the experience of slavery. A less extreme example of this position, conventionally phrased, is to be found in Johnson's work. Noting that the role of the mother is of "much greater importance than that in the more familiar American group," he goes on to state:

This has some explanation in the slave origins of these families. Children usually remained with the mother; the father was incidental and could very easily be sold away. The role of mother could be extended to that of "mammy" for the children of white families.[64]

Frazier has presented this point of view at greater length. One statement reads:

We have spoken of the mother as the mistress of the cabin and as the head of the family. . . . Not only did she have a more fundamental interest in her children than the father but, as a worker and a free agent, except where the master's will was concerned, she developed a spirit of independence and a keen sense of her personal rights.[65]

"In spite of the numerous separations," it is stated, "the slave mother and her children, especially those under ten, were treated as a group";[66] while, "because of the dependence of the children upon the mother it appears that the mother and smaller children were sold together."[67] To make the point, slave advertisements such as the following are cited:

A Wench, complete cook, washer and ironer, and her four children— a Boy 12, another 9, a Girl 5 that sews; and a Girl about 4 years old. Another family—a Wench, complete washer and ironer, and her Daughter, 14 years old, accustomed to the house.[68]

These citations are not made to suggest that due attention has not been paid to the place of the father in the slave family, though it is undoubtedly true that he has received less study than has the mother in research into the derivation of present-day family types among the Negroes. The fact of the matter, however, is that the roles of both parents were individually determined, varying not only from region to region and plantation to plantation, but also being affected by the reactions of individual personalities on one another. Not only was the father a significant factor during slaving, but a reading of the documents will reveal how the selling of children—even very young children—away from their mothers is stressed again and again as one of the most anguishing aspects of the slave trade. Whether in the case of newly arrived Negroes sold from the slave ships or of slaves born in this country and sold from the plantations, there was not the slightest guarantee than a mother would not be separated from her children. The impression obtained from the contemporary accounts, indeed, is that the chances were perhaps more than even that separation would occur. This means, therefore, that, though the mechanism ordinarily envisaged in establishing this "maternal" family was operative to some degree, the role of slavery cannot be considered as having been quite as important as has been assumed.

The total economic situation of the Negro was another active force in establishing and maintaining the "maternal" family type. No considerable amount of data are available as to the inner economic organization of Negro families, but the forms of Negro family life themselves suggest that the female members of such families, and especially the elderly women, exercise appreciable control over economic resources. That the economic role of the women not only makes of them managers but also contributors whose earnings are important assets is likewise apparent. This economic aspect of their position is described by Johnson in the following terms:

The situation of economic dependence of women in cities is reversed in this community, and is reflected rather strikingly in the economic independence on the part of the Negro women in the country. Their earning power is not very much less than that of the men, and for those who do not plan independent work there is greater security in their own family organization where many hands contribute to the raising of cotton and of food than there is for them alone with a young and inexperienced husband.[69]

In Mississippi the following obtains in plantation families:

In many cases the woman is the sole breadwinner. Often there is no man in the household at all. In a number of instances, elderly women in their seventies and their middle-aged daughters with or without children and often without husbands, form one household with the old woman as head.[70]

It is to be expected that such a situation will be reflected in property ownership:

In this town of a little more than three thousand inhabitants, . . . 202 colored people own property. The assessed value for the majority of these holdings ranges from $300 to $600. Of the 202 owners, 100 are men, owning property valued at $61,250, and 93 are women, with holdings valued at $57,460. Nine men and women own jointly property totaling $3280 in value. Among the Whites also, about half the owners are women. When White women are owners, it usually means that a man has put his property in his wife's name so that it cannot be touched if he gets into difficulty. Among the Negroes, many women bought the property themselves, with their own earnings.[71]

Of the high proportion of holdings by men in the more favored socio-economic group of Negroes, it is stated, "if more property were owned by Negroes in the lower strata, there would probably be a higher percentage of female ownership." Yet as it is, the percentage would seem to be sufficiently high in terms of current American economic patterns, especially since, as stated, Negro women actually bought and hold their property for themselves rather than for their husbands, as is the common case among the whites.

The absence of any reference to African background in the citations concerning Negro families headed by women is merely another instance of the tendency to overlook the fact that the Negro was the carrier of a preslavery tradition. It is in the writings dealing with this aspect of Negro life that we find truncated history in its most positive expression, since in this field the existence of an African past has been recognized only in terms of such denials of its vitality as were cited in the opening pages of this work. Yet the aspects of Negro family which diverge most strikingly from patterns of the white majority are seen to deviate in the direction of resemblances to West African family life.

It cannot be regarded only as coincidence that such specialized features of Negro family life in the United States as the role of women in focusing the sentiment that gives the family unit its psychological coherence, or their place in maintaining the economic stability essential to survival, correspond closely to similar facets of West African social structure. And this becomes the more apparent when we investigate the inner aspects of the family structure of Negroes in the New World outside the United States. Though everywhere the father has his place, the tradition of paternal control and the function of the father as sole or principal provider essential to the European pattern is deviated from. In the coastal region of the Guianas, for example, the mother and grandmother are essentially the mainstays of the primary relationship group. A man obtains his soul from his father, but his affections and his place in society are derived from his mother; a person's home is his mother's, and though matings often endure, a man's primary affiliation is to the maternal line. In Trinidad, Jamaica, the Virgin Islands, or elsewhere in the Caribbean, should parents separate, the children characteristically remain with their mother, visiting their father from time to time if they stay on good terms with him.

The woman here is likewise an important factor in the economic scene. The open-air market is the effective agent in the retail distributive process, and business, as in West Africa, is principally in the hands of women. It is customary for them to handle the family resources, and their economic independence as traders makes for their personal independence, something which, within the family, gives them power such as is denied to women who, in accordance with the prevalent European custom, are dependent upon their husbands for support. In both West Africa and the West Indies the women, holding their economic destinies in their own hands, are fully capable of going their own ways if their husbands displease them; not being hampered by any conception of marriage as an ultimate commitment, separation is easily effected and a consequent fluidity in family personnel such as has been noted in the preceding pages of this section results. Now if to this complex is added the tradition of a sentimental attachment to the mother, derived from the situation within the polygynous households of West Africa, ample justification appears for holding that the derivations given for Negro family life by most students of the Negro family in the United States present serious gaps.

As in the case of most other aspects of Negro life, the problem becomes one of evaluating multiple forces rather than placing reliance on simpler explanations. From the point of view of the search for Africanisms, the status of the Negro family at present is thus to be regarded as the result of the play of various forces in the New World experience of the Negro, projected against a background of aboriginal tradition. Slavery did not cause the "maternal" family; but it tended to continue certain elements in the cultural endowment brought to the New World by the Negroes. The feeling between mother and children was reinforced when the father was sold away from the rest of the family; where he was not, he continued life in a way that tended to consolidate the obligations assumed by him in the integrated societies of Africa as these obligations were reshaped to fit the monogamic, paternalistic pattern of the white masters. That the plantation system did not differentiate beween the sexes in exploiting slave labor tended, again, to reinforce the tradition of the part played by women in the tribal economics.

Furthermore, these African sanctions have been encouraged by the position of the Negro since freedom. As underprivileged members of society, it has been necessary for Negroes to continue calling on all the labor resources in their families if the group was to survive; and this strengthened woman's economic independence. In a society fashioned like that of the United States, economic independence for women means sexual independence, as is evidenced by the personal lives of white women from the upper socio-economic levels of society. This convention thus fed back into the tradition of the family organized about and headed by women, continuing and reinforcing it as time went on. And it is for these reasons that those aspects of Negro family life that depart from majority patterns are to be regarded as residues of African custom. Families of this kind are not African, it is true; they are, however, important as comprehending certain African survivals. For they not only illustrate the tenacity of the traditions of Africa under the changed conditions of New World life, but also in larger perspective indicate

how, in the acculturative situation, elements new to aboriginal custom can reinforce old traditions, while at the same time helping to
accommodate a people to a setting far different from that of their
original milieu.

It will be recalled that at the outset of this section it was stated
that other survivals than those to which attention has been given
thus far are betokened by certain facts mentioned more or less in
passing in the literature. One of these concerns the size of the relationship group. The African immediate family, consisting of a
father, his wives, and their children, is but a part of a larger unit.
This immediate family is generally recognized by Africanists as
belonging to a local relationship group termed the "extended family," while a series of these extended families, in turn, comprise the
the matrilineal or patrilineal sibs, often totemic in sanction, which
are the effective agents in administering the controls of the ancestral
cult.

That such larger relationship groupings might actually exist in
the United States was indicated during the course of a study of the
physical anthropology of Mississippi Negroes, where, because of the
emphasis placed on the genetic aspects of the problem being studied,
entire families were measured wherever possible.[72] In the town of
Amory (Monroe County) and its surrounding country, 639 persons
representing 171 families were studied, the word "family" in this
context signifying those standing in primary biological relationship
—parents, children, and grandchildren, but not collateral relatives.
How large the kinship units of wider scope are found to be in this
area, however, is indicated by one group of related immediate families which comprised 141 individuals actually measured. Such matters as how many more persons this particular unit includes and
its sociological implications cannot be stated, since no opportunity
to probe its cultural significance has presented itself. The mere fact
that a feeling of kinship as widespread as this exists among a group
whose ancestors were carriers of a tradition wherein the larger relationship units are as important as in Africa does, however, give this
case importance as a lead for future investigation.

Instances of similarly extensive relationship groupings are occasionally encountered in the literature. A description of one of these
corresponds in almost every detail to the pattern of the extended
family in West African patrilineal tribes:

The other community, composed of black families who boast of pure
African ancestry, grew out of a family of five brothers, former slaves,

and is known as "Blacktown," after the name of the family. Although the traditions of this community do not go back as far as those of White-town, the group has exhibited considerable pride in its heritage and has developed as an exclusive community under the discipline of the oldest male in the family. The founder of the community, the father of our informant, was reared in the house of his master. . . . The boundaries of the present community are practically the same as those of the old plantation, a part of which is rented. . . . But most of the land is owned by this Negro family. The oldest of the five brothers was, until his death fifteen years ago, the acknowledged head of the settlement. At present the next oldest brother is recognized as the head of the community. His two sons, one of whom was our informant, have never divided their 138 acres. He and his three brothers, with their children numbering between forty and fifty and their numerous grandchildren, are living in the settlement. Twelve of their children have left the county, and three are living in a near-by town. Our informant left the community thirty-four years ago and worked at a hotel in Boston and as a longshoreman in Philadelphia, but returned after five years away because he was needed by the old folks and longed for the association of his people. One of the sons of the five brothers who founded the settlement is both the teacher of the school and pastor of the church which serve the needs of the settlement.[73]

This passage is to be compared with the account of the formation and later constitution of the Dahomean "collectivity" and extended family.[74] In such matters as the inheritance of headship from the eldest sibling to his next in line, in the retained identity of the family land as a part of the mechanism making for retention of identity by the relationship group itself, and in the relatively small proportion of members who leave their group, immediate correspondences will be discerned.

Like the neighboring "Whitetown"—both these terms are fictitious, but the communities are presumably located in Virginia—sanctions and controls are to be seen such as mark off the African extended family group, succession from elder brother to younger being especially striking in this regard. This kind of "extended" family is also found among the racially mixed stock who, descended from freed Negroes, comprise the population of Whitetown:

At present there are in the settlement ten children and thirty grand-children of our informant. His brother, who also lives in the settlement, has six children and one grandchild. Working under the control and direction of the head of the settlement, the children and grandchildren raise cotton, corn, peanuts, peas and tobacco. In this isolated community with its own school this family has lived for over a century. . . . These closely knit families have been kept under the rigorous discipline of the older members and still have scarcely any intercourse with the black people in the county.[75]

Botume writes of the strangeness to her, a white northerner, of this tradition of extended familial affiliation in the Sea Islands during the Civil War:

> It was months before I learned their family relations. The terms "bubber" for brother, and "titty" for sister, with "nanna" for mother and "mother" for grandmother, and father for all leaders in church and society, were so generally used, I was forced to believe that they all belonged to one immense family.[76]

It is not unreasonable to suppose that this passage is indicative of survival, on the islands, of the classificatory terminology so widely employed in West Africa, though this, as well as the entire problem of the wider ramifications of kinship among Negroes in the United States, remains for future research. On the basis of such data as have been cited, however, African tradition must in the meantime be held as prominent among those forces which made for the existence of a sense of kinship among Negroes that is active over a far wider range of relationship than among whites.

What vestiges of totemic belief have persisted in the United States cannot be said. Certainly no relationship groups among Negroes claiming descent from some animal, plant, or natural phenomenon, in the classic manner of this institution, have been noted in the literature. But what may be termed the "feel" given by certain attitudes toward food may perhaps be indicative of a certain degree of retention of this African concept. Firsthand inquiry among Negroes has brought to light a surprising number of cases where a certain kind of meat—veal, pork, and lamb among others—is not eaten by a given person. Inquiry usually elicits the response, "It doesn't agree with me," and only in one or two instances did the inhibition seem to extend to relatives. Yet this fact that violation of a personal food taboo derived from the totemic animal in West Africa and in Dutch Guiana is held to bring on illness, especially skin eruptions, strikes one immediately as at least an interesting coincidence and perhaps as a hint toward a survival deriving from this element in African social organization, since it is so completely foreign to European patterns. Puckett records a statement published by Bergen in 1899 that, "Some Negroes will not eat lamb because the lamb represents Christ";[77] and this may be an instance of that syncretism which is so fundamental a mechanism in the acculturative process undergone by New World Negroes. Systematic inquiry concerning kinds of foods not eaten by given persons, the reasons or rationalizations which explain these avoidances, and particularly whether or not such taboos are held by entire families and if so, how they are transmitted, are badly needed. Such data, when available, should provide information which will tell whether or not this one aspect of an important African belief has had the strength to survive, in no matter how distorted a form, even where contact with European custom has been greatest and retention of aboriginal custom made most difficult.

Before considering other survivals of African culture, a point which touches upon certain practical implications of the materials dealt with in this section may be mentioned. At the outset of this discussion, it was noted that stress on values peculiar to Euro-American tradition has tended seriously to derogate the customary usages of Negroes which depart from the modes of life accepted by the majority. It was also pointed out that when the logical conclusions to be drawn from the position taken are accepted by Negroes themselves, this tends to destroy such sanctions as the Negroes may have developed, and injects certain added psychological difficulties into a situation that is at best difficult enough. Comment along these lines becomes especially pertinent when one encounters a passage such as the following, where the disavowal of a cultural heritage is emphasized by the assumptions mirrored in its phrases:

These settlements . . . of . . . higher economic status . . . and . . . deeply rooted patriarchal family traditions . . . represent the highest development of a moral order and a sacred society among the rural Negro population. This development has been possible because economic conditions have permitted . . . germs of culture, which have been picked up by Negro families, to take root and grow.[78]

The community referred to does not matter; it is the use of a figure which envisages a people "picking up" "germs of culture," to name but one such to be found in these lines, that gives us pause. To accept as "moral" only those values held moral by the whites, to regard as "culture" only those practices that have the sanctions of a European past is a contributory factor in the process of devaluation, if only because to draw continually such conclusions has so cumulative an effect. A people without a past are a people who lack an anchor in the present. And recognition of this is essential if the psychological foundations of the interracial situation in this country are to be probed for their fullest significance, and proper and effective correctives for its stresses are to be achieved.

6

Numerous beliefs, attitudes, and modes of behavior centering about children that have been reported from the United States point to African counterparts. But it must be made clear that such general matters as the great desire of Negroes for children and the affection which eventuates on occasion in the greatest sacrifices for the young of their households are outside the range of such counterparts. For in all human societies well-recognized biological drives are everywhere rationalized into active desire for offspring, and everywhere there must at least be a benevolent tolerance of the young if the group is to survive. As a matter of fact, such statements should never have required mention were it not that echoes are still heard of the polemics between supporters of the slave system and its opponents, wherein the former on occasion maintained that the Negro was a creature without sentiment toward his young. The need for serious consideration of such assertions is past; their historic role once recognized, they can be dismissed with mere statement.

That both prestige and economic advantage go with a large family, and that the desire for children in these terms is not generalized but definitely channeled, is important in terms of our major concern. In the New World everywhere, as in West Africa, situations entirely comparable to those indicated in the following passage are to be encountered:

In a system which requires the labor of the entire family to earn a living, children of a certain age are regarded as an economic asset. They come fast, and there is little conscious birth control. The coming of children is the "Lord's will." . . . There is pride in large families. "Good breeders" are regarded with admiration. One woman quoted a doctor as explaining that she was "sickly" because she "needed to breed." For men the size of the family is a test of virility and for the women fecundity has tremendous weight in their valuation as mates.[79]

In most parts of the area from which the slaves came a woman without children is socially handicapped. And while the system of polygyny does not place on a single woman the burden of providing the large family that will give a man prestige in this world and security of position in the next, and hence births per woman are perhaps lower than would otherwise be the case, regard for children as testimony of a man's virility and as a valuable economic asset are deep-rooted African tenets.

Adoption as a means of enlarging the family is widespread in Africa and the New World. Johnson explains the validations for the tradition in these terms:

> Children after a certain age are . . . an economic asset. Childless couples, for whatever reason, have not the social standing in the community of families with children. The breaking-up of families, through desertion or migration, results in the turning-over of children to relatives or friends, and since little distinction of treatment enters, they soon are indistinguishable from the natural children, and assist them by dividing the load of heavy families. Moreover, adoption is related to illegitimacy, and frequently the children in families which are referred to as adopted are really the illegitimate offspring of one's own daughter or neighbor's daughter. The child of an unmarried daughter becomes another addition to the children of the parents of the girl with all the obligations. Discipline is in the hands of the original parents and the young mother's relationship to her son is in most respects the same as her relationship to her younger brother. These children call her by her first name and refer to their natural grandparents as "mamma" and "papa." It has happened that men have adopted into their legitimate families extra-legal children by other women, and with no apparent distinction that would make them unfavorably conspicuous among the other children. Again, children orphaned by any circumstances are spontaneously taken into childless families.[80]

The same writer further comments on the phenomenon:

> Adoption . . . is commonly a convenience for children without the protection of a family organization of their own. A motherly old woman said: "These chillun here, they mother in Plaza. They father somewhere 'bout near here. They all got the same mother but different fathers. The two oldest ones was born 'fore they mother married. I tuk them all soon atta they was born." Older families, and especially old and widowed women, look upon adoption as more of a privilege than a burden: "Lord, I almost like to not be able to raise me that child; he was so sickly at first." The sentiment is sometimes carried to the point of surrounding the child with an importance which many children in normal families lack.[81]

In Mississippi, a similar incidence and importance of adoption has been reported:

> It has been remarked that the adopted and illegitimate children included in so many Negro households are considered full members of the family. Adoption is practically never made legal, and is referred to

as "giving" the children away. One of the several reasons for so frequently giving children away is the repeated breaking up of families
and the inability or unwillingness of the remaining mate to care for
them. Because of the strong desire most people have for children, there
is always someone ready to take them in. . . . Except in the small
upper class, a child practically always calls the woman who adopts him
"mother." This is done even when the real mother is one of the household, which would occur chiefly in cases of adoption by a grandmother.
. . . Whatever the motivation of the adoption, there is no attempt to
conceal their origin from adopted children. Even if the attempt took
place in early infancy, they usually know they have been given away,
and adults have no hesitation in talking about it before them. No stigma
attached to giving a child, it is an accepted procedure. Nor is it ordinarily considered a misfortune to be a "gift child." As a rule no difference is made between them and the children of the house, although a
case has been quoted in which a woman felt that she had been made to
work harder than her aunt's own children. The children seldom evince
any sense of being outsiders.[82]

That the pattern of adoption in these Negro communities differs
from the conventions concerning adoption operating in white groups
in this country is apparent without further analysis. The problem
thus once again becomes that of accounting for the distinctive quality of Negro custom. The data in hand are unfortunately neither
sufficient nor effectively enough placed in their cultural matrix to
permit conclusions to be drawn without further field research into
the ethnology of at least a few Negro communities in the United
States. Yet on the basis of comparative background materials, even
such general statements as have been quoted make it clear that the
principle of multiple causation is to be employed if a realistic analysis is to result. Slavery and the present economic and social scene,
while effective forces, again preserved and continued the force of
aboriginal tradition in this as in other aspects of Negro social life.

Reports of procedures in connection with childbirth[83] consist
mainly of scattered references to isolated items of folk custom. No
account of the birth of a child in a Negro village, where only the
midwives and other elderly women available were in attendance, has
been published in its full context, but only fragments of total procedure, principally "beliefs" of one kind or another. Many of these,
it should be said at once, seem of themselves to present a blend of
European and African elements of folk belief such as might be
expected under contact of two cultures having a common sub-
stratum. Such measures as placing iron under the bed at parturition
so as to ease birth pangs, however, or refraining from sweeping out

ashes until some time after the child has been born are the counterparts of procedures recorded in various portions of West Africa.
The use of cobwebs as a means of stopping hemorrhage is found
in Africa, where dressings of this material are commonly used both
there and in the New World tropics to stop bleeding. The care
used in disposing of the placenta and the treatment of the navel cord
are also largely African.[84]

Certain Negro attitudes reported from the United States toward
abnormal births are highly specific in their African reference. Twins,
the child after twins, children born with teeth or with a caul or
other peculiarities are, among African folk, regarded as special types
of personalities whose spiritual potency calls for special treatment.[85]
Equally widespread is the African belief that special measures must
be taken against malevolent spirits believed to cause a woman to
have a series of miscarriages or stillbirths, or consistently over a
period of time to bear infants who die one after the other. Among
the Geechee Negroes of Georgia,[86] it is believed that, "if you cannot
raise your children, bury on its face the last one to die and those
coming after will live." A technique of tricking the malevolent
spirits, described as occurring among these Georgia Negroes, is
equally African: "If you wish to raise your newborn child, sell it
to someone for 10 or 25 cents and your child will live." A case is
cited to illustrate the custom:

> A woman, the mother of 16 children, lost the first 10. The tenth one
> was buried on its face, and the other six, as they were born, were raised
> without difficulty. This woman's daughter lost her first two children, but
> the third was sold, and it lived.[87]

Puckett, who has also included this case in his discussion of Negro
folk beliefs, has recognized its African character from a passage he
quotes from Talbot in support of his contention.[88] Customs of this
nature are, however, spread much more widely than just in the
Niger Delta area, being found far to the east and west of that
region.

The African concept that anomalous births indicate the future
powers of a child is also a living belief in this country. Parsons
states:

> One born "foot fo'mos'" or a twin cannot be kept in bonds. "You
> kyan' put um down in de pail, come right out." If you tie him, he will
> "cross his feet, sleep, rise right up an' go 'way; take out his han' an'
> feet, rope don' go loose. He stay dere as long as he not aworried. In
> confusion (trouble) de oder twin loose him, my gran'moder say, an' de

sperit loose him dat born foot fo'mos'.'' I heard of one remarkable child born foot foremost and "in double caul."[89]

Steiner reported a Georgia Negro who, having been born with a caul, attributed to this fact his possession of two spirits, one that remained in his body and one that went about aiding him,[90] this being also reminiscent of the African belief in multiple souls. Puckett gives a further list of traits which at birth indicate the baby's fate or future powers,[91] which are likewise of African derivation and are to be encountered throughout the Negro West Indies as well as in the United States.

Names are of great importance in West Africa. Names are given at stated periods in an individual's life, and, as among all folk where magic is important, the identification of a "real" name with the personality of its bearer is held to be so complete that this "real" name, usually the one given him at birth by a particular relative, must be kept secret lest it come into the hands of someone who might use it in working evil magic against him. That is why, among Africans, a person's name may in so many instances change with time, a new designation being assumed on the occasion of some striking occurrence in his life, or when he goes through one of the rites marking a new stage in his development.

No great amount of information is to be found concerning the circumstances under which names are given Negro children in this country, but the available data indicate that African ceremonials in name-giving have by no means been lost. Parsons reports as follows from the Sea Islands:

A baby is named on the ninth day. At this time, or when she first gets up, a mother will carry the baby around the house, "walk right 'roun' de house."—The mother or some friend will give the name, probably a family name—"keep de name right in de fahmbly."[92]

Puckett gives an account of a Mississippi naming custom which is in the same tradition as that just cited, though it emphasizes different elements in the aboriginal complex:

An old Mississippi slave says that the child will die if you name him before he is a month old—seeming to indicate the fact that the spirit should have a chance to familiarize itself with this locality before it is pegged down. This conjecture is strengthened by the fact that when the child is a month old he is taken all around the house and back in the front door, then given a thimbleful of water. The meaning of this practice has been forgotten although one informant claims that the thimbleful of water is to keep the baby from slobbering.[93]

This rite of taking the infant about the house closely resembles the Haitian custom of circling the habitation on any important ritual occasion; taking a child to those places which will be of importance to him—"introducing" him to them, in a sense—in a manner to be encountered in many parts of West Africa.

How sturdily African traditions concerning names and naming have resisted European encroachment can be made clearer if the preceding passages, and the data to be adduced in paragraphs to follow, are compared with materials describing analogous rituals and beliefs found in the Gold Coast or Dahomey. The elaborate ceremonies that mark the birth of a child and the events of his life, the numerous categories of names that are given the infant, especially in Dahomey, to reflect specific circumstances held to mark his conception, or indicating the manner of his birth or certain physical characteristics manifested at that time, and the like, all demonstrate how meticulously these folk follow regulations concerning these matters that have been laid down in accordance with their beliefs.[94] Nor are these two peoples of West Africa unique. They are cited merely because the most complete data are from them; there is, however, enough material in reports from other parts of West Africa and the Congo to demonstrate that the patterns of which they represent so great an elaboration are everywhere present, and hold a place important enough that their survival in the New World, even under intense acculturation, is readily to be understood.

Puckett and Turner have made the most extensive collections of Negro names in the United States.[95] Puckett's findings are based on the analysis of designations found in documents of the slave period, and on lists obtained from present day Negro college students; Turner's data are derived from field work in the Gulla Islands. Puckett suggests that among the factors making for the retention of African names operative during the eighteenth and nineteenth centuries may have been the prestige associated with African designations:

. . . Cobb, in mentioning four native Africans, named Capity, Saminy, Quominy, and Quor, who were slaves in Georgia, states that they had facial tattooing and "were treated with marked respect by all the other Negroes for miles and miles around." This suggests that the cultural value of American names may not have been the same with the slave as with the modern immigrant. African captions may even have conferred a certain amount of distinction among the slaves, and thus have continued where the master allowed it. In fact, freedom from control of white owners, in addition to a slowly forming family tradition, may have been one reason why the free Negroes of 1830 seem to have possessed a larger assortment of African names than did the slaves of that period.[96]

The list of African slave names of the eighteenth century provided us is replete with designations whose provenience is evident. Abanna, Abnabea, Abra, Ankque, Annika, Bamba, Bayna, Bilah, Binah, Boohum, Braboo, Bumbo, Bungoh, Comba, Cudah, Cumba, Curiarah, Demeca, Ducko, Fantee, Gumba, Lango, Monimea, Mowoorie, Ocra, Ocrague, Ocrasan, Ocreka, Oessah, Pattoe, Quack, Quaco, Quamana, Quamno, Quash, Quoney, Samba, Sena, Simbo, Simboh, Tanoe, Temba, Warrah, Yamboo, Yaumah, Yearie, Yonaha, and Yono Cish,[97] despite the quaint spellings, are equivalents not only of the Gold Coast "day names" such as are found today in the Gulla Islands and elsewhere in the New World, but also of place names and terms commonly employed as personal names in the Niger Delta and the Congo. Turner, who in his mimeographed preliminary report, "West African Survivals in the Vocabulary of Gullah," identifies seventy names as African, mainly Mende, has later indicated[98] that it was only on close acquaintance that he was able to obtain the many African designations he has since recorded. For among the Gullah, "basket names" are used only within the family and among close acquaintances; and it is Turner's conviction that without proper entree and the support of adequate knowledge of African data, a student could go long without suspecting, much less recording materials of this type. Negro nomenclature diverges in no respect more from white practice than in its great diversity. Turner's comment on his experience in collecting personal names[99] is to the point as concerns the origin of this trait:

Even though the Gullahs may not know the meaning of many African words they use for proper names, in their use of English words they follow a custom common in West Africa of giving their children names which suggest the time of birth, or the conditions surrounding it, or the temperament or appearance of the child. All twelve months of the year and the seven days of the week are used freely. In some cases the name indicates the time of day at which the birth occurs. In addition to the names of the months and days, the following are typical: Earthy (born during an earthquake), Blossom (born when flowers were in bloom), Wind, Hail, Storm, Freeze, Morning, Cotton (born during cotton-picking time), Peanut, Demri (born during potato-digging time), Hardtime, Badboy, Easter, Harvest, etc. Names suggestive of the West African totems or clan names are Rat (female), Boy Rat (male), Toad, etc.

Another element of the naming complex is the ease with which a Negro may assume one name after another, especially in dealing with whites. The truth of the matter is that a name given a Negro by an outsider is something of the order of a nickname, worn even more lightly than are the nicknames of whites, which are seldom bestowed more than once on a given person, and are often retained through life. Experience in Dutch Guiana was enlightening in this regard. Here a man who had been known for some time, when first asked for in his own village by a name regularly used for him, could not be located. His people used quite a different name for him, but even this name proved not to be his "real" one, which had been given him at birth and was held a close secret within the family circle. It is thus not only possible, but quite probable that Puckett's list of slave designations actually represents but a portion of the African names employed. In accordance with a pattern operative in West Africa, the West Indies, and Guiana, names given by the slave-owners were most likely regarded as but an added designation to which one responded. They were likewise very possibly thought of as names to be employed by fellow slaves in the presence of whites, being accepted with the reservation that different, "real" names were to be used in the cabin or on other occasions when none but fellow slaves were present.

Botume, who, it will be remembered, worked with the freedmen of St. Helena Island immediately after their emancipation, has set down her bewilderment concerning the use of names. When placed at the side of Turner's findings on the "basket name" and the wide-spread Negro tradition of accepting additional names, her remarks tend to document the point just made so as to bring it out of the realm of conjecture:

In time I began to get acquainted with some of their faces. I could remember that "Cornhouse" yesterday was "Primus" today. That "Quash" was "Bryan." He was already denying the old sobriquet, and threatening to "mash you mouf in," to anyone who called him Quash. I reproved the boys for teasing him. "Oh, us jes' call him so," with a little chuckle, as if he ought to see the fun. The older people told me these were "basket names." "Nem'seys (namesakes) gives folks different names." . . . It was hopeless trying to understand their titles. There were two half-brothers in school. One was called Dick, and the other Richard. In one family there were nine brothers and half-brothers, and each took a different title. One took Hamilton, and another Singleton, and another Baker, and others Smith, Simmons, etc. Their father was

"Jimmy of the Battery," or "Jimmy Black." I asked why his title was Black. "Oh, him *look* so. Him one very black man," they said.[100]

That such confusion could never be tolerated within a society is self-evident; in this case, our bewildered author was merely attempting to cope with a chaos that existed only for those outside the group, within which such ephemeral designations merely represented a play on names over a stable reality of correct appellations.

African influences in customs concerning Negro children are also found in the isolated items that have been published having to do with the training and later care of the child. Puckett,[101] particularly, has made available numerous "superstitions" which suggest how deep-seated in African traditions are certain sanctions which determine folk behavior bearing on elements in child development. A passage may be cited as an example:

In the Sea Islands and in Mississippi, according to one informant, when a child is slow to walk you should bury him naked in the earth to his waist, first tying a string around his ankle. The same informants also speak of carrying a child to the doctor to have his tongue clipped when he is slow to talk. While sweeping is sometimes used beneficially, one should never sweep the room while the child is asleep. The idea is that you will sweep him away, and this seems to be possibly a half-remembered notion of the African "dream-soul" which leaves the body during sleep.[102]

Parsons recounts a related belief from the Sea Islands:

If you have to "go a distance wid de chil'," you notify de speret, call, "Come, baby!" Unless you called back in this way, wherever "you stop dat night, you wouldn' get any res' at all, 'cause de speret lef' behin'. Call him at eve'y cross-road you come to."[103]

To "call" the soul of a child before going on a journey is routine in West Africa, and elaborate care must be taken on numerous other occasions to ensure that it stay with its owner and continue to exercise benevolence toward him. Among the Yoruba, and in Dahomey, well-recognized rituals exist in which a person pays homage to his soul, while in the Gold Coast the patrilineal soul line is of equal importance with the matrilineal descent line. The correspondence of the material given in the passage just quoted, however, is most striking when analyzed with reference to a situation encountered in Dutch Guiana. Here a young woman informant, who had been ill for some years after she had moved to Paramaribo with her family from another town, recovered her health when, at the instructions of a diviner, she went through a ceremony calculated to return her soul to her. It had remained behind in the town of her birth because her mother had neglected at the time ritually to inform it where the family was moving, and it had thus failed to accompany its owner.[104]

That those concerned with education and health have been content to formulate projects of vast proportion without regard to their relationship to folk custom in child rearing and child care can only be regarded as a commentary on procedures in initiating and carrying through such enterprises. Quite without reference to the African background, the fact that Mrs. Cameron, working to a considerable extent in urban centers, was able to document the "high positions" which the practitioners of folk medicine and magic "possess in their respective communities," north and south, is eloquent of the shortness of the perspective under which good works are too frequently undertaken:

> Their hold must be very strong to allow them to maintain their ground in the face of such powerful interferences as the State Boards of Health, free dispensaries and free education. But the mould for the reception of these beliefs is set from babyhood in many families and the traditions surrounding these practitioners seem to still retain enormous force.[105]

In Trinidad, Haiti, and Dahomey appropriate rituals mark the appearance of the permanent teeth; the essence of one such rite is to throw the first deciduous tooth to fall out on the roof of the mother's house or into some near-by place, asking that the new teeth be strong and beautiful. Parsons reports from the Sea Islands that: "When a 'chil' sheddin' teet', take an' put 'em in a corn-cob, an' fling it right over de house.' This practice was referred to as 'callin' de new teet' back.' "[106] That its provenience is other than the English custom cited by Puckett[107] in connection with the Negro belief that deciduous teeth must be protected from dogs, which "requires the dog to eat the tooth," is apparent when its African counterpart is pointed out. As in so many other instances of strained ascription of origin, the difficulty in this case has been that the precise African correspondence had never been recorded, and was thus not available to the comparative folklorist.

The importance of whipping among American Negroes as a technique of training the young has been frequently remarked. An example of this is the following:

> A woman in her late fifties said: "Today parents don't make children mind enough. We used to take and whip them." She went on to tell that she grew up in a small rural community, and "when I was young, every

woman in the place was my mother. If I did wrong and one of them saw me she'd whip me, and then she'd tell my mother and I'd get another whipping. Today parents don't whip their children enough and the children are getting worse."[108]

Attempts to account for this phenomenon, which again diverges from common practice among whites, are usually couched in historical or psychological terms referring to the experience of the Negroes under slavery. In a passage which follows the one quoted, this explanation is given:

Formerly whipping served both Whites and Negroes as an accepted form of discipline and as a convenient outlet for sadism. The grandparents of the present young colored parents were themselves whipped by their white masters. The majority of old Negroes, in contrasting the present with the past, bring up the point of corporal punishment, saying: "They can't whip us now like they used to." The slaves adopted whipping as the approved way of correcting and punishing faults. Moreover, they had no means of retaliating for their own beatings, unless on their own children. . . . Although whipping was a pattern taken over from the masters, and still survives among their descendants, today the failure of Negro parents to whip their children may be criticized as "aping the Whites." A woman of sixty made that accusation against a young mother of the upper class, who always tries to explain things to her children and never beats them at all. It is of course true that reluctance to whip children is a newer white pattern which is gradually displacing the old.[109]

This attempt to account for beating is appealing because of its logic, but in the light of the facts it is not only poor history but poor psychology, since it completely disregards the fact that the outstanding method of correction in Africa itself and elsewhere among New World Negroes, whether of children or of adults, is whipping. In point of fact, the literature of slavery gives no indication that slaves did beat their children to "take out" their own humiliation on those who were as impotent before them as they themselves were under the lash of the master. Finally, it is not easy to understand just why sadistic tendencies should have taken this particular form among a people whom observers almost never characterize by this term.

When we turn to the data from Negro cultures concerning whipping as a form of correction, we find a great deal of material to confirm an assumption of historical relationship to New World practice. In Dahomey and among the Yoruba, flogging of an order of severity almost unknown in Europe, except as a penal device, was the rule. Children were likewise flogged—not so severely, it is true, but severely in terms of comparable modes of applying this form of discipline in white societies. Whipping is considered an integral part of West African pedagogical method; indeed, no better expression of the theory behind it could be given than the statement quoted by Powdermaker in the first of the two citations from her work, for this matches expression of opinion heard several times in West Africa itself when the training of children was under discussion.

In Haiti, to shift to the New World, or in Guiana or Trinidad or Jamaica, the cries of young boys and girls being whipped for misdeeds are heard even by the casual visitor. The right of any elder to whip an erring younger member of his family is vested in all Haitians, and on occasion a grown man will kneel before his father or uncle to receive the strokes that have been decreed as a punishment. Again, the comment given by Powdermaker as to the right of any woman of a community to whip a girl has specific correspondences both in Dutch Guiana and in Dahomey. In the latter, a boy or girl is whipped by any aunt, who thus makes it less likely that the father will obtain a poor impression of the child when hearing the outcry, and favor another wife's offspring. And in Dutch Guiana, a young woman calls old women of her village by the term for co-wife—"*kambosa,* she who makes trouble for me," the explanation of the practice being that every elderly woman is on the lookout for misbehavior. The old women are thought of as interfering unduly in the life of the younger women, making their escapades more difficult and assuring punishment on discovery.

7

The principle that life must have a proper ending as well as a well-protected beginning is the fundamental reason for the great importance of the funeral in all Negro societies. This results from several causes, among the most important being the widespread African belief in the power of the ancestors to affect the life of their descendants. The place of this belief in the total African world view is in keeping with its significance for the people. For the dead are everywhere regarded as close to the forces that govern the universe, and are believed to influence the well-being of their descendants who properly serve them. The worship of the ancestors thus supports all social institutions based on kinship, giving them that measure of stability and integration that has been so frequently remarked by those who have had firsthand contact with African tribes.

In West Africa, the ceremonial richness of the ancestral cult is enhanced because of the greater resources of the tribes of this region when compared to other areas, yet the feeling of the ever-present care afforded by these relatives in the world of the spirit is essentially the same among all African folk. The ritual for the ancestors begins with the death of a person, who must have a funeral in keeping with his position in the community if he is to take his rightful place in the afterworld. As far as surviving relatives are concerned, two drives cause them to provide proper funeral rites. The positive urge derives from the prestige that accrues to a family that has provided a fine funeral for a dead member; negative considerations arise out of the belief that the resentment of a neglected dead person will rebound on the heads of surviving members of his family when neglect makes of him a spirit of the kind more to be feared than any other—a discontented, restless, vengeful ghost.

The ancestral cult resolves itself into a few essentials—the importance of the funeral, the need to assure the benevolence of the dead, and, in order to implement these points, concern with descent and kinship. As illustrative of how these essentials have persisted, even where acculturation to white patterns has been most far-reaching, we may turn to the description of a family reunion of a group who, as the descendants of a free mulatto couple, are in their customary behavior as far removed as possible from the behavior of such Africans as may be included in their ancestry:

This family has had family reunions for fifty years or more. When the family reunion took place in 1930 there were grandchildren, great-grandchildren, and four great-great-grandchildren in the ancestral homestead to pay respect to the memory of the founder of the family, who was born in 1814 and died in 1892, and his wife, who died in 1895 at the age of seventy-one. His only living son, eighty-four years old, who was the secretary-treasurer of the family organization, was unable to attend because of illness. The founder of the family had inherited the homestead from his father, who was listed among the free Negroes in 1830. A minister, who had founded a school in the community in 1885 and knew him intimately, described him as "an old Puritan in his morals and manners and the only advocate of temperance in the county" when he came there to work.

The meeting was opened with a hymn, chosen because of its theme, "leaning on the Everlasting Arm." The widow of the son of the founder of the family spoke of the necessity of the children's "walking in the straight path" that the founder "had cut out." Her daughter, a recent

Master of Arts from Columbia University and the vice-principal of a colored high school in a large eastern city, had returned to the family reunion. Another granddaughter read, as was customary, a paper embodying the history of the achievements of the family and a eulogy of their ancestors. The program included a prayer service after which dinner was served. The ceremony was ended by a visit to the family burying-ground where there is a tombstone bearing the names of the founder and his wife and the date of their birth and death.[110]

This passage clearly indicates that, though the conversion of New World Negroes to Christianity in its varying forms has obliterated overt manifestations of the ancestral cult to the extent that European religious beliefs have been taken over, the extinction of the cult does not mean that its spirit has disappeared or that its sanctions have not persisted. Family reunions are common enough in this country, but it is somewhat doubtful whether at many white family reunions the day is ended with a visit to the tombstone of the founders; whether eulogies of the "ancestors"—this family was founded in 1814, it will be remembered—are included in the festivities; or whether such a strong religious tone is given the proceedings. One must look for these elsewhere than in custom governing affairs of this sort common to whites and Negroes. If the more detailed accounts of ancestral rites are consulted, such as have been recorded, for example, by Rattray for the Ashanti and for Dahomey in the work already cited, indication will be found of the provenience of the intangible validations which have made for self-consciousness on the part of this particular "extended family," and have shaped its family rituals.

The range of variation implied in resemblances between survivals in the practices of a group sophisticated in terms of Euro-American behavior, and the full-blown rituals of Africa itself or, for another region of the New World, Dutch Guiana[111] is thus seen to be great. With a realization of the various acculturative steps represented in other New World instances lying between the two extremes in customary usage where the dead and their souls are concerned,[112] we may therefore turn to a consideration of other Africanisms in death customs, funeral practices, and belief in ghosts that have been recorded for the United States.

Odum, in an early work, recognized the important place accorded death in the mores of the Negro community:

It is a great consolation to the Negro to know that he will be buried with proper ceremonies and his grave properly marked . . . there are few greater events than the burial, and none which brings the community together in more characteristic attitude. The funeral is a social event, for which the lodge appropriates the necessary expenses. Here the religious trend of the Negro is magnified and with praise of the dead and hopes for the future he mingles religious fervor with morbid curiosity and love of display.[113]

In the Mississippi community studied by Powdermaker, we learn that:

Burial insurance is usually the first to be taken out and the last to be relinquished when times grow hard. It is considered more important by the very poor than sickness or accident insurance, although the latter is becoming more popular. No Negro in Cottonville can live content unless he is assured of a fine funeral when he dies. Fifteen cents a week and five cents extra for each member of the family will guarantee a hundred-dollar funeral, in which the company agent plays an active part.[114]

In a later passage, the importance of providing for adequate burial is emphasized:

There are certain expenses besides taxes which must be paid in cash. One of these is insurance. In the dilapidated shacks of undernourished families, whose very subsistence depends upon government relief, the insurance envelope is almost invariably to be seen hanging on the wall. Even when sickness and accident insurance are allowed to lapse, the burial insurance is kept up.[115]

Johnson's report is to the same effect:

The tradition of the burial society hangs on in the mutual organizations which, though concerned chiefly with death benefits, build up and hold their membership on the strength of the social features. In a situation under which families were losing such insurance as they had, the burial societies were gaining in strength.[116]

Societies of this sort are ubiquitous among New World Negroes as the most widespread and institutionalized survivals of the African desire for proper burial. As is often the case, drives of this sort are illuminated by negative examples, one of which can be given in terms of an incident that occurred in the Trinidad village of Toco during the summer of 1939. An extremely poor man, whose wife and children no longer lived with him, was found dead in the shack he inhabited. Since he had no relatives and belonged to no insurance society, his burial was left to the officials charged with the care of paupers. In the tropics, a corpse is ordinarily buried in early morning or late afternoon, and during the day following death Public

Works carpenters could be heard hammering on "de box" they had been hired to make. After they had finished, the young men who had made the crude coffin placed it on their shoulders, and, with no concern to form a procession, walked down the road with it to the cemetery, laid it on the ground until the grave was dug, and then, lowering it, refilled the hole and went their way.

Indignation was voiced on every hand, and pity. Expressions of opinion were heard not only from members of the village of pure Negro descent, but those of mixed blood as well. One minor official, a mulatto of upper-class status, said: "It wasn't right to put him in the hole just like he wasn't human, it wasn't right of the ministers to stay away, and it wasn't right nobody laid him out." No one was surprised when one noonday, shortly afterward, some children on their way home from school, gathering fruit beneath a tree that stood in front of his hut, ran with fear as, glancing into the branches, they "saw" him glowering at them. And the door of his poor hut, blown open by the wind, remained unshut as folk sedulously avoided what must be a residence haunted by an angry, dissatisfied, vengeful spirit.

On the southern plantations, the feeling of the slaves that proper attention be paid the requirements of the dead was in some measure respected, as is shown by contemporary testimony on slave funerals. This, however, meant keeping alive the African tradition that the principal ritual take place some time after the actual interment, separating this, so to speak, from the funeral as such. The practice was encouraged by economic and social conditions under slavery; but it must be remembered that here, as in other forms of behavior previously considered, this situation merely tended to rework a tradition which, in such a manifestation as the Dahomean partial and definitive burials,[117] is found widely spread throughout West Africa and is today encountered in the New World where imposed regulation does not require immediate burial. The following passage shows how in outline the entire African funeral complex, including the delayed interment, was continued among the slaves:

There was one thing which the Negro greatly insisted upon, and which not even the most hard-hearted masters were ever quite willing to deny them. They could never bear that their dead could be put away without a funeral. Not that they expected, at the time of burial, to have the funeral service. Indeed, they did not desire it, and it was never according to their notions. A funeral to them was a pageant. It was a thing to be arranged for a long time ahead. It was to be marked by the gathering of kindred and friends from far and near. It was not satisfactory unless there was a vast and excitable crowd. It usually meant an all-day meeting, and often a meeting in a grove, and it drew white and black alike, sometimes almost in equal numbers. Another demand in this case—for the slaves knew how to make their demands—was that the Negro preacher "should preach the funeral" as they called it. In things like this, the wishes of the slaves usually prevailed. "The funeral" loomed up weeks in advance, and although marked by sable garments, mournful manners and sorrowful outcries it had about it hints of an elaborate social function with festive accompaniments.[118]

Another version of this same manner of honoring the dead by the slaves reads as follows:

One of the big days among our people was, when a funeral was held. A person from New Jersey who was not acquainted with our customs, heard it announced that: "next Sunday two weeks the funeral of Janet Anderson will be preached." "Well," said the stranger, "how do they know that she will be dead?" The fact was, she was already dead, and had been for some time. But, according to our custom, a custom growing out of necessity, we did not hold the funeral when the person was buried. The relatives—and friends—could not leave their work to attend funerals. Often persons would be buried at night after working hours. If the deceased was a free person, and the immediate family could attend a week-day funeral, there might be others, both friends and relatives who could not attend, hence, the custom became general.[119]

That the custom, noted likewise by Puckett for recent times,[120] has by no means died out is illustrated by the recent experience, in two instances, of having Negroes leave jobs to return south in order to attend delayed funerals, in one instance, "of my mother who died last spring." As in earlier days, the explanation of the principals was in terms of the need to make proper preparations, and the difficulty of gathering the family on short notice. Yet one may well ask why such delayed funerals are not found among other underprivileged groups in the population—immigrants, for example, whose need for delay in terms of their inability to leave jobs on short notice is quite as great as that of the Negroes. This is made the more evident when it is pointed out that the explanation for this custom given by Negroes, while in line with the practical requirements of their life, happens to be very similar to the explanation given by Dahomeans for their aboriginal form of the institution. For when asked why they permit time to elapse between the "partial" and the "definitive" burial of their dead, they likewise point out their need for time to effect necessary preparations if the rites are to be carried out in proper style. Whatever the rationalization, the provenience of the tradition as found in the United States is clear; the light it throws on attitudes toward death and burial among Negroes in this country is merely further testimony of the vitality of the entire complex of attitudes and rituals toward death that have carried over in however changed outer form.[121]

More Africanisms are found in some of the details of Negro funeral procedure. Here, as elsewhere, it is to be regretted that no consecutive account of the rituals of death are to be had for analysis,[122] yet such data as have been published unambiguously include many African correspondences. The importance of proper mourning, by which is meant public vocal expression of grief, finds many counterparts in the ancestral continent.[123] Crape is worn by members of the family, and not only placed on the door of the house where the dead lived, but is even reported as being tied on "every living thing that comes in the house after the body has been taken out—even to dogs and chickens." As an "attempt to pacify an avenging spirit which was the cause of death,"[124] this likewise reflects African procedure and belief. The extension of separating burial and funeral rites into the holding of multiple funerals for a person of status in the several communities or several organizations he served[125] is similarly non-European. The great need that a funeral proceed smoothly, as shown in the belief that "if the procession should stop another death will soon follow, mishap on the way probably indicating that the corpse is dissatisfied and regrets having to leave this world,"[126] can be readily matched in Africa.

Parsons gives further hints of direct Africanisms in connection with the funeral itself:

When an Odd-Fellow dies, "de body cover up, nobody mus' touch. Six men come to bade an' dress de body." Similarly, on the death of a Good Samaritan, "de body cover up, no one can touch de body 'til de Sisters come. Sen' to de Wordy (Worthy) Chief. Fo' Sisters come wash de body an' lay out. Nobody can look at de face widout de Sister say so. Say, 'Can I look at de face?'—'Yes.' Each Sister has to watch de body fo' one hour." [127]

The correspondence of this complex to other New World Negro customs and those of West Africa is immediate, especially that part wherein it is forbidden for anyone to touch the body until the members of the society to which the dead belonged have prepared it—to say nothing of the further secret rites which future research may perhaps reveal. To refer again to Dahomey, the body of a member of a religious cult group there may not be touched by relatives until the priest and the surviving members, employing elaborate secret rites, come and "take the spirit" from the head of the body.[128] In Trinidad, secret society members, most notably in the case of Masons, gather in the room where the body of a dead "brother" lies to perform secret rites and prepare it for burial. In Haiti, a person with a "spirit in his head" must have it removed before the more common rituals are performed.[129]

Other African aspects of the funeral appear as we continue our search. At the funeral of one Jesse Harding as described by Johnson, it was hard to arouse the congregation; he had been a good man, but not a type sympathetic with the easygoing ways of the group among whom he lived. One of those present, called upon to speak, did the best he could:

> The chill of the audience bore down upon him, and he admitted, almost bargainwise: "Brother Jesse had his faults, like you and me. I talked with him at home and at the hospital." He excused himself for not visiting at the hospital oftener: "They had to ask me not to come to the hospital so much, 'cause there was so many sick folks just like Brother Jesse." Everybody knew the deceased's forthrightness and it could be mentioned again.[130]

To evaluate frankly at a funeral the characteristics of the dead, to expose in direct address the differences he may have had with those in contact with him during his life, as though the spirit could hear what is said; all these characterize West African rituals. At the funeral of a Liberian Kru in Chicago some years ago, attended by the men of the African "colony" of that city, all of them spoke in this manner to the body of the dead, so that the corpse would bear them no resentment that would interfere with the tranquillity of his spirit existence and cause his return to trouble them.

Puckett is of the belief that "in a general sort of way those practices up to actual burial are European, while grave decoration and avoidance of the spirit are more African in type."[131] It does not seem likely, however, that this analysis will be proved valid when full accounts are written of the entire cycle of death rites performed in a considerable number of West African tribes, in various Negro communities over the New World, and particularly in the United States, especially if these accounts are presented with coherent analyses of conceptions as to the causes of death and the role of the dead in the world. It has already been indicated how, in the Negro funeral as found in the United States not only many of the elements in its ritual but also its underlying motivations and its setting in the matrix of custom reflect an impressive retention of African traits.

The function of nature deities in West African pantheons is to punish those who have transgressed accepted codes, and of these forms of punishment death by lightning is one of the most widely recognized. One of the elderly informants queried by Johnson as to conditions of slavery, said this:

My master's brother's wife was so mean tel the Lord sent a peal of lightenin' and put her to death. She was too mean ter let you go ter the well and git a drink of water, and God come 'long and "squashed" her head open.[132]

Puckett also points out that other beliefs as to the relationship between lightning and death are operative when he states that "If it rains while a man is dying, or if the lightning strikes near his house, the devil has come for his soul."[133]

What may be regarded as a generalized pattern of formal leave-taking of the dead by all his relatives and close friends, with varied rites during the process, is deeply rooted in West African funeral rituals. The custom of passing young children over the coffin has not been reported for West Africa, but something closely related to it has been witnessed among the Bush Negroes of Dutch Guiana.[134] Parsons, for the Sea Islands, quotes an informant as follows:

"Dead moder will hant de baby, worry him in his sleep. Dat's de reason, when moder die, dey will han' a little baby 'cross de box (according to others, across the grave) same time dey fixin' to leave de house, befo' dey put um in de wagon."[135]

Puckett says:

In another case in South Carolina the children march around the father's casket singing a hymn, after which the youngest is passed first over and then under the casket and the casket is taken out on and run upon the shoulders of two men.[136]

We also learn from this same source that fruit trees in an orchard are sometimes notified of the death of their owner, "lest all decay";[137] and that at wakes the body, lying on a "coolin'-board," is addressed by the mourners as they take their farewell of the dead.[138] The wake is as important in Africa as it is in the West Indies and the United States. It is reasonable, however, to suppose that as found in the New World it is an example of the process of mutual reinforcement experienced when similar cultural impulses from two sources come into contact.

The importance of exercising caution when dealing with the spirits of the dead is fundamental also in West African belief. The statement of Puckett that, "It is thought to be bad for any one to work around a dead person until he is tired, i.e., in a weakened condition where spiritual harm might result"[139] is to be met with everywhere in West Africa and among New World Negroes. Among the Bush Negroes, to dig a grave requires several days, because of the danger that a worker might perspire and allow a drop of sweat to fall in the excavation. The ghost could then utilize this to take with him the soul of the one who had labored too hard. The conception that a man has "two ghosts, an evil ghost, derived from the body and a 'Holy Ghost' derived 'frum de insides' "[140] is to be referred to the multiple soul concept of West Africa, which elsewhere in the New World takes the form of ascribing to a person a dual soul, one inside the body and the other manifested as the shadow.

The spirits of the dead are held to be dangerous if death occurred in some strange or terrible manner. They are headstrong if wishes they expressed while living are not followed, vengeful if their relatives are not respectful or if spouses marry too soon; and various devices must be employed to ensure that their bodies will remain quiet, such as fastening their feet together or weighting them down.[141] The dead may on occasion return to the scenes they knew when alive, and in such instances a feast may be provided for them.[142] Or, again, offerings may be placed in the coffin, or in the form of coins on a plate near the coffin to be used by the family of the dead, or on the grave.[143] And while all these customs, as found in the United States, probably represent syncretisms of African and European belief, they are to be encountered in many parts of West Africa, and everywhere among the Negroes of the New World outside this country.

Chapter VII

THE CONTEMPORARY SCENE: AFRICANISMS IN RELIGIOUS LIFE

The prominent place held by religion in the life of the Negro in the United States, and the special forms assumed by Negro versions of Christian dogma and ritual, are customarily explained as compensatory devices to meet the social and economic frustration experienced by Negroes during slavery and after emancipation. Such explanations have the partial validity we have already seen them to hold for various phases of Negro secular life but, as must be emphasized again, cannot be regarded as telling the entire causal tale. For underlying the life of the American Negro is a deep religious bent that is but the manifestation here of the similar drive that, everywhere in Negro societies, makes the supernatural a major focus of interest.

The tenability of this position is apparent when it is considered how, in an age marked by skepticism, the Negro has held fast to belief. Religion is vital, meaningful, and understandable to the Negroes of this country because, as in the West Indies and West Africa, it is not removed from life, but has been deeply integrated into the daily round. It is because of this, indeed, that everywhere compensation in terms of the supernatural is so immediately acceptable to this underprivileged folk—and causes them, in contrast to other underprivileged groups elsewhere in the world, to turn to religion rather than to political action or other outlets for their frustration. It must therefore be assumed that not only in particular aspects of Negro religious life to be pointed out in this chapter, but in the very foundations of Negro religion, the African past plays full part. And we must hold this in mind as we turn to a review of those manifestations of Negro religion which, like its fundamental sanctions, can be traced to a pre-American past.

2

We may begin by treating the organizations that comprise the institutionalized forms of Negro religion. From the earliest times of slavery, it has been the less inhibited, more humble denominations which have attracted Negroes in the United States. Perhaps because this is so striking, a formula which explains it in terms of simplicity, naïveté, and emotionalism has attained a certain currency among students. Thus:

> The worship of the Negro is of the simplest sort. He has no appreciation of elaborate rituals, of services consisting of forms and ceremonies. Hence the great mass of colored races have united with either the Methodist or Baptist Churches. These churches have the simplest, least complicated forms of church services, and the Negro naturally gravitated toward them.[1]

The simplicity assumed in this citation, however, is but one of those questionable generalizations encountered again and again in this analysis. Actually, Negro propensity for ritual, as evidenced in aboriginal cultures where no contact with whites has to be taken into account, is quite the equal in intricacy of any series of European rites. Nor must it be forgotten that when the New World is considered as a whole, the Negroes who adhere to Catholicism, with its elaborate ceremonialism, far outnumber those who are affiliated with Protestant sects having simpler rituals.

Dollard turns to an historical and psychological explanation:

> It is impossible to say from census materials what percentage of Negroes and whites are members of religious bodies in our community. We do know for the county that about half the adult Negroes are church members and of these, four-fifths are Baptists. We do not know how far these proportions hold for Southerntown and county but Southerntowners say that if a Negro is not a Baptist someone has been tampering with him. Apparently the Baptists and Methodists were most energetic in their early measures to capture Negro allegiance by means of their itinerant preachers. Furthermore, the religious behavior of these denominations was less formalized and stereotyped than that of the Presbyterian or Episcopal churches, and the evangelical mode of preaching seemed to have a spontaneous appeal to the Negroes; perhaps they were disposed toward emotionally toned group meetings by their African background. They seemed to have a marked selectivity for the tensity and emotionalism of the Baptist and Methodist preaching. . . .[2]

The question of why "less formalized and stereotyped" rituals should appeal to the Negroes may be put aside for the moment. That the social and economic status of the communicants was an effective cause that operated in the case of Negroes and whites alike is apparent, however, and must be taken into full consideration.

Jackson, informing us that in Virginia, "in every instance we note that the church established [by Negroes] was a Baptist church," goes on to say, "it is to be noted also that through Virginia generally the servant class leaned to the Baptist connection rather than to the other churches."[3] He attempts to account for the lack of appeal of the more sober sects in the following terms:

The greatest handicap in the ministrations of the Established Church, however, was its lack of emotionalism and a spirit to fire the masses. The functionaries of this body, clinging to European conceptions of religion, were unable to sense the nascent evangelism of the American people with its insistence on the sinfulness and depravity of man, a condition which in turn called for this thorough regeneration. To develop this new feeling a special technique was needed. Such a technique was found in the revival.[4]

Members of this denomination themselves recognized the need for adaptation to a more congenial pattern:

Episcopalians in Virginia under Bishops Meade, Johns and others became evangelical to a degree approximating Baptists and Methodists. They then accepted the revival and preached the gospel and became disciplinary on matters of amusement and public entertainments.[5]

As concerns the particular drives which made for Negro affiliation to the Baptist Church, this sociological explanation is offered:

The Baptist church by reason of its policy is par excellence the church of the masses. It is the religious organization to which the underprivileged class, more so than to any other denomination, is likely to turn. This church is extremely democratic and is characterized by a local autonomy which makes each church practically a law unto itself. The man who is, therefore, passed over in every-day secular affairs turns to an organization in which he can find that very expression which is otherwise denied him.[6]

Furthermore, we learn that

. . . there was a strong attraction of the slaves for the Baptist church because they were given greater participation in religious exercises. . . . There was also greater liberality among the Baptists in giving Negroes permission to preach while also in addition the Baptist method of administering communion was not calculated to discriminate against them. Finally the mode of baptism among the Baptists satisfied the desire of the Negro for the spectacular.[7]

Certainly some of the reasons why Negroes were not attracted to the Established Church are implied in the following statement:

As a general rule, the Episcopal minister went to the family mansion, the Methodist minister preached to the Negroes and dined with the overseer at his House.[8]

There is no question of the popularity of the Baptist and Methodist churches among Negroes at the present time. Johnson's community today is "predominantly Baptist," with Methodists next in number—of "612 families, 439 were Baptists and 147 Methodists."[9] C. C. Jones[10] and Jackson[11] give data which clearly show how the situation today is merely the continuation of an earlier tradition. The need in their religion for emotional release was understood by the Negroes, as is apparent in this comment of Jones, whose concern with the conversion of the slaves makes his writings especially to the point:

True religion they are inclined to place in profession, in forms and ordinances, and in excited states of feeling. And true conversion, in dreams, visions, trances, voices—all bearing a perfect or striking resemblance to some form or type which has been handed down for generations, or which has been originated in the wild fancy of some religious teacher among them. These dreams and visions they will offer to church-sessions, as *evidences* of conversion, if encouraged to do so, or if their better instruction be neglected.[12]

Independent testimony regarding the force of the drive for emotional expression among the slaves is contained in an account given by an ex-slave of conditions known to her:

Referring to a plantation located in Louisiana, Mrs. Channel says: "On this plantation there were about one hundred and fifty slaves. Of this number, only about ten were Christians. We can easily account for this, for religious services among the slaves were strictly forbidden. But the slaves would steal away into the woods at night and hold services. They would form a circle on their knees around the speaker who would also be on his knees. He would bend forward and speak into or over a vessel of water to drown the sound. If anyone became animated and cried out, the others would quickly stop the noise by placing their hands over the offender's mouth."[13]

The importance of the Negro preacher in furthering this patterned emotionalism has often been pointed out,[14] while cases have been recorded where his influence has been felt by whites, even during the period of slavery.[15] The fact that differences between denominations are unimportant in the minds of members of the various churches in the town studied by Powdermaker would further indicate that it is the expression of religious feeling that is essential, not the label.[16]

In analyzing Negro religious institutions, those autonomous groups not affiliated with denominations whose primary membership is drawn from whites must also receive adequate treatment. These are the "shouting" sects, which play a large part in Negro religious life. Such sects, termed "cults" in the passage which follows, are in it compared to and differentiated from the evangelistic churches:

. . . the following general characteristics seem common to both groups: (1) primary emphasis upon "preaching the 'Word' "; (2) salvation by faith; (3) worship as fellowship; and (4) vernacular singing. In addition to these, certain other features observed particularly in connection with the cults appear more or less common to evangelistic churches also. They are (1) lengthy exhortations and sermons punctuated by stereotyped phrases such as, "Amen!" "Glory to God!" "Praise His Name!" "Hallelujah!" and so forth; (2) sermons featuring polemics against the so-called "sins of the flesh," in contrast to the "blessings of the Spirit" and the "rewards of the hereafter"; and (3) the dogmatic assertion by each of its monopoly on the "only true gospel" of Jesus. Although the cults and evangelistic churches seem to have the above features in common, certain others appear to be more especially distinctive of the religious cults only. These may be listed as follows:

1. A leadership that is magnetic to an almost hypnotic degree and virtually dictatorial in its control over the cult devotees.
2. Frenzied overt emotional expression, such as shouting, running, jumping, screaming, and jerking as a regular feature of the worship services.
3. Frequent repetition of hymns transformed into jazzy swingtime and accompanied with hand-clapping, tapping of feet and swaying of bodies.
4. Testimonies given in rapid succession and certifying to the reception of "miracles," healings, messages, visions, etc.[17]

Within the cults this author distinguishes groups whose "entire program seemed designed to magnify the personality of the leader of the cults"; those marked by " 'spirit-possession,' a type of highly emotionalized religious and ecstatic experience commonly designated by such terms as 'filled with the Holy Ghost,' 'lost in the spirit,' 'speaking in tongues,' and 'rolling' "; and those to be considered as

"utopian, communal or fraternal." However, despite the distinctions
between these cults, they are sufficiently alike that they may be differentiated "from all other institutions, religious, fraternal, civic,
or otherwise." The essential traits that define them all are, "first
'spirit-possession'; and second, the mass hypnotic effect of the group
gatherings."[18]

The emotional displays to be witnessed in Negro churches have
been recounted so often that it is hardly necessary to quote any of
the numerous detailed accounts that have been published. An early
report, which shows how firmly the pattern had set at the time of
its writing, has been given by Bremer.[19] R. J. Jones describes at
some length a number of religious meetings where the quasi-hyster-
ical quality was prominent.[20] Daniels,[21] telling of his visits to the
Boston Church of God, Saints of Christ, gives details of services
where possession hysteria occurred. Odum[22] presents a generalized
version of typical behavior at "shouting" services; while Puckett,[23]
Dollard[24] and Powdermaker[25] describe various rites and incidents
at services witnessed or recounted in the literature. One example,
from yet another source, will be sufficient to indicate details of the
pattern:

The company has long been swaying back and forth in the rhythm
of the preacher's chant, and now and then there has come a shout of
assent to the oft repeated text. Each time the preacher's almost inco-
herent talk becomes articulate in a shout, "I have trod de wine-press,"
there are cries of "Yes!" "Praise de Lawd!" and "Glory!" from the
Amen corner, where sit the "praying brethren," and from the Hallelujah corner, where sit the "agonizing sisteren." In the earlier demonstration the men rather lead, but from the time when Aunt Melinda
cries out, "Nebbah mind de wite folks! My soul's happy! Hallelujah!"
and leaps into the air, the men are left behind. Women go off into
trances, roll under benches, or go spinning down the aisle with eyes
closed and with arms outstretched. Each shout of the preacher is a
signal for someone else to start; and, strange to say, though there are
two posts in the aisle, and the women go spinning down like tops, I
never saw one strike a post. I have seen the pastor on a day when the
house would not contain the multitude cause the seats to be turned and
take his own position in the door with a third of the audience inside
and the rest without. . . . I have seen the minister in grave danger of
being dragged out of the pulpit by some of the shouters who in their
ecstasy laid hold upon him. I have seen an old man stand in the aisle and
jump eighty-nine times after I began to count, and without moving a
muscle of his thin, parchment face, and without disturbing the meeting.[26]

This account may be compared with still another description of a

service at the Damascus Baptist Church in Macon County given
by Johnson. Most of this is devoted to excerpts from the sermon
and the reproduction of prayers; yet through its fragments one can
sense the emotional stresses that play on the audience until that
point in the services is reached when,

> The shouting has begun with sudden sharp groans of spiritual tor-
> ture, then screams of exultation. Three or four persons are expressing
> themselves with shouts accompanied by a variety of physical demon-
> strations, while most of the audience responds in low accents.[27]

For the great majority of Negroes in the United States, therefore, whether they worship in churches that are part of organizations including white congregations as well as their own or in purely
or predominantly Negro denominations of humbler physical resources, the essence of their belief is its intimate relation to life,
the full participation of the communicants, and the emotional release
that finds expression in the hysteria of possession. In its purely institutionalized aspects, Negro religion is marked by a disproportionate importance of its leadership in comparison with whites, and in
the extent to which each unit—each church group—preserves its
autonomy.

It is to be noted that this summary excepts such denominations
as the Catholic, Episcopalian, and Presbyterian, which from an absolute point of view have no inconsiderable number of Negro communicants, and where the behavior of Negro worshipers, in so far
as present data permit any generalization, is indistinguishable from
that of their white fellow members. Whether a study of the religious life of Catholic Negroes in the United States elsewhere than
in Louisiana would reveal syncretisms not in accord with official
theology and ritual cannot be said until such a study has been made.
Similarly, differences between Negroes and whites who belong to
these more restrained churches in the minutiae of belief and ritual
practices are not known—and need not be studied, indeed, until far
more materials are in hand concerning the churches that represent
greater deviations from majority practice.

3

At this point it is essential to summarize in greater detail than
in our earlier discussion the forms of belief and ritual that exist in
West Africa, and to follow this summary with a brief outline of
the transmutation these forms have experienced in the New World.

It will be well to bear two points in mind when considering this summary. In the first place, the generalizations made concerning the essential aspects of the religion of most Negroes of the United States are used as points of comparative reference. Secondly, stress is laid on the outer forms of religious expression rather than on inner values and beliefs. For, as will be seen, while Christian doctrine by no means escaped change as it passed into Negro hands, the most striking and recognizable survivals of African religion are in those behavioristic aspects that, given overt expression, are susceptible of reinterpretation in terms of a new theology while retaining their older established forms.

In the region of Africa from which the slaves were principally drawn, the outstanding aspect of religion, noted by every writer who has dealt with these peoples, is its intimate relation to the daily round. The forces of the universe, whether they work good or evil, are ever at hand to be consulted in time of doubt, to be informed when crucial steps are to be taken, and to be asked for help when protection or aid is needed. Thus, while it is quite incorrect to describe the religion of the African as essentially based on fear, as has often been done, the very nearness of the spirits means that their requirements must be cared for as continuously and as conscientiously as the other practical needs of life. Cult practices, therefore, have their humblest expression in individual worship. Sacred localities do exist, and priests have their social and religious functions to perform, but in the final analysis the rapport between a person and the invisible powers of the world are his own immediate concern, to be given over into the hands of an outsider only in times of special need.

These less formal modes of worship are, however, no more than a beginning, for everywhere organized groups exist which, because of the special training given their members, are regarded as vowed to the service of particular spirits or deities. Such groups ordinarily include leaders—priests, that is—and followers, whose competence varies with the degree to which they are permitted acquaintance with the esoteric knowledge needed to give adequate service to the god who is the object of devotion. The group may be a family affair, and the god may in reality be an ancestor so important that his worship has been taken over by the community at large. Ritual may be strictly followed or may be more or less improvised; the priest may exercise the closest control over his followers or his position may depend on their pleasure; membership itself may be fixed or fluctuating; a given devotee may be vowed to the exclusive service of a single spirit or may worship a number of these. But everywhere the group is essentially local in character, and the organization of religion in a tribe is never so tightly knit that the control of the principal, or eldest, or spiritually most potent priest acknowledged as the head of a given cult extends beyond the reach of his personal influence; rarely, indeed, beyond control of his own particular group.

Ritual is based on worship that expresses itself in song and dance, with possession by the god as the supreme religious experience. Under possession the worshiper, who is either one of the initiate or is possessed by a deity who thus is believed to express a desire to have this individual as a servitor, merges his identity in that of the god, losing control of his conscious faculties and knowing nothing of what he does until he comes to himself. This phenomenon, the outstanding manifestation of West African religion, is, for all its hysterical quality, by no means undisciplined. On the contrary, in every culture definite rules govern the situations under which it is to be experienced, the behavior of the possessed person while under the spell, the manner in which he is controlled by those in authority while possessed, and how he is to be cared for as he comes out of his seizure.

Possession is everywhere a social phenomenon; it is in this, indeed, that it differs most strikingly from the possession of European holy men, whose visitation by holy spirits, a "miracle" and thus something outside common religious experience, customarily occurs when they are alone. Among the Africans, such "private" possession is unknown. A given rhythm of the drum, the sound of a rattle, singing and handclapping of a chorus are almost invariably essential if possession is to ensue, and the devotee of unstable emotional qualities who, by himself, may become unsettled and go into a possession presents an unusual case. As a rule, possession comes on at some ceremony where a follower of a god is moved by the singing, dancing, and drumming of a group of which he is a member; the god "comes to his head," he loses consciousness, becomes the deity, and until his release dances or performs after the fashion of the spirit who has taken possession of him.

In those parts of the slaving area where possession has been studied, the motor behavior of those possessed is consistent to a remarkable degree. Whether a person is merely a devotee who has been experiencing a generalized feeling of restlessness for some time preceding the ceremony and is thus ripe for the "visit of the god" or has been designated by the leader of his group as the recipient

of this attention from the deity, the worshiper to be possessed begins by clapping his hands, nodding his head, and patting his feet
in time to the rhythm of the drums. In this his behavior resembles
that of the others present, but he soon is to be distinguished by
the vigor of his movements and the fixity and remoteness of his
gaze. His motions become more and more emphatic, until, still
in his place, his head is thrown from side to side and his arms
thresh about him. Finally he dashes into the center of the cleared
space, where he gives way to the call of his god in the most violent
movements conceivable—running, rolling, falling, jumping, spinning, climbing, and later "talking in tongues," and prophesying.
As time goes on and he feels the ministrations of the one in charge
of the ritual take effect, he subsides and joins the dancers, who
always move about the dancing circle in a counterclockwise direction. In this case, his release from the spell is gradual; sometimes,
however, his frenzy continues unabated until he falls in a faint, is
removed by those about him, and eventually returns to the dancing
space to resume his role as spectator. In every case, however, the
drummers must continue to beat the rhythm of the god until all
those under the spell have come to themselves; otherwise, their own
spirits might not return and the consequences would be disastrous.
Furthermore, were drumming to stop abruptly, the often dangerous
positions in which those possessed find themselves, high in a tree
or atop a roof, for example, would cause them to suffer harm.

In all this region, persons worship gods they have inherited, or
to whom they have been vowed at birth, or who have expressed a
desire for them in a dream or by actual possession. In all cases,
however, it is necessary to have adequate training in order properly
to worship a spirit. The person under possession for the first time
moves awkwardly in comparison to the trained dancing of the
initiate; the novice is overwhelmed by his emotional experience and
only with time attains the complete release that comes to the
seasoned cult member. Correct procedures of all kinds, such as knowing the songs to sing for one's god and the dances to dance in his
honor, and how to cope with others possessed by the god, as well
as more esoteric facts concerning the deity and his associated divin-
ities are taught a candidate in the training he receives before he
can become an active member of local religious groups. It is during
this initiation period that he also learns the meaning of those strange
syllables, akin to "speaking in tongues," that a devotee utters under
possession and which, when interpreted, turn out to be a prophecy,

or a new cure, or how to cope with magic, or any of those other matters which concern the gods when they come to earth.

In some parts of the area—in Dahomey, among the Yoruba and other Nigerian tribes and, to a certain measure, on the Gold Coast—the new devotee undergoes actual seclusion during his training. Whether this is the rule in the Congo and to the north cannot be said on the basis of available data. Yet teaching there must be, whether formal or informal, for it is as dangerous for a man in Africa to become possessed by his god without proper knowledge of how to cope with him as it is difficult to be a full-fledged member of a Trinidad "shouting" church without having gone through the "mourning" period and the rite of baptism.

Certain instances of possession may be cited out of firsthand experience—some of them as yet unpublished, and all comparable in that they represent the findings of the same observers. The first concerns a ritual witnessed among the Ashanti of the Gold Coast, performed to summon the gods to discover certain evil magic troubling the people of the remote village where it was held. The crowd assembled to watch the ceremony was so large that it almost completely enclosed the rectangular dancing space wherein those who were possessed moved as the spirits directed. At one side were the drummers and singers. The seven drums, rattles, and other percussion devices kept up a steady beat that set the tempo for the singers gathered near, who also accompanied their singing by handclapping that matched the basic rhythms. From time to time, one person or another would "get the god," jump from his seat and run to the center of the circle. One woman acted the cripple, at the outset moving with the greatest difficulty, though always in time with the beat of the drums. As the afternoon wore on, her ability to get about gradually improved—first with the aid of a crutch, then with a stick, until, as the dancing became more and more ardent, she threw even this away, and, with a shout, danced violently without any support.

Various persons came to those possessed, kneeling before them. In some cases infants were lifted that the spirits who had come to the heads of the dancers might bless them. The attitude of the spectators was of concerned interest—but as always during rituals to African gods, the sanctimonious behavior that is associated with European religious exercises was quite absent. There was a task to be accomplished and the gods were being summoned by the proper specialists to perform their work. Spectators were therefore free to enjoy themselves or, where the opportunity offered, to profit from the presence of a spirit by having a request transmitted to it.

Tension heightened; more persons joined the corps of singers, more possessed dancers were in the circle reserved for them. The chief priest himself became possessed, and stalked about speaking unintelligibly. Suddenly, after several hours, a number of the participants, including the principal figure, dashed at full speed through the crowd and through the village until they reached a point on the bank of a stream a short distance outside it. There they began to dig, and, soon after, a shout went up from those watching them. The evil that had been dogging the community had been discovered; now steps could be taken against it, especially since the powerful gods that had located it could be called on to nullify its capacity to work harm. As those who had made the discovery returned to the dancing circle, the possessed devotees danced even more vigorously; a sign that the spirits were pleased at what had been accomplished. The drumming and singing continued, but the climax had been reached; one after another the gods "departed" as their devotees subsided, coming to themselves gradually as their dancing stopped, or going into the patterned faint which marked the end of their possession.

We may now turn to the New World to fill in the steps by which this worship of the African gods, with drum and rattle as well as song, and without the ritual accouterments of Christian churches, were transmuted into the forms of Negro religious practice found today in the United States. We may first consider Dutch Guiana, where worship both among the more African Bush Negroes and among the urban group, long in contact with European culture, has been described. Reference may be made to the published descriptions of worship by the former people[28] without repeating those descriptions here, since the physical setting of the bush, and the freedom of the people to indulge their religious emotions without interference from the whites whenever the occasion calls for it, makes their practice essentially that to be encountered in West Africa itself.

In Paramaribo, however, regulations of the colonial government have made difficulties for the followers of non-Christian cults. African-like ceremonies are to be witnessed, though this is permitted with some reluctance and is possible only at certain seasons. A description of the manner of possession by one of the gods who "came" to such a ceremony may be quoted:

The next *winti* called was the deity of the cross-roads, *Leba*. As the drums played and the singing began anew, several persons, who were seated, began to tremble. Their trembling began with the agitation of the lower limbs, after which the knees began to shake. This was followed by the quivering of the hands, the twitching of the shoulders, and the head. The facial expression was that of a person in a trance. Their eyes were either shut or they stared blankly, and the muscles were set and tense. As the drumming and singing continued, the heads of those who were experiencing possession began to shake agitatedly and to roll from side to side, and in this state they raised themselves from their seats, and sank back again. As the twitching and trembling and rolling of the head became more and more violent, a friend or relative seated beside the ones who were becoming possessed straightened the head-kerchiefs which were by now askew, if the persons were women, and helped them back to their seats. From time to time an exclamation issued from their lips, a shout, a groan, or words spoken rapidly and unintelligibly. They were speaking the secret language of the *winti*. As their movements increased in violence, the arms were thrown about so that anyone sitting next to a possessed man or woman was struck. The jerking movements of the head were repeated with greater and greater frequency, until the head seemed to be rolling about on the shoulders. When the one who was going through these movements of possession was not in the front row, room was made so that there would be no obstacle in his way when he rushed forward into the dance-clearing.[29]

In the coastal area of Guiana, the behavior of the drummers and singers who accompany the possessed dancers is almost identical with that witnessed in West Africa. The same relaxed movements of the fingers as the drummers sometimes even play rhythms identical with West African beats on the drumheads, the same swaying of the bodies by the singers that makes of their singing itself a dance, and the same cupped hands with which the clapping is done, all testify to the manner in which these descendants of Africa are but repeating motor habits current in the homeland of their ancestors. There is likewise little difference between the two regions—or, for that matter, between these two and what is found in the United States "shouting" churches—in the meaning of such a rite for the participants. Curing, the solution of practical difficulties, protection from the forces of evil operative here and now; the immediacy of the ends reflected in the words of songs and in the supplications to the gods might be the attitudes shown in prayers and sermons heard in Negro churches of this country.

Yet in Paramaribo, where the dance described in part in the excerpt quoted was observed, Christianity is a functioning element in the life of the Negroes. A large proportion of those who were in attendance at this ceremony were professing Christians, baptized members of the Moravian or Lutheran or other sects, and as often as not, frequent attendants at church. The Negroes of Paramaribo

recognize that it is important to have such an affiliation, since a
baptismal certificate is often requested when a job is sought, while
it is easier for a child who has been baptized to be accepted in a
school than one who has not thus embraced Christianity. We shall
encounter the same phenomenon elsewhere in the New World as
other instances of the pliability shown by the Negro in the face of
situations beyond his control are given; a pliability which, as we
have seen, is manifested in Africa itself when the gods of other
tribes are taken over and incorporated into a system already an integrated whole. In Paramaribo this tradition is ready at hand
whereby folk continue to worship ancestral gods while belonging to
Christian churches for practical reasons. In other Protestant New
World countries where the proselytizing drive has been more in-
sistent, Christianity has prevailed to a greater extent; but even in
such localities only at a cost of substantial concessions to African
forms of worship and of reinterpretations of belief within the framework of Christian theology and ritual.

The reconciliation of pagan African and Christian belief approaches equality only in Catholic countries, in those cults, such as
the *vodun* of Haiti, carried on outside the church. Catholic theology
and ritual are too fixed to give rise to the variation characteristic of
the type of Negro Christianity engendered by Protestantism; in so
far as Negroes participate in the activities of the Catholic Church,
they must conform to standard practice. But in those Catholic countries where adequate reports are available,[30] especially Haiti, Cuba,
and Brazil, it is plain that official Catholicism only partially satisfies
the heritors of African religious traditions, just as the type of
Protestantism practiced by the whites in Dutch Guiana or the West
Indies or the United States has required adaptation to serve their
needs.[31] The difference recognized by the people themselves between
the Catholic Church and these sects is, however, not matched in the
United States and other Protestant countries, where Baptist or
Methodist churches, whatever their local habits of worship, are
Baptist or Methodist, so that not until a group gives over the name
itself does it become something distinct.

The numerous resemblances to be discerned between Brazilian
practices and those of all other parts of the New World and West
Africa are exemplified in the photographs reproduced by Ramos of
a *filha do santo*, a "daughter of the saint,"[32] as an initiate of the
fetish cult is called. The very term used in Brazil for such a person
constitutes an important correspondence with West Africa, on the
one hand, as illustrated by the designation *vodunsi*, "wife of the

god," applied to a cult initiate in Dahomey, and with the United
States, on the other, as seen in relationship between the "sanctified"
and their God in such a sect as the Church of God in Christ. The
women depicted in these photographs, in physical type and manner
of dress, so closely resemble persons to be encountered in United
States "shouting" churches that they must be seen to obtain the full
effect of the comparison. The motor behavior depicted in these reproductions, furthermore, links West Africa, on the one hand, and spirit
possession in North American Negro churches, on the other, in
unmistakable fashion.

Because of the overwhelming adherence to Protestantism of
Negroes in this country, however, the significance of what is found
in these Catholic countries is not as great for comparative purposes
where Negro religious behavior in the United States is to be assessed as are the data from such Protestant regions as Dutch Guiana
or Jamaica or Trinidad. Trinidad is especially important in this connection, since in this island the gamut runs from cults as African in
their forms of worship and theology as anything to be encountered
in Guiana or Haiti or Brazil to a formal Protestantism among the
Negroes that is as "correct" in its observances as Negro Episcopal
or Presbyterian or other more restrained churches in the United
States. Citations to these data in published form cannot be made at
this time, inasmuch as the field material was only gathered in 1939.[33]
Specifically, religious custom in Trinidad varies from the completely
African Shango cult through the Baptist "shouters" (who are, in a
sense, an "underground" movement, since they are proscribed by
government ordinance), to the European-like groups affiliated with
Moravian, Presbyterian, Seventh-Day Adventist, Church of England, and Catholic denominations.

It is unnecessary to describe the Shango cult procedures in any
detail, since, except for certain relatively minor aspects, they dupli-
cate corresponding rites that have been sketched as found in Africa,
Guiana, Haiti, and Brazil. The importance of the local group is as
apparent in this cult as in these other areas, and the role of the priest
is that of the leader in Africa. Drums and rattles and song bring on
violent possession of the classical type, accompanied by the same
magnificent dancing that marks the worship of African gods wherever they "mount" their devotees. In one respect this cult leans more
toward the Negro practices in Catholic countries than in Protestant,
for these folk make the same identifications between African gods
and Catholic saints that occur elsewhere in regions where the Church

hierarchy of spiritual beings is reinterpreted in terms of African nature gods.

The most revealing segment of Trinidad Negro religious life is that of the Baptist "shouters" who, on casual inspection, would be regarded merely as more individualistic adherents of that Christian sect. The "shouters" themselves distinguish two types of Baptists, however, the "carnal" group, wherein "shouting" is not countenanced and a greater degree of decorum exists, it may be said, than in Negro Baptist churches in the United States; and their own group, the "spiritual" Baptists. They were outlawed by an ordinance in 1917, ostensibly because of the disturbances these groups created in their fervor, but probably in more realistic terms because of the understandable need felt by the more conventional denominations to counteract the inroads these "shouters" were making into their following. They strikingly resemble the early Christians in their communal cooperativeness, in the measures they take to exact discipline and morality within their own groups, and in the gentle nonresistance with which they persist in carrying on despite the edicts against them and what they regard as constant persecution resulting from enforcement of the law which makes them subject to frequent raids and fines or jail sentences.

When initially visited, meetings of this sect seem to differ but little from services in churches of the more decorous denominations, the outstanding thing about their ritual being the devotion of the communicants to the "Sankeys," as they term songs from the Sankey and Moody hymnal, which they know in enormous numbers, with every verse to each song memorized. Yet even at first sight certain aspects of their humble meeting places are apparent that, differing from what is found in more conventional Christian churches, at once strike the eye of the Africanist. Markings in white chalk on the floor, at the doors, and around the center pole are reminiscent of the so-called "verver" designs found in Haitian *vodun* rituals. The presence of a large bell, the ritual importance of the central post, and other elements in the building complex comprise further of these deviations from customary practice in the direction of West African ritual.

As one becomes better known to the membership, more variants are permitted to come to light. A period of initiation for neophytes, called "mourning" here as in the Negro churches of the United States, suggests the seclusion of novitiates in Africa and the period of probation undergone by candidates for membership in a Haitian *vodun* cult group. The incidence and character of the visions "seen" by these people are as reminiscent of African as of European tradition, but the manner in which baptismal rituals, begun as decorous Baptist meetings, turn into "shouts" is not at all European. The sequence characterizing this process of injecting legally tabooed Africanisms into approved Christian procedure is condensed in a recording, made in Trinidad, of the singing of the "Sankey": "Jesus, Lover of my Soul." The song begins in its conventional form, sung, if anything, with accent on the lugubrious measured quality that marks hymns of this type. After two or three repetitions, however, the tempo quickens, the rhythm changes, and the tune is converted into a song typically African in its accompaniment of clapping hands and foot-patting, and in its singing style. All that is left of the original hymn is the basic melody which, as a constant undercurrent to the variations that play about it, constitutes the unifying element in this amazingly illuminating music.

The change from Baptist ritual to the African-like "shout" during a given service is gradual, for, as is often the case in Africa itself, even the leader does not know when the spirit will come and possession will occur. Restraint, in the European sense, may reign for an hour or two after the beginning of a Sunday night ceremony, as actually was the case in at least several services visited. But sooner or later the restraint is broken—unless, that is, the service is one where no "shouting" can be indulged in because of danger from the police—and then the scene turns into one entirely comparable to those witnessed in West Africa or in the New World wherever African patterns of worship have been preserved. Drums and rattles, forbidden in Christian rite, are naturally absent, but the deficiency is compensated for by handclapping and the improvisations of rhythm taking the form of a vocal "rum-a-tiddy-pum-pum" sung in the bass by men who have the power needed to make their contribution heard above the blanket of choral singing. Possession is present in full vigor, with only the African element of the dance lacking, though on occasion even this is represented in the manner in which "patting" the foot is done by the person possessed. Shoes are removed because of Biblical precept; quite unrealized by those who practice this custom, it is also in accordance with the African canons of good form in dancing. Numerous other details which indicate how this sect affords insight into the way in which African and European practices have been reconciled could be given were space to permit. These, however, must await the publication of the complete data, when it will be demonstrated how this Baptist "shouting" sect is a direct reinterpretation of the Shango cult, and thus leads immediately into relationship with a full-blown African religious custom.

4

As manifestations of African religion are thus systematically traced, the neglect of so many students to allow for the African past in the explanations they offer of aberrant elements in Negro religious behavior in the United States is seen to make a sorry chapter in the history of scholarly procedure. Not all students have refrained from taking Africa into consideration, however, and Puckett's analysis of Negro religious beliefs is outstanding in this respect.[34] It is possible that the difficulty has in many cases been a semantic one for, as this author remarks:

The mere fact that a people *profess* to be Christians does not necessarily mean that their Christianity is of the same type as our own. The way in which a people interpret Christian doctrines depends largely upon their secular customs and their traditions of the past. There is an infinite difference between the Christianity of the North and South in America, between that of city and country, between that of whites and colored, due in the main to their different modes of life and social backgrounds. Most of the time the Negro outwardly accepts the doctrines of Christianity and goes on living according to his own conflicting secular mores, but sometimes he enlarges upon the activities of God to explain certain phenomena not specifically dealt with in the Holy Scriptures.[35]

This confusion, when added to a reluctance to admit the presence of African elements in Negro behavior in the United States, has made for the uncritical acceptance of a label as a substitute for investigation, and has resulted either in a tendency to overlook the deviant types of behavior manifest in Negro churches or to refer them to white influence. That the African past must be included under the rubric "traditions of the past," whether these traditions are held overtly or not, becomes apparent when the religious habits of Negroes in the Caribbean and South America are anchored to both ends of the scale whose central portion they comprise—to Africa, the aboriginal home of all these varieties of religious experience, on the one hand, and to the United States, on the other, where the greatest degree of acculturation to European norms has taken place.

The importance of extending our conception of the traditional sanctions of Negro behavior so as to include the African past may be documented by indicating one or two points where ascriptions of origin are open to serious challenge, just because of failure to realize the strength of this tradition. In our first instance, indeed, not only have Negro patterns from Africa been perpetuated, but there is a strong probability that these patterns were themselves of importance in giving to the whites just that tradition which, among Negroes, is customarily ascribed to white influence! The problem of derivation referred to concerns American revivalism. Are Negro "shouts" due to the exposure of these people to the white revivalist movement? Or is white revivalism a reflex of those Africanisms in Negro behavior which, in a particular kind of social setting, take the form of hysteria?

A reservation of some importance must be made in our present consideration. While the impulses of a traditional past play their part in influencing such a phenomenon as revivalism, the special orientation of the local scene is likewise important in developing aspects of the institution which, resulting from contact, are unlike anything in the original cultures. Hence, whether Negroes borrowed from whites or whites from Negroes, in this or any other aspect of culture, it must always be remembered that the borrowing was never achieved without resultant change in whatever was borrowed, and, in addition, without incorporating elements which originated in the new habitat that, as much as anything else, give the new form its distinctive quality.

We may now turn to the analysis of certain statements made to explain the religious hysteria of Negroes in the United States, first considering the conclusions of R. J. Jones, whose data on Negro cults have already been cited at some length in preceding pages:

Two significant observations seem fairly definitely established as a result of this study. They are first, religious cult behavior, commonly designated as particularly Negroid, cannot be construed, either in nature or function, in spite of its prevalence, as a racial characteristic. And second, as long as any group of people, irrespective of race, continues to labor under conditions of economic, social, and cultural disadvantage, sufficiently acute to necessitate emotionally compensatory forms of reaction on a comparatively large scale, manifestations of religious cult behavior, as have been revealed in this study, will continue to exist as perhaps a negative element in the context of our contemporary American culture.[36]

The first proposition in this quotation is a good example of that confusion concerning biological and cultural causation which, as has been pointed out at the opening of this work,[37] has seriously handi-

capped investigations into the derivation of American Negro behavior. The validity of this proposition is obvious to anyone whose
grasp on the nature of culture is based on acquaintance with the
many data demonstrating the independence of tradition from physical
type, and the resulting untenability of any assumption of a causal
connection between the two. The second proposition, a quasi-
Freudian interpretation of the socio-economic situation of the Negro,
is by now familiar as one of those explanations which, something
less than the whole truth, yet has sufficient validity to be persuasive
by and of itself.

Proceeding with his argument, this same student continues:

Notwithstanding the numerous analogies pointed out, this contemporary religious cult behavior cannot justly be considered as a survival
of primitive religious manifestations either among Negroes or any other
group. For whatever may be said regarding the cultural carryover which
the African slaves might have brought along with them to this country
over three centuries ago, they have, for the most part, been submerged
completely, if not virtually eliminated from Negro life as a result of the
long and thorough-going processes of acculturation and Americaniza-
tion to which they have been subjected since their arrival on the American continent. It seems hardly possible, therefore, to posit any significant direct connection between the contemporary primitivism and the
contemporary scene.[38]

The assumption here that the "Negroes" have been in the United
States for three centuries is one of those points which, in fact, is only
as true as its related but untenable implication that no impulses from
Africa have been received since. Large numbers of Negroes
from Africa were legally received in this country to the first decade of
the nineteenth century, and later as contraband until the outbreak
of the Civil War. That the African element had at least a mechanism
for survival in this constant recruitment from the Old World is
completely lost sight of in statements such as the above. But even
more serious objections are to be registered against the conclusions
of the study from which these two quotations have been taken. For
the method, typical of most of those who discuss Africanisms in
Negro life in the United States, is based on what may almost be
termed a positive disregard of materials from Africa itself, to say
nothing of materials from other Negro communities in the New
World. What Jones does is to compare the hysteria in the American
Negro cult groups with similar religious phenomena among various
primitive folk over the world, with particular reference to the Es-
kimo! The reasoning is clear. It has been charged that the religious

behavior of the Negro in the United States is primitive. The Africans from whom the Negroes are descended are primitives. Eskimos are also a primitive people. The Eskimos manifest religious hysteria. But this hysteria is unlike that of the American Negroes. Hence the American Negro religious excitement is not primitive. Hence it is not African.

From where, then, has this intense emotionalism come? With African tradition ruled out as a causal factor, the search must be bent toward finding "more direct social and cultural influences that might be responsible." And this, it turns out, is the camp meeting of the whites. The matter is put in these terms, which comprise as good an example as any other that might be adduced[39] to indicate the conventional, simplistic explanation of Negro religion:

Granted that the American Negro, as a former slave, received enough of a basic pattern through the observance of white camp meetings to imitate and introduce it, with slight modifications, into his plantation church; assuming that this was definitely adopted from the whites, it cannot be denied that the situation under which the Negro was brought from Africa to this country and the conditions to which he was exposed after his arrival, laid the basis for his being particularly psychologically susceptible to the reception and exaggeration of certain patterns of religious behavior he observed among the white majority.[40]

Here we again meet the familiar theme—the Negro as a naked savage, whose exposure to European patterns destroyed what little endowment of culture he brought with him; the Negro as a cultureless man, with his entire traditional baggage limited to the fragments he has been able to pick up from his white masters and, because of innate temperamental qualities, to "exaggerate" them into exuberant and exotic counterparts.

Powdermaker is somewhat more realistic and hence more tentative in her approach when she considers the problem of derivations. She observes:

Many features now common in Negro meetings, especially in rural districts—"jerks," the "singing ecstasy," the "falling exercise," visions —were exhibited in white religious revivals of the eighteenth century, and are still to be found today, though far less generally, among certain Whites. But just as the Negro has metamorphosed white hymns and folk tunes into spirituals that are different enough to be considered creations rather than modifications, so has he made of Christianity something very much his own. Only against the historical background which has been sketched can it be appreciated how much his own, in content as in administration, the church has become.[41]

What are the differences between Negro and white revivals as observed by this student? They are indicated as follows:

In contrast to colored revivals, a large portion of this audience remains unaffected throughout, merely looking curious. Participation was confined to the few most active. Among the Negroes also there are a few who come just to look on, but the general feeling is that the audience are also actors. There is also a contrast in appeal: the fear of damnation as opposed to the hope of salvation held out before. A further, less definable difference seems due to an impression of greater rhythm and spontaneity in the Negro revival, not wholly accounted for by the greater participation of the audience. The rhythm of the white minister's speech was more halting than that of the Negro minister, and shaped to a less vigorous melodic line. The movements of the white congregation were more convulsive and jerky than those of the Negroes. This general contrast corresponds to the popular feeling that Negroes have greater sense of rhythm and greater freedom in bodily movement than white people. Such motor differences do not necessarily arise from differences in physical makeup, but may be to a large extent socially conditioned.[42]

The explanation of these differences is found to be difficult—and is difficult if their historic depth is neglected except for the short historic past usually conceded Negroes. The quandary is well described by Powdermaker herself:

. . . one can only speculate about why the Negroes respond with such marked readiness to the opportunity for this form of display. There is much to be said for the theory that the repressions caused by the interracial situation find relief in unrestrained religious behavior. Such an explanation is partial, however. Other factors, unrevealed by this study, are still to be sought.[43]

The nature of these other "unrevealed" factors is made apparent when it is recalled that the work from which these sentences are taken is entitled *After Freedom*; for one cannot, indeed, explain deep-rooted phenomena of this order by reference to the last moments of Negro experience. But it may be suggested, in the light of the preceding discussion of African and New World Negro religious conventions, that their causes are not as mysterious as is implied. Comparative study does reveal them as the manifestations of African tradition; a tradition which was strong enough not only to hold its own in contact with white religious custom, but also in all likelihood to make its contribution to white religion as well.

Certain forms of white revivalism had undoubted independent origin. Davenport, whose work of several decades ago is perhaps still the best discussion of this question, points out that as early as 1734 the "Great Awakening," begun in New England under Jonathan Edwards, spread from Maine to Georgia during the decade between 1740 and 1750, and was the inspiration of the Scotch-Irish revival in Ulster of a slightly later period. The presence of a sect of "jumpers" in Wales in 1740, and the appearance in England of a sect of French Prophets in the early eighteenth century, after they "had been driven out of France and had already spread the well-known phenomena of nervous instability through Germany and Holland," likewise testifies to the presence of revivalism and hysteria in regions removed from African influence.[44] But these earlier revivalist movements also differed in many respects from the camp-meeting frenzies of a later period which, probably more than the Great Awakening, laid the foundations for the revivals that played so characteristic and colorful a part in the early history of the United States.

Before considering the possibility of an African contribution to this movement, one element in its setting that has been consistently overlooked when derivations are discussed must be mentioned—the influence of American Indian custom. The many "revivalist" movements among the Indians that have occurred in historic times, at least indicate the need to bear in mind the fact that among these people a pre-Columbian messianic complex existed that facilitated the rise of the various movements recorded since white contact. That these latter movements, such as the Ghost Dance, have many Christian elements is to be expected.[45] Yet the hysterical seizures that mark many Indian cult practices, and the dancing and singing that are integral parts of this worship, make it permissible to ask whether a relationship does not exist between these indigenous movements and both white and Negro religious developments in this country. The fact that this is a matter for future analysis does not in any way lessen the importance of the problem; that it has been neglected by students of Negro religion is merely another point to be regarded as documenting the prevalent attitude concerning the sources of anything distinctive in present-day Negro custom, namely, that such traits must be referred to the influence of white practice.

The later, most characteristic form of American revivalism, the camp meeting, dates from the beginning of the nineteenth century. One of its most famous early manifestations was in Kentucky about 1800. During the five days of this "Gasper River" meeting:

The preaching, praying and singing continued almost without cessation save for a few hours in the early morning. It was not until Saturday evening, however, that any special outbreak of overwrought nature manifested itself. Then two women became greatly excited, and their fervor was communicated by contagion through the whole multitude. The camp became a battle-ground of sobs and cries, and ministers spent nearly the whole night in passing from group to group of the "slain."[46]

This type of evangelism began four years previous to the meeting, "with the Rev. James Crawford and others from the Carolinas and Virginia." It stressed the nature of the conversion experience and the need of knowing when and where the "new birth" had taken place. At one of the meetings described, the hysteria was so intense and widespread that some of the participants lay unable to speak or move; some spoke though they could not move; some beat the floor with their heels; some, "shrieking in agony, bounded about like a live fish out of water"; while some rolled over and over for hours, and others plunged into the forest.[47]

A number of items in the preceding description have important bearing on our problem. For one thing, it is significant that those who stimulated this new movement, which was characterized by a far greater incidence of hysteria than its precursor, came from southern or border states where contact with Negroes was continuous. The social setting of the hysteria that we have seen to be so fundamental in African and New World Negro practice is equally important in the possession hysteria of these whites, while the rolling, bounding, shrieking, and running about are all common to possession by the gods in Africa. Puckett gives numerous examples of identical behavior of whites and Negroes under possession or religious seizure,[48] though in a later paper he comes nearer to the conventional point of view by laying emphasis on white influence in shaping its forms.[49]

The contrast drawn by Davenport between the seizures characteristic of the United States and the Scotch-Irish revivalist pattern is enlightening in that it indicates that the two forms, though of a single historical origin, represented derivations from dissimilar habit patterns:

I wish in closing to call attention to the difference in type of the automatisms of Kentucky and Ulster. In Kentucky the motor automatisms, the voluntary muscles in violent action, were the prevailing type, although there were many of the sensory. On the other hand, in Ulster the sensory automatisms, trance, vision, the physical disability and the sinking of muscular energy were the prevailing type, although there were many of the motor. I do not mean that I can explain it. It may be that as the Charcot and Nancy schools of hypnosis brought out by chance, each in its own field, different kinds of hypnotic phenomena which, when known, spread by imitation in the respective localities and under the respective influences, so in Kentucky and the north of Ireland by chance there appeared different types of physical manifestation which were then imitated in the respective countries.[50]

Yet that here, as in an earlier citation, the pre-American tradition of the Negroes is an additional factor seems not unlikely. It is just in the forms of motor behavior remarked on as characteristic of the "automatisms" of the (white) Kentucky revivals that aboriginal modes of African worship are to be marked off from those of Europe. In the New World, exposure of the whites to Negro practices as well as of Negroes to European forms of worship could not but have had an influence on both groups, however prone students may be to ascribe a single direction to the process from whites to Negroes alone. Certain details of Negro religious behavior taken over by whites have actually been remarked, as when Puckett[51] concludes that modes of clapping the hands and patting the feet found among whites are to be ascribed to Negro influence. Whites had opportunity to learn other motor habits from the Negroes:

The slaves attended these [early white camp] meetings in large numbers. . . . The time of meeting was the interval in the late summer between the laying by and gathering of the main crops (exactly the period most in use today for rural white and colored revival meetings) and the general pattern of service, even to the mourner's bench at the front of the auditorium, was remarkably like that followed by modern rural Negroes and mountain whites.[52]

What, then, are the reasons to justify support for an hypothesis that the white camp meeting was influenced by contact with Negro religious practices? For one thing, as has been said, the camp meeting-revivalist tradition most characteristic of this country originated and had its greatest vogue in the southern and border states, where Negroes participated together with the whites. Again, the tradition of violent possession associated with these meetings is far more African than European, and hence there is reason to hold that, in part at least, it was inspired in the whites by this contact with Negroes. Finally, in so far as Negroes are concerned, the differences between their revival meetings and those of the whites today in the manifestation of ecstasy and hysteria, in the form of the services, and in the attitudes of communicants toward these rites underscore the differences between the worship characteristic of the cultures from which the ancestors of these two groups were derived.

Thus we see how, in assessing the forces that have given Negro religious hysteria its present-day forms, and the extent to which these forces have shaped corresponding modes of behavior among whites, a problem must be faced that is far more complex than is ordinarily recognized. The same conclusion must be reached when another problem of derivation lying in the field of religion is considered. This has to do with the popularity among Negroes of the Baptist Church, which has been stressed by all students of Negro religion. Explanations of this fact, it will be remembered, are couched in terms of the greater democracy of the Baptist Church organization, the greater emotionalism permitted in the services of this church, and that the services of this denomination are closer to the requirements of humbler folk than those of other churches. That the first two of these reasons is congenial to African religious patterns has already been pointed out. Yet neither this fact nor an explanation in terms of the socio-economic situation of the Negroes under slavery and in postslavery days is of much aid in helping the student understand why the Baptist Church, rather than autonomous "cults," should have had such a great appeal to Negroes, or why denominations other than the Baptist did not attract comparable numbers of followers.

For an answer to this question we must turn to baptism by total immersion, indispensable for affiliation with the Baptist Church. It will be remembered how, earlier in our discussion of the religious patterns of West Africa, the importance of the river cults was stressed. It was pointed out that the river spirits are among the most powerful of those inhabiting the supernatural world, and that priests of this cult are among the most powerful members of tribal priestly groups. It will be further recalled how, in the process of conquest which accompanied the spread of the Dahomean kingdom, at least (there being no data on this particular point from any other folk of West Africa), the intransigeance of the priests of the river cult was so marked that, more than any other group of holy men, they were sold into slavery to rid the conquerors of troublesome leaders. In all those parts of the New World where African religious beliefs have persisted, moreover, the river cult or, in broader terms, the cult of water spirits, holds an important place. All this testifies to the vitality of this element in African religion, and supports the conclusion, to be drawn from the hint in the Dahomean data, as to the possible influence such priests wielded even as slaves.

In the New World, where the aggressive proselytizing activities of Protestantism made the retention of the inner forms of African religion as difficult as its outer manifestations, the most logical adaptation for the slaves to make to the new situation, and the simplest, was to give their adherence to that Christian sect which in its ritualism most resembled the types of worship known to them. As we have seen, the Baptist churches had an autonomous organization that was in line with the tradition of local self-direction congenial to African practice. In these churches the slaves were also permitted less restrained behavior than in the more sedate denominations. And such factors only tended to reinforce an initial predisposition of these Africans toward a cult which, in emphasizing baptism by total immersion, made possible the worship of the new supernatural powers in ways that at least contained elements not entirely unfamiliar.

The importance of the association of water with African ritual may be further documented to indicate its fundamental character. In ceremony after ceremony witnessed among the Yoruba, the Ashanti, and in Dahomey, one invariable element was a visit to the river or some other body of "living" water, such as the ocean, for the purpose of obtaining the liquid indispensable for the rites. Often it was necessary to go some distance to reach the particular stream from which water having the necessary sacred quality must be drawn; in one instance, at Abeokuta, a bedecked procession of worshipers left a shrine atop a high hill, followed a long path to the riverside over two miles away, and returned before the ceremonies could be carried out at the shrine of the god. On one occasion, in Dahomey, the bed of a sacred stream run dry was "filled" from near-by wells so that this water could be ritually redrawn for use in an especially important ceremony.

Among the Ashanti, pilgrimages to Lake Bosumtwe and other sacred bodies of water regularly occur. And it is on such occasions that the spirit of the river or lake or sea manifests itself, by "entering the head" of a devotee and causing him to fling himself, possessed, into the water. The same kind of possession occurs in the Guiana bush, where the rites of various African tribes for their water spirits impel the one possessed to leap into the river with the strength necessary to swim even against the swift currents of the rapids. Possession by the river spirits in Haiti, or by spirits of snakes that inhabit the water, bring the devotee threshing into the stream near which the rituals are held, and where the deity is thought to reside.

But in the United States, where neither Bosumtwe nor *watra mama* nor Damballa is worshiped, Negro Baptists do not run into the water under possession by African gods. Their water rituals are those of baptism. Yet it is significant that, as the novitiate whose revelation has brought him to the running stream or the tidal cove is immersed, the spirit descends on him at that moment if at all, and a possession hysteria develops that in its outward appearance, at least, is almost indistinguishable from the possession brought on by the African water deities.[53] The importance of the Biblical concept of "crossing the river Jordan" in the religious imagery of the Negroes, and as a symbol of what comes after death, is a further part of this complex. For, like baptism, the river Jordan embodies a concept in Christianity that any African would find readily understandable. In the transmutation of belief and behavior under acculturation, it furnished one of the least difficult transitions to a new form of belief.

The slaves, then, came to the United States with a tradition which found worship involving immersion in a body of water understandable, and encountered this belief among those whose churches and manner of worship were least strange to them. When, in addition, they found in this group those whites who tended to be closest to the lowly, and thus tended to be the least formidable persons in their new setting, they understandably affiliated with it and initiated a tradition which holds to the present time. The favorable influence of the traditional past, and the new socio-economic setting were not, however, the only forces that furthered this particular process of reinterpretation. It is not generally recognized that the Cherokee Indians, a tribe with whom Negroes were perhaps more in contact during the days of slavery than with any other except the Seminoles and Creeks, themselves had a well-developed river cult.[54] It was neither African nor Christian, but its mere presence would act to strengthen any river cult foreign to the newly arrived Africans. This Indian rite included total immersion at each recurring new moon. It required fasting before immersion, something which in spirit is not too far removed from the restraints laid on the novitiate of any cult in Africa, or in the rites of certain "shouting" sects where new members "go to mournin'" before baptism. Certainly its presence in the Negro milieu reemphasizes, if this is now necessary, the complexity of the elements that determined the present-day forms of Negro religion wherein baptism plays so prominent a part, and the fact that membership in the church which gives the rite of immersion so large a place in its ritual is the most popular single denomination among Negroes.

5

Little is known of the theory of magic held by Negroes in the United States. Dollard has defined it as "a means of accommodating to life when it is not arranged according to one's wishes,"[55] which in broad terms places these forms in the category of magic as generally drawn. His further discussion is more specifically to the point:

Of course, one can think of magical practices among the Negroes as lagging cultural patterns, which they are, but one can also think of them as forms of action in reference to current social life. Magic accepts the *status quo*; it takes the place of political activity, agitation, organization, solidarity, or any real moves to change status. It is interesting and harmless from the standpoint of the caste system and it probably has great private value to those who practice it. These psychological satisfactions are important, even if they do not alter the social structure and are mere substitutes for effective efforts to alter it. . . . Magic, in brief, is a control gesture, a comfort to the individual, an accommodation attitude to helplessness. There is no doubt that magic is actively believed in and practiced in Southerntown and county today.[56]

The problem of the derivation of Negro magic and folk belief in the United States involves reference to the concept of the Old World province[57] perhaps more than any other aspect of Negro culture. The importance in magic of the charm, or "fetish," properly speaking, is outstanding; and all those who have had occasion to study African magic in Africa or its manifestations in the New World will at once recognize the applicability to it of the "rag and a bone and a hank of hair" formula, representative of European magic. This is why in many parts of the New World (and, under contact, in West Africa itself) such printed "magic books" as the medieval *Albertus Magnus* have so wide an appeal, or that in these areas Chicago and New York mail-order concerns specializing in "magic" do so considerable a business. It is not strange, therefore, that amalgamation between magic and other types of folk belief of the two continents occurred; it is necessary, however, first to recognize that the process was one in which both parties participated, and then to seek out the elements essentially African in the magic practices actually found among Negroes in this country at the present time.

Puckett holds that most Negro beliefs of this order are African but that European parallels can be discovered for many of them. He cites such instances as burying a bag containing parings of a dead man's nails and hair under the threshold of a neighbor in order to afflict him with ague, or the belief that to sleep in the moonlight will result in insanity, or that going to bed hungry will cause a man to sin. These European examples are to the point, even though the materials that go into the Negro conjure placed under a doorstep are intended to achieve far more sinister ends, in both Africa and the New World, than the European parallel given by Puckett. Again, New World Negroes believe that to sleep in the moonlight results in paralysis of one side of the face, not insanity, and this stems directly from West Africa; while in West Africa and the New World where studies have been made, the belief which holds that it is harmful to go to bed hungry goes on to say that this is because a person will lie awake, and thus give an unfriendly spirit an opportunity to take away his soul, rather than because he will sin, which is the European concept.

Other folk beliefs may be noted which are specifically African. Cable is cited by Puckett as mentioning in several places the pouring out of "oblations of champagne and the casting upon the floor a little of whatever a person was eating or drinking to propitiate M. Assouquer (the voodoo imp of good fortune)."[58] This custom prevails in all West Africa; in Haiti, Trinidad, and Guiana what falls to the ground during a meal is not swept up that day, since the spirits (sometimes ancestors) must be permitted to come and eat what they have thus indicated they desire. Cable also points out that in New Orleans a red ribbon was worn about the neck of a devotee "in honor of Monsieur Agoussou."[59] In the pantheons of the West African tribes, various colors are favored by the several gods, of which red is always one. The reason for wearing the red ribbon given by Puckett, namely, the similarity of this color to blood, obtains no confirmation in the comparative data. The name of the "demon" honored by this color comes directly from Dahomey and is found in the voodoo cult of Haiti as well as in that of New Orleans.

In one of her novels, Julia Peterkin speaks of this belief among the Gullah Negroes:

All Kildee's life he had heard that to stir the earth on Green Thursday was a deadly sin. Fields plowed, or even hoed to-day would be struck by lightning and killed so they couldn't bear life again. God would send fire down from heaven to punish men who didn't respect this day.[60]

With this passage the following may be compared:

> Not every day of the Dahomean four-day week is devoted to work in the fields, for on Mioxi no farming is done. Violators of this custom incur the wrath of the Thunder gods who kill offenders with lightning. The story is told of a man whose house may still be seen in Abomey who, being ambitious, was cultivating his fields on this day, when a bolt of lightning struck and killed him.[61]

The historic connection between these two would seem to be inescapable. Other beliefs about lightning present similar correspondences. The reluctance of Negroes in this country to burn the wood of a tree struck by lightning[62] would seem to be a survival of the cult of the African thunder-gods, the most feared of all in those African pantheons where they figure. Whatever lightning strikes in Africa may no longer be the property of men, but is taken by the avenging deity who has thus grimly claimed it for his own.

The many references to beliefs involving the crossroads given by Puckett[63] indicate that its great importance in West Africa has been continued in this country. Among West Indian Negroes, the crossroads is the favorite locale for operations in black magic; and this is a New World development comparable to the fact that if a person in the United States wishes to become a practitioner of black magic, he must go to the crossroads and pray to the Devil for nine days and nine nights.[64] In West Africa, the trickster-god who guards the entrances to villages, households, and sacred shrines is referred to by natives when speaking to Europeans as the Devil; his importance, as Legba in Dahomey, or Elegbara in the Yoruba country, or as Lebba or Legba in Guiana and Haiti and Trinidad, is paramount. In these New World regions he is conceived as the god of the crossroads; that his American equivalent, the Devil, must be propitiated at the crossroads merely means that here, as elsewhere among the Negroes, he continues to control supernatural traffic.

In Atlanta, Mrs. Cameron was informed, "if a frizzled hen is kept in the yard she will scratch up and destroy all conjuration which will cause discomfort for the family."[65] The ascription of this function to these peculiar fowl has been recorded several times from the South;[66] hens of this kind are to be found performing the same task in West Indian Negro yards no less than in West African compounds. The belief that hair and nail parings must be carefully watched lest they fall into strange hands is widely spread in this country, for in the most general terms of sympathetic magic it follows that what is a part of a person represents that person, and that should such parts be obtained by an enemy he could and would make the most of them. Yet the following passage has a somewhat different reference:

> Some . . . old . . . people try to save every strand of hair and every finger and toe nail, because they say that when they die they will have to show them before they can get into heaven. These hair combings are sometimes kept in a paper sack and the teeth and nails in a small box, both of which are buried with the individual when he dies.[67]

In West Africa, and among at least the Negroes of the Guiana bush, hair and nail clippings are employed in place of the body itself when circumstances make it impossible to bring back a corpse to his family for burial. An instance of this African practice was had when on the death of the member of the Kru tribe resident in Chicago already mentioned, cuttings of the hair of his head, and his finger- and toe-nail parings, were returned to Liberia to be interred and thus ensure that his soul would remain at peace.

A final example can be taken from an account of the wealth of Negro beliefs concerning snakes:

> The most elaborate and entertaining of these snake beliefs concern marvels which the narrator generally has heard about from others but which are related with all gusto and relish of the eye-witness. Thus the coachwhip (sometimes assisted by its mate) wraps itself around its victim and flogs it to death with its tail, which is said to be plaited in four strands like a whip (the arrangement of scales actually resembles a whip), sticking its tongue up the person's nose to see if he is breathing. The hoopsnake, which can lure its victim with its human whistle, seizes its tail in its mouth and rolls over and over like a hoop until it overtakes the person and kills him with a thrust of the poisonous stinger at the end of its tail (which has a terminal spine). The glass snake or joint snake (really a degenerate lizard which can voluntarily snap off part or all of its tail and grow a new one) can come together after being broken up (unless one buries its head or fastens it in a split sapling). The milk snake sucks cows dry or makes them give bloody milk, poisonous to human beings, and the cow forms an attachment for it and dies of grief if the snake is killed. The black snake charms children—in fact, all snakes can charm birds, animals, and human beings with their gaze. The green snake is the doctor snake and the darning-needle is the snake-doctor, both of which cure injured snakes and even bring the dead back to life. A horse hair deposited in a watering trough will turn into a snake if left undisturbed for a period of six weeks (a belief originating, no doubt, in the fact that a parasitic worm, spending parts of its life cycle in a grasshopper, is often found in horse troughs).[68]

It is possible that not a single one of these nature tales could be duplicated in West Africa, or even in the West Indies, if for no further reason than because of the differences in the species of snakes there and in the southern part of the United States. Yet the preoccupation with snakes on the part of American Negroes is significant. In West Africa, and in those parts of the New World where aboriginal religious beliefs have been retained in relatively pure form, the serpent is a major figure among supernatural beings. This is illustrated by the importance of the Dahomean rainbow-serpents, Aido Hwedo and Damballa Hwedo, conceived as having been present at the creation of the world, and by their counterparts, the Dangbe serpent-spirit of Guiana and the Haitian Damballa. The richness and variety of Negro beliefs concerning snakes are thus carry-overs of African religious concepts salvaged in the face of the adoption of Christianity.

The principal forms of Negro folk belief found in the United States are differentiated by Puckett into "signs" and "hoodoo":

While hoodoo is possibly the most picturesque form of Negro occultism, yet an exact knowledge of its usages is restricted to a relatively small number of persons, chiefly men, although women are not entirely excluded. "Signs," on the other hand, constitute the largest body of Negro magical beliefs and number among their devotees mainly women, although men are by no means counted out. "Signs" are generally what a person is thinking of when he speaks of Negro superstitions, although the term as used by the Negro is somewhat more inclusive than the English term "omens," taking in not only omens but various small magical practices and taboos as well. The distinction between hoodoo and "signs" is not clear-cut even to a Negro. Perhaps it lies more in the number of adherents than in any inherent quality (hoodoo being the more exclusive), though as a general rule the hoodoo charm is more complex.[69]

This, in turn, accords with the distinction made by Mrs. Cameron regarding the kinds of specialists who administer folk remedies:

. . . two groups of practitioners are known and recognized not only by themselves but also by their particular clienteles, as distinct from each other. One deals in what may be termed "medicine," that is, roots, herbs, barks and teas, while the other is composed of those who work by means of magic. So clear cut is this feeling of difference between the members of these two groups that there is reason for deep insult if a practitioner of the medical type is mistaken for one of those who practices magic.[70]

These two types of practitioners are distinguished by their dress, especially since, despite the fact that there is no hard and fast rule, "medical practitioners are predominantly women, while those who practice by magical means are, for the most part, men"; a statement that again is reminiscent of Puckett's findings. Even disregarding the sex line, the dress of a practitioner is not that of ordinary persons, though this tendency is becoming less pronounced.

The ways in which persons attain these callings, detailed in this unpublished account, have not received the attention they deserve. While "there are, of course, no established institutions . . . to which these novices may go for a definite period of time in which they may complete a defined amount of work," the methods of obtaining status as competent practitioners are nonetheless well recognized. Among the "medical group," status may be achieved in three ways:

The first . . . is to be especially endowed with supernatural power. This most often takes the form of seizures similar in type to cataleptic ones. This, as is explained, comes "like a thunderbolt from a clear sky." . . . By whatever means the message comes it instantly makes the recipient qualified for the task, and he possesses all the techniques of the craft. These healers so strongly believe in their ability to perform cures that they not only become deeply insulted when one expresses disbelief in their method of diagnosis and treatment, but may say that this disbeliever can expect to be chastened by the supernatural power. . . . Human selection by personal initiative is another means. . . . In such cases a novitiate is apprenticed to a practitioner. An older doctor takes this novice under his care and gives him guidance, and in this way he learns by clinical contact with actual cases. In some instances the novice may possess a blood-tie with the trained practitioner, while in others there may be nothing more than a friendship which exists between the family of the former and that of the latter. . . . The third manner of entering the profession may be a combination of the other two. That is, one may set out of his own volition and then, as opportunities permit, he serves. Each case gives added experience until a good reputation is obtained. . . . Many practitioners state that in occasional crises they receive direct help from a voice which gives directions and tells of remedies that bring marvelous results which gain for the practitioner distinction and fame.[71]

The difference between the two types of healers goes deeper than mere outward appearance, as can be seen when the following passage is compared with the preceding one:

Unlike the medical practitioner, there is a great deal of secrecy surrounding both the preparation and the technique of the magician. Though it appears that the position may descend by inheritance, in some instances it is not only voluntary, but mandatory that the son follow the footsteps of his father, and a refusal to do so is punishable by bad luck or sickness. It cannot be gainsaid that the craft is free to all, for one may enter by choice or be selected by older magicians, provided, however, that the chosen one is a "seventh-month" child or if he is the seventh son of a father in which family no girl has been born, or if he is born with a caul over his face. . . . As contrasted with the inheritance of power we find that one may be put into apprenticeship for training, though this occurs very rarely. Even then the novitiate must have exhibited special ability to manipulate magic. The magician seems to be more dependent upon the spirits of the dead ancestry than upon God, as is the case with the herb-doctors, and their emphasis is principally upon ritual and ceremonial. In all probability all novices of both general groups receive some technical training from older practitioners, for otherwise the question arises as to how they could obtain all the knowledge of their profession which they must have. We rarely find cases where fees are paid for instruction unless services rendered during the time in which the novitiate is in training are considered in this light.[72]

These citations give unique information concerning a critical aspect of the magico-medical complex of the American Negro, indicating point after point at which African tradition has held fast. In West Africa those who deal in herbs, roots, and other curatives are invariably differentiated from those whose cures for illness and other less mundane evils come from their supernatural powers. Precisely the same distinction is made in Dutch Guiana, while in Haiti the difference between the *traitement* and the use of a *wanga* or *arrêt* makes the same point. That these two types of practitioners have been known since the days of slavery—it is not necessary to document the well-recognized fact that magic in some form or other has characterized Negro life since the earliest days of their presence in this country—is to be deduced from the many instances afforded in contemporary writings. Thus one may compare a case involving a healer, cited by Catterall,[73] with a passage from Douglass[74] wherein he describes how, in a fellow slave named Sandy, he found a "genuine African" who had inherited "some of the so-called magical powers" of his homeland, which he offered to call into use to help protect Douglass from the wrath of his masters.

But this is only a beginning of the correspondences between African and American Negro practices to be found in Mrs. Cameron's analysis. The importance of revelation in giving remedies to mankind is a fundamental West African tenet; the role of the *azizan*, or forest spirits, or of Legba, the trickster, in giving to qualified persons in Dahomey the information that makes of them healers[75] corresponds in many points with the more generalized concept of spiritual inspiration given the healer in this country. The fact that practitioners of magic are "dependent upon the spirits of the dead ancestry" is likewise purely African, and in accordance with the general pattern of the cult of the dead. The mandatory nature of the magician's calling, and the fact that circumstance of birth may determine his selection, are again familiar Africanisms, as is the fact that the use of herbs involves supernatural sanction no less than does the practice of magic.

The pragmatic test which all faith must meet has, of course, been continuously applied by the Negroes to magic and folk belief over the years, and this means that the devices employed by these specialists have fulfilled their function to the satisfaction of their clients. That belief has held fast even in the light of the relative inapplicability of these magical devices where whites are concerned is due to a process of reasoning which itself came to the New World from Africa. The tradition that a certain kind of magic is only efficient in the case of a certain type of people is to be met with widely. "White man's magic isn't black man's magic," a succinct statement of this principle, was heard on a number of occasions in the interior of Dutch Guiana. For the Negroes realize that, lacking belief, supernatural powers cannot work effectively. Where belief is held, however—and the matter of belief is not the result of individual volition, but of early training and affiliation—the power of these forces can operate in all its strength. It has remained for a novelist to depict the psychology of magic as it operates in the mind of a man against whom its power has been turned. In William March's *Come in at the Door*, the reaction of the mulatto teacher to the charm set against him goes far, as described, to explain how powerful magic can be for one whose later experience could not erase the sense of inevitability that seizes him when he believes himself assailed by these intangible forces.

The conviction held by American Negroes that no dichotomy exists between good and evil in the realm of the supernatural, but that both are attributes of the same powers in terms of predisposition and control, is characteristically African. As concerns the type of magic found in this country, the matter has been well put by Puckett:

It is a long lane from heart-winning to "cow cuds," but in almost every practical episode of life along the way conjure is operative. Evidently we have here a force, second in utility only to West African religion itself, by which mankind can (or thinks he can) achieve almost every desired end—a force which is closely interwoven with his daily life and one which deserves his earnest attention. But this power, like African religion, is not moral, but is capable of indifferently working harm as well as benefit. Thus it behooves its troopers to look to their armor as well as to their arms . . . Fortunately this matter of armament is simplified in that for the most part one and the same substance serves alike for shield and sword.[76]

Of similar import is what Puckett terms "turnin' de trick," a "trick" being the charm that has been laid against an enemy:

If the person desires, the trick may now be turned against the person who planted it. Ed Murphy did this by laying the trick he had discovered in a piece of paper, sprinkling quicksilver over it, and setting the paper on fire. The trick exploded and made a hole in the ground a foot deep as it burned up—his enemy soon died. "It is said that if any one tricks you and you discover the trick and put that into the fire, you burn your enemy, or if you throw it into the running water you drown him."[77]

The immediate African parallel to this point of view, striking because it contrasts vividly with the European habit of separating good and evil so strongly that the concept of the two as obverse and reverse of the same coin is almost nonexistent, is to be seen from the following discussion of attitudes held by the Dahomeans toward their *gbo*, or charms:

One point which emerges from a consideration of the *gbo* . . . is that good and bad magic are merely reflections of two aspects of the same principle. . . . The character of the *gbo* is such that while one of these charms helps its owner, giving its aid to protect him from the evil intentions or deeds of enemies, it also possesses the power to do harm to the one who would do such evil to its possessor. Thus when a man leaves a house for a few days, he places a *nguneme* charm—so common as hardly even to be thought of as a *gbo*—to protect his belongings. . . . The power of this . . . is such as to harm those who violate it. . . . It is believed that one who did violate property guarded by such a leaf, if a man, would become impotent, or, if a woman, would become barren, while children of such violators would meet early death. . . . Thus it can be seen that the Dahomean is relating what is, to him, an obvious fact when he says that good and bad magic are basically the same.[78]

Some years ago Miss Mary Owen divided magic charms into four categories—good tricks, bad tricks, all that pertains to the body, and commanded things consisting of "such things as sand, or wax from a new beehive—things neither lucky nor unlucky in themselves, but made so by commands." Puckett's reservations to this system, taken on the ground that the classifications so merge with one another in the charms used by the Mississippi Negroes studied by him as to be indistinguishable, are to the point, but perhaps merely reflect the general process of blurring which seems to be a concomitant of acculturation among New World Negroes.[79] Certainly the fact that charms of Missouri Negroes can be classified in this way does not vitiate the principle enunciated by Puckett concerning the dual function of magic among the Negroes.

No systematic treatment of the categories of charms used by African folk has been made except for Dahomey, but this analysis, based on detailed materials regarding the manufacture and employment of a series of over forty charms, gives a comparative basis for such classifications.[80] In Dahomey *gbo* are classed both according to function and as to the materials that go into them. These classes cut across each other just as do the first and the last two items in Miss Owen's set of categories, since in Dahomey eighteen types of *gbo* are distinguished on functional lines, while six categories are couched in terms of the materials employed in their making. Hence both in what may be termed the theory of the operation of magic charms and in ways of differentiating them, immediate correspondences to data found in at least one tribal group in West Africa are discernible.

It would be impossible to give in any detail the bewildering variety of Negro magical devices employed to achieve ends desired for oneself, or to bring harm to others, or as preventatives when the needs of one person bring him into conflict with another. Fortunately, it is quite unnecessary to give such a catalogue here, since this aspect of Negro life has been treated more exhaustively than almost any other. Aside from the innumerable "tricks" named and described by Puckett, and the full-length works of Owen[81] and Hurston,[82] one finds great wealth of materials in the appropriate *Memoirs* of the *American Folk-Lore Society* and in the *Journal of American Folk-Lore*. Especially in the earlier numbers of this *Journal*, one comes on detailed signed reports of various cases involving magic and descriptions of the charms used,[83] while the editors of that period were also alert to abstract accounts appearing in other scholarly journals and newspapers bearing on the subject.[84]

These numerous data demonstrate that the broad principles of sympathetic magic that function in Africa and the West Indies have lost none of their appeal to Negroes of the United States. Instance after instance proves again that the concept of magic applies to results obtained from what in scientific parlance would be termed "physical causation" no less than supernatural; that identification of poison with black magic exists in the United States as it does in the West Indies or West Africa. There are even suggestions of carry-overs in specific details of African life, reinterpreted but nevertheless immediately recognizable as, for example, a remedy, cited in the first of Miss Moore's papers listed above, which included drawing in the sand a design similar to that used everywhere by the Yoruba as a decorative and quasi-religious motif.

6

Most of the materials considered in the preceding section refer to Negro communities in the southern states east of the Mississippi, and to those northern communities whose Negro populations are derived from this area. The part of the South that lies west of the Mississippi River, especially Louisiana and Missouri, represents something of a special case. Particularly is this true of the region about New Orleans, which is the locality where those aspects of African tradition peculiar to this specialized region have reached their greatest development. The reason for the distinctiveness of its customs, and for the degree to which particular kinds of Africanisms not found elsewhere have been preserved, is to be found in its historical background and the kind of European culture to which the Negroes of the region had to accommodate themselves. The white population was French and Catholic, the early affiliations of this area pointed to the French rather than to the British West Indies, and later impulses resulted from the migration of Haitian planters with their slaves to Louisiana. These circumstances have given to its present-day Negro population, no less than to its white, the qualities which set off the region from other parts of the United States.

The outstanding aspect of the Negro culture in this area is sub-sumed under the term "voodoo." The uniqueness of the cult as found in New Orleans in earlier days, at least, is due to those cir-cumstances that have been mentioned—the differences in the French as against the English plantation system, and the fact that exposure to Catholicism caused accommodation to take on different forms than contact with Protestantism. To what extent the voodoo cult—in the patois of the area, *vodun,* as in Haiti—has persisted to the present time cannot be said. The testimony of Hurston, who has re-ported on her field-work among the cult-heads in New Orleans at some length, would seem to indicate that the former well-integrated system of ritual and belief has degenerated considerably, and taken the protective cloak of spiritualism,[85] although many direct correspondences will be found to exist between the spirit-possession dances described in the final pages of another discussion by this author,[86] Haitian *vodun* practices, and Dahomean cult rituals. The description given by Saxon of a voodoo rite,[87] despite the heightened tone of its treatment, likewise suggests elements in accord with Haitian and Dahomean procedures. Until these customs and beliefs are studied in such a manner as to present the life of the people without undue weighting of the sensational and esoteric phases of their life, however, any discussion of Negro culture in this area must be fragmentary.

The survival of Africanisms to very recent times is apparent in practically every work dealing with the region, and this comment is the more impressive because the writers in most cases were entirely innocent either of concern with correspondences or of knowledge of African life. It was customarily taken for granted that those traits of New Orleans and Louisiana folk life that could not be accounted for by reference to French traditions must have come from somewhere else, and that that somewhere was Africa; but this is incidental in such writings, which customarily attempted only to describe the "quaint" customs that characterized their subjects.

One of the richest stores of data pertaining to Negro custom is the writing of George Cable, whose articles on New Orleans life, and particularly whose novel describing this life in preslavery days, *The Grandissimes,* hold special significance for research into the ethnography of United States Negroes. Based on intimate knowledge of the locality and its history, it must be accepted as a valid document if only on the basis of comparative findings. It is thus a real contribution to our knowledge of life in this area during the time of slavery, and a book which investigations into present-day custom should take into careful account.

The names of several deities which figure in the *vodun* cults of Haiti and Dahomey are mentioned in Cable's novel. Papa Lébat[88] "who keeps the invisible keys of all the doors that admit suitors," is the Papa Legba of Haiti and the Dahomean trickster of this same name, who has already been referred to. Danny[89] is the Dahomean serpent-god Dan, the Haitian Damballa, who in his West African manifestation is the god of good fortune. Agoussou,[90] whose color is red, has already been considered.[91] M. Assouquer[92] is a deity whose name would seem to be a conglomerate of the designations of several West African gods, unidentifiable in this form because so little is told of his functions except that he is "an imp of good fortune." The familiar pouring of a libation, whose place in a wider pattern in connection with magic has already been mentioned, is encountered, as where we read how one character, in distress, "has recourse to a very familiar, we may say time-honored prescription—rum. He did not use it after the vodou fashion; the vodous pour it on the ground. Agricola was anti-vodou."[93] The concept of the *zombi* as spirit,[94] of the magic charm embodied in the term *"ouangan"* (*wanga*) and the importance, in the syncretism of the region, of the *P'tit Albert*[95]—that book of medieval European magic so feared in Haiti that its importation is prohibited by the law of the country—all these are familiar aspects of Haitian terminology and important elements in Haitian no less than West African life. As for broader aspects of custom, the descriptions of the manner in which the charms to be set against various characters at various points in the action of this novel are made, and of their effect on those against whom they have been "set," give vivid insight into the working of magic in this area.

One of the last recorded *vodun* ceremonies was that of June 24, 1897, which has been given in abstract.[96] A more complete account, of an earlier rite, was taken from a report of the trial, held in August, 1863, of some of the important dignitaries of the cult.[97] The ceremony was of a type held once annually. The account speaks of a "witches' brew" in a vase at the center of the cleared space where the ceremony occurred; more significantly, it also mentions three snakes that "lifted their heads nonchalantly" when the police entered, and hundreds of lighted candles about the central sacred spot.

The role of the serpent in the Haitian *vodun* cult and in Dahomey points to a certain validity for the claims of those who give serpent worship a prominent place in the cult of New Orleans. Hurston, telling the story of Marie Laveau, the *vodun* priestess, as recounted to her by the "hoodoo doctor" Turner, gives an important place to the rattlesnake that "came to her bedroom and spoke to her," presumably calling her to membership in the cult. This serpent remained with her—"the rattlesnake that had come to her a little one when she was also young was very huge." Turner's tale continues:

He piled great upon his altar and took nothing from the food set before him. One night he sang and Marie Laveau called me from my sleep to look at him and see. "Look well, Turner," she told me. "No one shall hear and see such as this for many centuries." She went to her Great Altar and made a great ceremony. The snake finished his song and seemed to sleep. She drove me back to my bed and went again to her Altar. The next morning, the snake was not at his altar.

His hide was before the Great Altar stuffed with spices and things of power. Never did I know what became of his flesh. It is said that the snake went off to the woods alone after the death of Marie Laveau, but they don't know. This is his skin that I wear about my shoulders whenever I reach for power.[98]

It would be revealing to know more of Marie Laveau's story, especially the price she paid for the power her serpent brought her. For in West African and Haitian and Guiana and Brazilian and Jamaican and Trinidad Negro belief, nothing is to be had without adequate price, and compacting with the supernatural is expensive in any terms. The tale of this serpent resembles other stories of men and women who also had a serpent or a spirit as familiar, and who over the years paid again and again with the souls of those beloved by them until at last there were no more souls, and they themselves paid the ultimate price. The versions of such affairs in all these related cultures are too similar to this one not to make of it a point to be probed by some student who, equipped with the requisite comparative knowledge of these phases of African religion, may in the future work among the believers of the Louisiana *vodun* cults, where the traditions centering about the name of this most famous of priestesses are still living.

There is much in Hurston's descriptions of the initiations she experienced into various cult groups that can be referred to recurrent practices in West Africa, and in the Catholic New World countries where pagan beliefs of Africa have persisted. The stress she lays on certain aspects of the initiation—seclusion as a novitiate, fasting, wearing of special clothing, dancing and possession, sacrifices—all these would be given prominent mention in describing the induction of a novitiate into a Dahomean religious cult or into a Haitian *vodun* group. An arresting correspondence concerns the sacrifice of nine chickens, the uneven number itself being characteristic of Negro sacred rites:

The terrified chickens flopped and fluttered frantically in the dim firelight. I had been told to keep up the chant of the victim's name in rhythm and to beat the ground with a stick. This I did with fervor and Turner danced on. One by one the chickens were seized and killed by having their heads pulled off. . . .[99]

This ritual method of killing chickens has been witnessed in many rites attended in West Africa and the New World; that it should have been continued in present-day voodoo of New Orleans is another indication of how minutiae can persist after the broader lines of ritual procedure and their underlying rationalizations have been lost.

During her initiation into this same group, Hurston is told she must come "to the spirit across running water"; she is given a new name because the priest sees her "conquering and accomplishing with the lightning and making her road with thunder."[100] The African reference of these allusions will be clear in the light of our earlier discussions of the religious significance of running streams and of lightning, and are further consistent with the tradition of renaming novitiates subsequent to the initiatory experience. Her own possession experience[101] is likewise strictly in African form, especially the manner in which, when taken in charge by the priestess, the spirit was immediately transmitted to her:

When the fourth dancer had finished and lay upon the floor retching in every muscle, Kitty was taken. The call had come for her. I could not get upon the floor quickly enough for the others and was hurled before the altar. It got me there and I danced, I don't know how, but at any rate, when we sat about the table later, all agreed that Mother Kitty had done well to take me.[102]

This may be compared with an account of how first a *vodun* priest and later a possessed woman brought on possession to others taking part in the *service* preceding a Haitian ceremony,[103] or how possession is regulated in Dahomey and the nature of the possession experience;[104] such a comparison will establish identities of the most precise nature.

Outstanding in the manner in which traits of African religion are carried over in those New World countries where Catholicism is predominant are the syncretisms between African and Christian sacred beings, especially the manner in which Catholic practices are incorporated into the African rituals of the Negro cult groups. The existence of this phenomenon has been proved for Brazil by Ramos, who in numerous works has provided full documentation, naming the saints that correspond to African gods, and depicting and analyzing the ritual altars on which crucifix and chromolithographs of the saints jostle African offerings and wood carvings in the African manner.[105] The Cuban form of this syncretism has been described by Ortiz[106] and has been reported several times from Haiti.[107]

The same phenomenon has been reported for the New Orleans cults; though, as might be expected, the correspondences are less specific and less numerous than in these other countries. The following may be analyzed:

In the first séance she tells of a white girl calling upon a Negro voodoo-woman to obtain help in winning the man she loves. At the meeting, the girl is allowed to wear nothing black, and is forced to remove the hairpins from her hair, lest some of them be accidentally crossed, thus spoiling the charm. In the room were paintings of the various Catholic saints, and an altar before which was a saucer containing white sand, quicksilver, and molasses, apexed with a blue candle burning for Saint Joseph (Veriquete). All the way through, there is this strange mixture of Catholicism and voodooism. The "Madam" kneels at the girl's feet and intones the "Hail Mary" of the Church, there is a song to *Liba* (voodoo term for St. Peter) and another to *Blanc Dani* (St. Michael). The money collected at the séance is put in front of the altar with the sign of the cross.[108]

In this passage certain identifications of first importance are encountered. The identity of Legba (Liba in the above) and St. Peter follows in principle the syncretism of Haiti; here, indeed, this identification is the more logical, since Liba, guardian of gate and crossroads, is conceived as St. Peter, guardian of the keys. It is likewise significant that he is the first of the two voodoo spirits called after the "Hail Mary," for this confirms our assumption of his identity with the Legba of Haiti and West Africa, where this god is likewise the first called. Why Blanc Dani, Cable's "Danny" and the counterpart of the Dahomean serpent-deity Dan, whose color is white, is identified with St. Michael cannot be said; it is apparent that the principle of identification is the same as that which in Haiti identifies the serpent-deity with St. Patrick (for understandable reasons, apparently overlooked in New Orleans). Other instances of the working of the principle of syncretism given by Puckett[109] need not be repeated here; it is of interest that, according to Miss Owen's account cited by him,[110] the same process is found among the voodoos of Missouri.

The pronounced African character of voodooism makes of the localities in which it occurs some of the most promising spots in which to seek other less dramatic but equally deep-seated survivals. For there is no more reason to believe that other, nonreligious Africanisms have died out than there is, in the light of Saxon and Hurston's accounts, to credit the assertions of those who state that the last voodoo performance took place in 1897. Nor, on the other hand, is it to be supposed that voodoo is by any means restricted to the region about New Orleans. Its influence has traveled far, and the voodoo doctor, as distinguished from the hoodoo doctor (a subject likewise a matter for future detailed field analysis) is found practicing his trade well outside the limits of the Catholic, sometime French, area of Louisiana. Powdermaker reports "four famous voodoo doctors" living and practicing within a radius of fifty miles of the community she studied,[111] while Puckett, who gives in some detail the distribution of the cult, states that in 1885 it was estimated that perhaps a hundred old men and women followed it as a profession in Atlanta, and that similar cults, reinforced by West Indian migration, have taken great hold in recent times in the Negro district of New York.[112]

In view of the manner in which the type of worship and magical control represented by voodoo drives deep into the traditional beliefs of the Negro, it should not be surprising if future study shows that much more of the cult has persisted than is customarily held. Certainly, in any analysis of African survivals in the United States, this Louisiana enclave, where special historical circumstances have made for the perpetuation of this African cult and for the preservation of more numerous and more specific African practices than in any other portion of the country excepting the Sea Islands, should receive far greater and systematic attention than has been given it.

7

Certain Negro beliefs which cannot be systematized in terms of even so rough a set of categories as has been employed in marshaling the facts thus far presented in this chapter, remain to be discussed. In a well-integrated system of belief and ritual this would not be true, but where contact has distorted values and changed modes of expression, phenomena of the kind now to be discussed lose contact with such forms of belief and behavior as have been previously considered. They are no less important for the study of African survivals because of the position they take in the ordering of Negro life, however, and require careful exposition and adequate analysis.

The conception of the Devil held by Negroes in the United States may be taken as the first of these points. In the religious system of the whites, this character holds secondary importance, except perhaps in those evangelical churches where the punishment of the damned as focused in the personality of Satan is stressed. Yet the Devil as conceived by the Negroes is a different Satan, even from that of the Protestant groups that preach the doctrine of damnation most vigorously; which means that the problem again arises of analyzing the elements contributing to a belief which, differing in its present form from its counterpart among whites, represents a combination of cultural impulses.

The matter has been put in the following terms:

The Africans cling to their tendency to worship the malevolent even after they have heard of Christianity. One bishop asked them why they persisted in worshiping the devil instead of God. The reply was, "God is good, God is love and don't hurt anybody—do as you please, God don't hurt you; but do bad and the devil will get you sure! We need not bother about God, but we try to keep on the good side of the devil." The Southern Negro likewise gives the devil as a personage considerably more attention than is paid him by the present whites, though in the past both in Britain and in the Early Colonies as well as in other parts of the world this personage was greatly feared if not actually respected, seeming to show that the Africans were not alone in their emphasis of the malevolent element in religion.[113]

Other comments on Negro concepts of the Devil may be cited to clarify the position further:

Satan . . . is a familiar figure in negro songs. It is to be noted that while he is a very real and terrible personage, there is always a lively, almost mirthful suggestion in the mention of his name. . . . The personality of Satan is, therefore, at once a terror and a source of enjoyment to the negro. The place he holds in negro theology is not unlike that which he occupied in the miracle plays of the middle ages. There seems an inherent tendency to insincerity in negro demonology. Satan is a decided convenience. It is always possible to load upon him what else must be a weight upon the conscience. That Satan holds the sinner responsible for this has its compensation again in the fact that Satan himself is to be dethroned.[114]

Hurston, who discusses the Negro point of view with the intimacy of inside knowledge, describes the character of the Devil as conceived by Florida Negroes in these terms:

The devil is not the terror that he is in European folk-lore. He is a powerful trickster who often competes successfully with God. There is a strong suspicion that the devil is an extension of the story-makers while God is the supposedly impregnable white masters, who are nevertheless defeated by the Negroes.[115]

That this Devil is far from the fallen angel of European dogma, the avenger who presides over the terrors of hell and holds the souls of the damned to their penalties, is apparent. So different is this tricksterlike creature from Satan as generally conceived, indeed, that he is almost a different being. To account for the difference, therefore, we turn again to that character in Dahomean-Yoruba mythology, the divine trickster and the god of accident known as "Legba"; the deity who wields his great power because of his ability to outwit his fellow gods. His importance in the daily life of these West Africans has already been discussed. We know beyond dispute that it has carried over into the New World in the evidence cited from Brazil, Guiana, Trinidad, Haiti, and New Orleans. Here under various names—Lebba, Legba, Elegbara, Liba—he rules the crossroads and, as an extension of his powers and duties in West Africa, "opens the gate" for the other gods at all rituals. It is of some importance to note that, in West Africa, this deity is identified with the Devil by missionaries; and considered from the point of view of their world concept, tuned as it is to a dichotomy between good and bad, this celestial trickster who balks the gods with his cunning could easily be interpreted in this fashion. It is thus understandable how, in the New World, where Protestantism placed special emphasis upon the difference between good and evil, the reinterpretation of this deity as the Devil was especially logical.

Yet reinterpretation was more verbal than otherwise; in no sense did it involve a wholehearted acceptance either of the Devil's personality as depicted in Christian theology or his function as the representative of evil in the universe. The tradition in African thought which holds nothing to be entirely good or entirely bad goes so deep that it is hard to see how it could be given over for the less realistic European penchant for concepts phrased in terms of blacks and whites. We have already seen this African point of view operative in the United States in such matters as assigning to charms or other supernatural devices powers at once good and evil. This is of a psychological piece with the pliancy of the Negro in social situations that makes of him the diplomat who has been able to weather some of the hardest times known to any group of human beings.

How much closer the Devil of American Negroes is to the character of the African trickster-god than to the bearer of his name among non-Negro peoples can be seen from the following description of one of the Devil's West African counterparts:

Legba is essentially a trickster; but like all supernatural Dahomean forces, he can be beneficent as well as malevolent. More than any other deity . . . he must be worshiped by all regardless of cult affiliation, for as messenger of all the gods and their spokesman, he is the one to be propitiated if a request is to be granted by a supernatural force; he alone has the power to set aside certain misadventures in the destiny of a person, and the power, also, to add to them.[116]

It is thus apparent that, while Christianity gave to the Negroes in the New World much of its own world view, in the United States, where the perpetuation of African gods under their own names was impossible, the process of readjustment permitted the deity to survive under a different designation. That he survived thus disguised does not matter, for disguise, itself a technique of survival, is highly congenial to the African habits of thought that, as we can see, were never entirely given over by the Negroes in their new setting.

Another Negro belief that may well be the survival of a concept having wide distribution in West Africa concerns what may in broadest terms be called "little people." It has never been formulated carefully by those who have studied the folk-beliefs of Negroes in this country, but here and there a sentence or a paragraph in the literature is highly suggestive of the possibilities of systematic inquiry into the nature and functioning of such beings.

One of the earliest accounts in which creatures of this type figure is a discussion of the folk-tales of Georgia, wherein a version of the well known tar-baby story is being discussed:

As I heard it in one of the southernmost counties of the State, the tar-baby was by no means a mere manufactured, lifeless snare, but a living creature whose body, through some mysterious freak of nature, was composed of tar, and whose black lips were ever parted in an ugly grin. This monster tar-baby, which haunted the woods and lonely places about the plantation, was represented as wholly vicious in character, ever bent upon ensnaring little folks into its yielding, though vice-like embrace. Well do I remember the dread of encountering the ogre-like creature in some remote spot, where I should be unable to withstand its fascinations; for it was said to be impossible to pass the tar-baby without striking it, so provoking was its grin and so insulting its behavior generally,—and when once you had struck it, you were lost. I was always on the lookout for it, but, it is needless to say, I never encountered it, except in dream-land, where again and again was suffered the unspeakable horror of being caught and held stuck fast in its tarry embrace.[117]

This document is interesting for many reasons other than the African tradition mirrored in it, and may first be compared to data gathered among the Negroes of the town of Paramaribo, Dutch Guiana, where the belief in the dwarfs called *bakru* causes mothers to instill in their children much the same fear that is expressed in the lines quoted above. These dwarfs, half wood, half flesh, are "given" by a practitioner of evil magic to a client who wishes wealth. They "work" for their owner; should someone try to strike them, they present the wooden side and then kill the one who has tried to harm them. Eventually:

After the owner of a pair of *bakru* dies, and there is no one to care for them, they disappear to live on the road. A favorite diversion of theirs is to mingle with children who are on their way home from school. They try to touch the children, to tease them, and to offer them a drink. It is death for a child to drink from the little bottle each *bakru* carries in his pocket. . . . A woman whose own aunt had had two such creatures, looked under the bed after her aunt died, caught a glimpse of the *bakru*, and fled. They were very black,—black hair, black skin, black eyes, . . . like Bush-Negro children.[118]

Several points of resemblance between the tar-baby of Georgia and the *bakru* of Guiana are to be discerned at once—their blackness, the appeal they have for children, and the danger that comes when children come into their power. The differences must also be considered, the most important for our purpose being that the tarbaby is reported as a huge figure, while the *bakru* are small and child size.

We may turn next to certain other citations which tell of what, on the surface, may seem to be beliefs appertaining to somewhat different beings. For the following citations we are indebted to Puckett's thorough search of the literature and his painstaking records gained after personal inquiry in the field. The first, from the literature, states:

In one Gullah district, Sabey, whom the Negroes feared because of his ability to throw spells, was a "queer, misshapen mulatto, almost an albino, with green eyes and yellow wool lighting and thatching a shrewd and twisted, though good-natured monkey face."[119]

An informant in Mississippi reported that:

Others say that hoodoomen, who always have long hair and beards, always carry a loaded cane with which they tell whether you are honest or not.[120]

In both these citations we encounter concepts that may be expressed in most generalized form as the association of abnormality with supernatural powers—in the case of the Sea Island creature, dwarfism and albinism; in that of the hoodooman, long hair and beard, physical traits not commonly encountered among Negroes. Two more quotations are pertinent at this point. One reads:

A Georgia convert told of God being a little white man two feet high with pretty hair, ending her testimony by mourning and singing, "Ain' dat pretty hair? Ain' dat pretty hair?"[121]

The other, from a Columbus, Mississippi, informant, is as follows:

Escape from an embarrassing predicament is another favorite theme, illustrated by the escape of "de widder-woman frum de Hairy Man. De Hairy Man had cotched her in de woods an' was fixin' ter kill her," but she asks for a few moments in which to pray. The "Hairy Man" didn't know what prayer was, so the woman took advantage of this spiritual ignorance to call her dogs, who ate up her monstrous assailant.[122]

We may now turn to West Africa for comparative materials which will bring these strands together. Rattray speaks of various kinds of supernatural beings which among the Ashanti empower magic charms. Of the *suman*, or charm, he states:

The power of suman comes from *mmoatia* (fairies), *Sasabonsam* (forest monster of that name), *saman bofuo* (ghosts of hunters), and *abayifo* (witches).[123]

Illustrations are given of wood carvings made by Ashanti artists of the first two types of creatures named above.[124] The *Sasabonsam* is large in contrast to the two *mmoatia*, the larger creature being distinguished by its long hair on head, face, and in the pubic region. The text implements the illustrations. Of the *mmoatia*, described by Rattray somewhat unfortunately in the citation above as "fairies," he says:

The most characteristic feature of these Ashanti "little folk"—the word *mmoatia* probably means "the little animals"—is their feet, which point backwards. They are said to be about a foot in stature, and to be of three distinct varieties: black, red, and white, and they converse by whistling. The black fairies are more or less innocuous, but the white and the red *mmoatia* are up to all kinds of mischief, such as stealing housewives' palm-wine and the food left over from the previous day. The light-coloured *mmoatia* are also versed in the making of all manner of *suman* which they may at times be persuaded to barter to mortals by means of the "silent trade." . . .[125]

The *Sasabonsam*, to which Rattray cites a parallel far to the east in the Niger Delta region from Miss Mary Kingsley's *West African Studies*,[126] are described as follows:

The *Sasabonsam* of the Gold Coast and Ashanti is a monster which is said to inhabit parts of the dense virgin forests. It is covered with long hair, has large blood-shot eyes, long legs, and feet pointing both ways. It sits on high branches of an *odum* or *onyina* tree and dangles its legs, with which at times it hooks up the unwary hunter. It is hostile to man, and is supposed to be especially at enmity with the real priestly class. Hunters who go into the forest and are never heard of again— as sometimes happens—are supposed to have been caught by *Sasabonsam*. All of them are in league with *abayifo* (witches), and with the *mmoatia*, in other words, with the workers in black magic. As we have seen, however, and will see again farther on, their power is sometimes solicited to add power to the *suman* (fetish), not necessarily with a view to employing that power for purposes of witchcraft, but rather the reverse.[127]

Directly comparable with these "little people" of the Ashanti are the *ijimere* of the Yoruba and the *azizan* of Dahomey.[128] These are creatures which, in the forest, accost hunters and give to them the knowledge of medicines and of magic that makes those who follow the occupation of hunter so powerful. Similarly comparable to the Ashanti *Sasabonsam* are the Dahomean *yehwe zogbanu*, the forest-dwelling, many-horned, fire-breathing monsters who are the subject of numerous tales in which the hunters, like the Mississippi woman caught in the "embarrassing predicament," are saved by some occurrence phrased in terms of the "Flight up the Tree" motif, wherein the one attacked devises some trick which enables him to call his dogs and thus outwit his pursuers.

The fascination which creatures of this sort hold for Africans cannot be underestimated. They figure in the thought and interests of the people, as reflected both in their everyday conversation and in their tales and myths. Their living quality can only be realized by those who have had firsthand contact with a functioning lore, that is not merely a collection of old wives' tales, representing happenings of the long ago. In the disintegration accompanying the acculturative process, these figures with their misshapen bodies, their hairiness, their supernatural powers, their whistling, and their fondness for trickery and destroying human life have understandably merged into one another, becoming blurred in outline and confused in attribute and function. Yet they are unmistakable in the quotations that have been cited. The *mmoatia* have lost their characteristic of backward-turned feet—at least, no such "little folk" have as yet been reported from the New World—but traits of the "white" and "red" varieties of *mmoatia* are to be discerned in the magic role assigned the Sea Island dwarf and the concept of God that came to the Georgia convert, while the half-wooden dwarfish *bakru* and the huge tar-baby likewise partake of other characteristics attributed to these Ashanti figures. Greater detail from other tribes of West

Africa is necessary if the variation in these creatures as envisaged over the entire area is to be grasped; in the New World, too, further search needs to be made for their survivals. When such data are in hand, however, we will be able to fill in our present rough outlines with precise knowledge of how, under acculturation, the merging of traits from various African tribes represented by the slaves worked out into generalized belief of the type embodied in these manifestations of the "little folk" concept.

Ghosts, witches, and vampires are as well known in Africa as in Europe, so that in this case the problem is to indicate the African aspects of the belief in these beings found among Negroes of the United States. Parsons speaks of old women being regarded as witches in the Sea Islands, and tells the preventive measures to be taken when they are thought to be about:

. . . there is the familiar belief about hags—women who shed their skins and victimize sleepers—"ol' haigs what ride people in de sleep." And the precautions to be taken are likewise familiar. "Say if you want to ketch dat haig, you scatter mustard-seed fo' de do', 'cause mustard-seed so fine, pick dat up 'til morning," or again, you must put salt and pepper in the discarded skin.[129]

A description of the powers of these old women as conceived by Missouri Negroes, though perhaps overdramatized, may likewise be set down:

Granny . . . knew a charmed child when she saw one, and was resolved to do what she could to relieve the unconscious victim. Oh! She knew Aunt Mymee, and so did the others. Although they visited and received her in turn, although she had lived in the cabin a few rods from Granny's for years, not one of them ever went to bed at night without hanging up a horse-shoe and a pair of wool-cards at the bed's head. Not one of them failed to pour a cup of mustard or turnip-seed on the doorstep and hearth, so that she would have to count all those seeds before she could go in at the door, or down the chimney to tie their hair into knots; to twist the feathers in their beds into balls as solid as stone; to pinch them with cramps and rheumatism; to ride on their chests, holding by their thumbs as by a bridle, while she spit fire at them till cock-crow. Not one of them had any doubt as to her ability to jump out of her skin whenever she pleased, and take the form of owl, black dog, cat, wolf, horse, or cow. Not one of them merely suspected, she *knew* Mymee could appear in two places at once, ride a broomstick or a bat like a charger, and bring sickness and bad luck of all sorts on whomever she pleased.[130]

Vampires also have their describable habits:

Vampires are not common, but one Negro tells of a young girl constantly declining while an old woman got better and better. This was because the harridan sucked young folks' blood while they slept. "De chillun dies, an' she keeps on alivin'."

Another Missouri "witcher-ooman has blood sucking children."[131] There are several ways of keeping these creatures from one's house:

Salt sprinkled thoroughly about the house and especially in the fireplace; black pepper or a knife about the person; or matches in the hair, all bring dire perturbation to these umbrageous visitors. . . . Ha'nts, like witches, may also be kept away by planting mustard seed under your doorstep, or by keeping a sifter under your head while asleep. Some say that ghosts will not budge from a foot with fern seed in the hollow, though one informant recommends fern seed or sulphur to keep spirits away.[132]

This same student gives attention to the African derivation of certain of these methods:

The greatest variance among the Negroes is to be found in the great number of methods used in avoiding or driving off witches. The most common legend in this regard is that of an old witch who took off her skin, hung it on the wall and went off to ride some one. While she was gone a man slipped in and sprinkled red pepper in the skin. The witch came back and tried to slip it on. "What de mattah, skin? Skinny, doan' you know me? Doan' you know me, skinny! Doan' you know me!" she cried in agony, hopping up and down until she was finally discovered and killed. In various forms this same plot exists all through the South—in Georgia, Missouri, Virginia, Louisiana, North Carolina and the Sea Islands, as well as in the Bahamas. The belief is too widespread to be an independent development; to the best of my knowledge it is not found in Europe; but in West Africa there is the widespread idea that the witch leaves her skin behind on going out, and among the Vais it is thought that salt and pepper will prevent her getting back into her hide.[133]

Yet these methods of discovering, holding, and punishing witches and vampires are present not only in the two parts of West Africa indicated by the references, but they have also been found, in the course of field-work, in Nigeria, Dahomey, and among the Ashanti, and, in the New World, in Guiana, Haiti, Jamaica, Barbados, and Trinidad. In Dahomey, there are those who will warn one to beware the old women with bloodshot eyes who sell in the market, since haggling with them will anger them and bring on their vengeance;

or, in Nigeria, to be on guard against old men and women, because they are capable of translating their love of evil-doing into action. What seems to be European in the citations given above—broomstick riding, the ability to turn oneself into an animal—is, however, African as well, and represents the strengthening of belief when comparable phenomena in the two cultures come into contact. How close European belief is to African, where such characteristics are concerned, is made plain by Rattray's tale of how witches were conceived by his Ashanti informant:

"The majority of witches are women," he continued, "but they need not necessarily be very old women. If an old witch wishes her daughter to become a witch she will bathe her repeatedly with 'medicine' at the . . . kitchen-midden. The great desire of a witch is to eat people, but she will not do this so that any one may see; they suck blood. Each witch has a part of the body of which she is particularly fond. . . . The person, on awakening, will complain of illness and die before nightfall. . . . Witches always try to obtain some object that belonged to the person whom they wish to kill, such as hair, nail-cuttings, or waist-beads; witches can transform themselves into birds, chiefly owls, crows, vultures and parrots; into house-flies and fire-flies, into hyenas, leopards, lions, elephants, *bongo* and all *sasa* animals, and also into snakes."[134]

It is apparent that this leaves such patent transformations to be accounted for as are dictated by life in the temperate zone—cows, horses, wolves, and the like—and indicates that the concept of the broomstick as a means of locomotion is European. But, by the same token, this passage significantly documents the manner in which beliefs derived from various portions of the Old World have reinforced each other, and indicates once again the caution needed when discussions are couched in terms that ascribe to any one area the role of absolute source of provenience for traits found both in Europe and in Africa.

Chapter VIII

THE CONTEMPORARY SCENE: LANGUAGE
AND THE ARTS

It has long been held that the principal contribution of the Negro to the culture of the Americas, and most particularly to the culture of the United States, lies in the expression of his musical gift.[1] Since the "discovery" of the spirituals shortly after the Civil War, and markedly in recent years with the spread of the "blues" and the development of jazz and swing, musicians have drawn freely on Negro folk melodies and rhythms. In some instances the borrowing has been direct, as where Negro religious songs have found their way into the hymnals of the white churches. "Swing Low, Sweet Chariot," for example, is as familiar to whites as to Negroes, and in all likelihood is sung more frequently by the former than by the latter. The tendency of white song writers to take over the stylistic values and melodic progressions of Negro music, as occurs in such popular songs as "Ole Man River" and "That's Why Darkies Were Born," is another case in point. A third influence of Negro music finds expression in the more serious works. Dvořák's symphony "From the New World" has set a fashion that has been increasingly followed, until today its manifestations range from immediate applications of Negro musical idiom, as in Gershwin's opera *Porgy and Bess*, through the reworking of this idiom in such compositions as Powell's "Southern Rhapsody," to its less conscious translation in recent works by such younger white American composers as Roy Harris and Aaron Copland.

In the light of the wide popularity of Negro folk music, and the inspiration it has been to composers, it is strange that it has not been subjected to a more extensive musicological analysis. Such studies as are found are too often marked by the undocumented assertion that has become familiar in the preceding chapters as characterizing Negro studies in general, so that, as in regard to other aspects of Negro culture, controversy as to derivations typically takes the form of mere statement and counterstatement. In the main, these discussions turn upon the extent to which Negro music reflects African patterns or is merely a revamping of European thematic materials borrowed from the whites. The supporting evidence is largely confined to studies of Negro religious songs—spirituals—as these are related to white religious music. Rare, indeed, are the available comparisons between the Negro music of the United States and Africa while, moreover, students have almost completely failed to recognize that the Negro songs of the United States are but part of a larger body of New World Negro music. Yet this larger body of song, as found in the Caribbean and Latin America, offers not only a sure approach to an understanding of the processes that have brought into being the various forms of Negro music we know today, but also affords a rare opportunity to study, under the unusual conditions of historic control already commented on, a problem of wide implications for an understanding of change in musical style wherever this occurs.

It is of some interest to trace the changes in point of view as to the origins of Negro music that have taken place in the United States. It was first assumed that, in essence, the songs of the Negroes represented a welling forth of the anguish experienced under slavery.[2] In time, however, opinion grew that, since this music differed from other forms of musical expression, Africa was to be looked to for an explanation of its essential characteristics; this point of view was most clearly and vigorously expressed in a volume by the musicologist and music critic, H. E. Krehbiel.[3] Krehbiel's special concern with Negro songs stemmed from his friendship with Lafcadio Hearn and George W. Cable, whose interest in the Negroes of Louisiana had resulted in the collection of a considerable number of voodoo cult songs from New Orleans, wherein non-European elements are pronounced. Krehbiel had heard some singing by African Negroes at the World's Fair in Chicago in 1893, and this undoubtedly influenced his approach. Yet, as with all later writers who treated of Negro songs, he made no detailed study of African musical style, but relied mainly on what he could glean from travelers' accounts and other nonmusical works.

No serious critique of Krehbiel's materials has ever been undertaken, but in 1926 a paper by the late Erich M. von Hornbostel, the distinguished musicologist of the University of Berlin, suggested a new line of thought[4] and initiated a countercurrent that has today become dominant among students. According to von Hornbostel, who based his conclusions on impressions of Negro singing during a brief visit to the United States, the outstanding aspects of the Negro spirituals are European, such characteristics as the pentatonic scale, the "Scotch snap," and a tendency to harmonize in thirds all being well-known traits of white folk music. Only one such feature is held to be of African derivation—"leading lines sung by a single voice, alternating with a refrain sung by the chorus." Hornbostel concluded that in the United States the Negroes have evolved a real folk music which, while neither European nor African, is an expression of the African musical genius for adaptation that has come out under contact with foreign musical values. "Had the Negro slaves been taken to China instead of to America, they would have developed folk-songs in Chinese style"; as it was, they devised "songs made . . . in European style." The purely African element in this music is the manner of singing these songs; in motor behavior alone has aboriginal habit persisted:

Not what he sings is so characteristic of his race, but the way he sings. This way of the Negro is identical in Africa and in America and is totally different from the way of any other race, but it is difficult, if not impossible, to describe or analyze it.[5]

Most students today lay emphasis upon the importance of what the Negroes borrowed from European melodies, a position typified by the analyses of Newman White,[6] George Pullen Jackson,[7] and G. B. Johnson.[8] Their position holds that whatever African elements may be present in the spirituals—they have not considered to any extent the musical structure of other types of Negro song found in the United States—the correspondence between them and the religious songs of the whites are so close and so numerous that one need search no further. Some retention of African elements is admitted to have been possible, but these are held to be of such slight incidence as to be almost negligible.[9]

Yet, we ask once again concerning this element of Negro life, can any analysis of affiliations based on the scrutiny of only one or two possible sources be regarded as valid from the point of musicological and historical scholarship? There is, indeed, a certain significant malaise concerning the point. Herzog,[10] who emphasizes the absence of African elements in the spirituals, but who has worked in Liberia (the music he collected there has not been published, nor has any analysis of the relationship between it and the songs of the Negroes in the United States been made available), stresses the need for considering the African element in the equation when reviewing works by Johnson and Jackson.[11] "All discussions of the

Negro Spirituals have suffered, and still suffer, through insufficient knowledge of African musical material," he writes; and he underscores the caution later in the same review when he states that, "it is definitely necessary to utilize the available findings on African music more seriously and painstakingly than has been done thus far." Johnson, in his 1931 paper already cited, also mentions the matter: "It may be objected that I have started at the wrong end of the investigation of the relation of the spirituals to African and to English music, that the African side should have come first." The reservation, however, is explained in terms of the special techniques needed for analysis of tribal music, and the paucity of the comparative African materials.

It is undeniable that above anything else more data are needed, for the major musicological problems of Negro songs can be solved only by intensive study of an adequate body of transcriptions of recorded melodies and rhythms. The range of variation in Negro music in scale value, general form, and rhythmic structure; whether the songs of a given tribe of Africa have a style restricted to the group or are a part of a more widespread pattern; what has happened to these various tribal and regional styles under contact with European music in the New World when the differences and similarities found between the generalized characteristics of Old World European and African music are considered—all these must await systematic recording, careful transcription, and concentrated analysis. Far greater attention must be given to nonreligious folk music of United States Negroes than has hitherto been accorded songs of this kind. They have attracted a certain degree of attention from those concerned with analyzing their words. But the actual melodies and rhythms of work songs, of songs of recrimination and ridicule, of prison songs, are needed to supplement the rather extensive collections of spirituals available for study.[12]

In the West Indies, except for Cuba, Haiti, Jamaica, and in Guiana, almost no published materials can be found;[13] even when unpublished recordings are taken into account, our resources represent only a fragment of the materials. Brazilian recordings are almost nonexistent, while but little has been published of Negro music from that country;[14] and this is the more true for those portions of Latin America wherein Negro populations may have retained something of their musical habits from Africa[15] and thereby influenced the songs of their European and Indian compatriots.[16]

It is essential to recognize, however, that transcriptions and analyses of recordings, no matter how carefully and minutely done,

can for all their importance never tell the entire story of the relationship of New World to African musical styles, nor of the differences between the music of various parts of the New World. For,
as von Hornbostel observed, the problem also involves the consideration of the intangibles of singing techniques and motor habit
accompanying song quite as much as the actual progressions that
may be copied down from recordings.[17] The matter has its analogy
in the playing of jazz and swing; in the difficulties which the New
York Symphony Society and Walter Damrosch, its conductor,
found in mastering Gershwin's "Rhapsody in Blue," or in the
trials of European bands when they attempt to imitate the swing
rhythms of the Americas. Even the style of white bands in the
Western Hemisphere is noticeably different from that of the Negro
organizations.

Among these intangibles is the close integration between song and
dance found everywhere in Africa. Motion pictures of African
choruses accompanying their soloists show that even handclapping
can become a dance, while in the New World this tendency to "dance
the song," whether it is religious or secular, is a commonplace. Im-
provisation is similarly a deeply rooted device of African singing.
With broad social implications, especially in the songs of recrimination so widely distributed in Negro cultures, its effect as a mechanism making for variation in the music of different peoples and in
developing individual style calls for careful study.

The pattern whereby the statement of a theme by a leader is repeated by a chorus, or a short choral phrase is balanced as a refrain
against a longer melodic line sung by the soloist, is fundamental,
and has been commented on by all who have heard Negroes sing in
Africa or elsewhere. The relationship of the melody to an accompanying rhythm—carried on by drums, rattles, sticks beaten one
against the other, handclapping, or short nonmusical cries—is also
of the closest. So prominent is the rhythmic element in Negro music
that this music as ordinarily conceived relegates the element of
melody to second place. This, to be sure, is only partically valid, as
is demonstrated by the performances of the Dahomean choruses
of chiefs' wives when they sing songs of the royal ancestral cult,
or by the long melodic line of some of the Shango cult songs from
the island of Trinidad, or by some of the Brazilian Negro melodies.
Yet the need to ornament an underlying rhythmic structure is fundamental, and when Negro music as a whole is considered this trait
must receive closest attention.

The broad approach conceived here as essential is to be thought of as part of the program of study, detailed in our opening chapter.[18] In its musicological phase, this program has resulted in the systematic collection of a large series of related Negro songs from various historically linked portions of the Negro world. The two field-trips to Dutch Guiana yielded a collection of 255 songs from the Bush Negroes and those of the coastal city of Paramaribo, together with some Haitian melodies obtained en route. These songs have been transcribed and analyzed by Dr. M. Kolinski, and in 1936 were made available in published form.[19] During field-work in West Africa, 464 melodies were recorded, principally from Dahomey and the Ashanti of the Gold Coast, though a few songs from the Yoruba of Nigeria and the coastal region of Togoland were obtained. These songs have been transcribed and analyzed by Dr. Kolinski, but as yet have not been published. A further collection of 300 songs, made in Haiti, has also been studied by Dr. Kolinski. All the above were recorded on cylinders, since the conditions of field work did not make the employment of the types of electrical apparatus developed to that time feasible.[20] The lightening of electrical field recording equipment in weight, and the available facilities in Trinidad did make it possible, however, to obtain disk recordings of the music of the Negroes of this island during field work there, and some 325 melodies were added to the collection.

These songs have been gathered with the constant objective of throwing as much light as possible on the problems of the results of cultural contact that are the focus of our discussion. In every case, all effort has been made to assure a fair cross section of the musical resources of each people studied; to gather, that is, as much information as could be obtained regarding the range of variation in their song types. For African tribes, this meant recording secular as well as religious songs, social as well as cult melodies, lullabies as well as dance tunes. Among New World folk, it meant not only collecting as many different kinds of songs from the same people as possible, but including a sampling of their entire repertory, without regard to derivation. Thus in Dutch Guiana, European nursery rhymes set to music were recorded as well as *winti* songs; in Haiti a French marching song used in the *vodun* cult was as welcome as African melodies in praise of *Ogun* or *Aida Wedo*; in Trinidad "Sankeys"—Baptist hymns—as well as possession songs of the Shango cult, which were accompanied by a full complement of African drums. It was, indeed, because of this insistence on the comprehensive approach that the representative recordings of these "Sankeys" are at hand not only to throw light on the development of the spirituals, but to clarify the entire question of the derivation of swing.

Certain results of the work done thus far under this program of research may be indicated. In music, the principle operative in other aspects of culture makes it important to recognize that in this special case few New World Negro songs, whether in Guiana or Haiti or Brazil or the United States, are without some mark of European influence. In the music of the Suriname bush, or in some Haitian cult songs, pure African melodies and rhythms may be encountered, but these are exceptions. On the other hand, it is rare to find a Negro song which, though quite European in melodic line, is not tinged by some Africanlike modulation, or is not given a subtle turn by the manner of its singing. In Trinidad and Brazil and Cuba, Iberian and African rhythms have combined with particular felicity, perhaps because of an earlier influence of African Moorish melodies on the music of Spain and Portugal. In Guiana and Jamaica and the United States, other combinations are present; but it must be realized that they are combinations, all components of which must be weighted if we are to sense the developments that marked the syncretizing process. Certainly the conclusion that the African musical tradition has in no case been entirely submerged is of primary significance; that no matter how intense or how long was its contact with European melodies, it has in some measure persisted.

It has also become apparent that we can speak of "African" music in about the same degree as we can of "European" music. Just as there are certain underlying patterns of folk musical style that can be discerned in the analysis of Western European songs, so there are similar least common denominators in the music of West Africa and the Congo. Some of these latter have been mentioned, while those of Europe have received sufficient attention so that it is unnecessary to restate them here. In some instances these general patterns approach each other, which complicates the problem where certain similarities of this nature in the two traditions have coalesced and reinforced one another in New World Negro music. Thus the tendency to sing in thirds, which von Hornbostel assumed to be a European trait, is rather widespread in West Africa itself. The Ashanti of the Gold Coast, for example, rarely sing otherwise, and instances of melodies were recorded where a beginning was made anew when one member of a group of singers was out of key. Nor is this tendency to harmonize attributable to the relatively slight contacts these people have had with the English; to assume this is to fall into the same error committed when in the United States only European sources are taken into account in studying the origins of Negro songs.

Dr. Kolinski's analyses of the groups of related materials he has studied may be sketched to document the difficulties with which research into the problems of Negro music bristle. The songs of the Guianas, when first investigated, were found to vary from almost purely African to almost completely European; yet when the recordings from Africa were available, it became necessary to revise this simplified conclusion in the light of the variations found in the large amount of African data at hand. A number of traits mark off Ashanti music from that of Dahomey, and these characteristics, in turn, differ from certain elements in the musical styles of Nigeria and Togoland. The Guiana Negroes, who are derived from all these territories, combined and recombined their local African styles in various ways while at the same time retaining examples of each in relatively undisturbed form. It is understandable how, when spirituals from the United States are compared with West African songs, this complexity becomes materially greater. For here not only must the inner combinations of varied West African types of music be taken into account, but a more far-reaching influence of various European styles as well. Yet even when the only available transcriptions, those published for general use, are employed, many Africanisms are to be recognized. From the songs appearing in several such volumes of spirituals,[21] thirty-six were found to have the same scales (tonal structures) as specific songs in the West African collection, while identical correspondences in melodic line were even found in a few instances. Thirty-four spirituals had the same rhythmic structure as some of the West African melodies, while the formal structure of fifty spirituals—their phrasing and time—were found to have African counterparts.

Just how the songs of the African gradually took on more and more European characteristics, as the Negroes experienced ever more contact with whites—our fundamental problem in studying this aspect of culture—is far from solved. The objective nature of the data obtainable in this field, however, gives it special importance for any attempt to throw light on the general inquiry of how the Negroes adapted themselves to the white patterns they encountered as slaves and as free men in the New World. It is necessary that the work to be done in recording Negro music be coordinated and extended; that it blanket the entire geographical region and include songs of all kinds sung by New World Negroes and their African forebears wherever found; and that it be so prosecuted that the musical resources which have in the past stimulated folk singers and their more sophisticated fellows, the trained musicians, be made even more available to all those who are responsive to musical beauty, in whatever form met.

2

The comparative study of African and New World Negro dances presents far more difficulties than does the study of music. For not only are the available data on the dance found in scattered literary descriptions of various occasions on which persons, usually untrained in the study of the dance, witnessed ceremonies of one kind or another, but no method has as yet been evolved to permit objective study of the dance. What we are reduced to, therefore, are statements of opinion of those who have witnessed Negro dancing in the New World and have found certain qualities in it that they feel resemble the African background more or less closely. Approaches to the study of the dance comparable to those worked out for music by such musicologists as von Hornbostel or by such psychologists as Seashore and Metfessel are entirely lacking; other than a general recognition that motion pictures should be useful, almost no scientific approach has been devised.

This does not mean that a beginning, albeit a small one, has not been made in studying the primitive dance. The manner in which the masked dances of the Dogon of French West Africa has been presented is one example of such an attempt.[22] After first filming the dances themselves, outline drawings were made of the principal figures, taken off a series of single frames. The movements of the dancers are thus presented in their bare essentials, which makes it simpler than any other means yet devised to compare these figures with others similarly treated, or for those interested in dancing to reproduce the dance figures. The method is the more interesting because of the inclusion, at the back of the book, of a small phonograph record, which has the appropriate series of outline drawings of the dance printed on its face, and reproduces the drum rhythms employed. As far as Africanisms in New World Negro dances are concerned, this particular study is too isolated, and deals with a tribe sufficiently outside the area of intensive slaving operations, to hold any great importance except as concerns its methodological suggestions. Nonetheless, even in this case certain sketches of the *sim* dancer[23] are strikingly reminiscent of steps executed by Negro dancers in the United States, particularly in some of the more vigorous dances where "footwork" produces the desired effect.

An attempt to begin the comparative study of dancing among Negro folk of the New World was made during 1936 by Miss Katharine Dunham in applying her training and experience as a dancer to the comparative study of Negro dancing in Jamaica, Martinique, Trinidad, and Haiti.[24] Motion pictures of various dances were taken by her, to make possible comparisons between these and the motion pictures of dances obtained in Dahomey, the Gold Coast, Nigeria, Guiana and Haiti during the field work on which has been based much of the approach to the comparative study of Negro cultures and survivals of Africanisms in the New World discussed in these pages. To the present time, the most important result of Miss Dunham's field investigations has been in her own creative dancing. The popular successes achieved by her reproductions of the dances she studied add weight to the testimony of numerous dancers and laymen as to the familiarity of her dances to them in the light of their own experience with Negro dancing in this country, or of these dance patterns as diffused to the whites. Such reactions, despite their impressionistic nature, are not without significance in terms of the search for African survivals, pointing to the rich returns to be gained from systematic scientific analysis, on the basis of comparative studies, of the tenacity of African dance styles and the effect of acculturation on New World Negro dancing.

That in its setting the dance presents the same kind of change in terms of fewer African elements as one proceeds in the New World from those areas where Africanisms have been preserved in greatest intensity in other aspects of culture to other localities where Africanisms are found in most dilute form is apparent. In Guiana, for example, and in the Haitian countryside the African character of the dancing is at once apparent to an observer who has witnessed West African dances. As one approaches the United States through the West Indies, however, the introduction of European dance patterns becomes more and more evident, until in the United States, as well as among the more acculturated upper socio-economic groups in the islands generally, pure African dancing is almost entirely lacking, except in certain subtleties of motor behavior.

It is interesting to note, however, that European dances—except for so-called "social" dances—have been taken over most completely, not in the United States but in the West Indies. The reel and the quadrille, for example, are so important in Trinidad that the first has become the dance par excellence which accompanies African types of healing rituals, while the quadrille has become a favorite among the repertory to be witnessed at rites for the dead, taking equal place with the *bongo* and other African-type dances performed on such occasions. What is most European in the dancing of Negroes in the United States and elsewhere in the New World is when a man and a woman dance with their arms about each other. In Africa and among those West Indian Negroes who are less sophisticated in terms of acquaintance with white behavior, this is regarded as nothing short of immoral. This reaction, it may be remarked, is exactly similar to that of Europeans who witness for the first time the manipulation of the muscles of hips and buttocks that are marks of good African dancing, or the simulation of motions of sexual intercourse also found in certain quasi-ritual African dances. Yet these latter are no more and no less lascivious to the Negroes than are ordinary "social dances" to white persons, where a man and woman dance touching each other.

Recognizable African dances in their full context are probably entirely lacking in the United States, except perhaps for the special area constituted by Louisiana; and they seem to have been absent for generations. At the time of which Cable wrote, however, the *calinda*, the *vodun* dances, the *congo*, the *bamboula* were all to be witnessed in New Orleans. The careful descriptions of these dances given by this observer are a notable contribution to our knowledge of how they were performed, being especially useful in linking them with related dances found at the present time in the West Indies.[25] What remains of this dance tradition cannot be said definitely, but certain descriptions given by Hurston[26] and Saxon[27] indicate that some of the dances described by Cable and other earlier visitors to this scene still survive, despite their having been driven underground.

African types of dancing elsewhere, as in Africa itself, are found in connection with various religious and secular situations. In the churches, the forms of spirit possession that have been described in preceding pages are essentially African, especially in so far as these include dancing as well as those more random, less organized motor expressions of hysteria such as "jerks" or bounding up and down. In the Gulla Islands, the secular dances where men and women dance opposite each other without touching are quite African; but in the main the marks of aboriginal lineage in secular dance forms are essentially in the dancing style to be seen in the movements of Negro "jitterbug" enthusiasts.

3

More attention has been paid to folklore than to any other aspect of New World Negro life. Not only is this true in the sense of the word in which it is interpreted to mean folk customs, but also in the special sense of signifying the literary aspects of folk life. Folk tales, proverbs, riddles, jokes, and other forms of Negro literary expression have been collected since the Civil War. Moreover, collectors have not failed to record these elements in the Negro cultures of the West Indies and in West Africa as well as in the United States, so that a large quantity of materials exist for comparative study.[28]

Though some writers have stressed European and Indian influences in Negro tales, there is little question of the retention of Africanisms. Materials of this kind are particularly susceptible to objective analysis, because of the many independent components which render assumptions of correspondence almost indisputable. A good example of how this operates is to be seen in the case of what is perhaps the best-known Negro story, *The Tar Baby*. It will be recalled that in essence this tale tells how a trickster-thief is himself tricked by the device of erecting in a field a figure made out of tar or some other sticky substance, to discover who is stealing the produce. Coming in the dark, the trickster speaks to the figure, and when it fails to reply, rebukes it for the lack of good manners it shows (a significant Africanism!). After an ineffectual reprimand, the trickster strikes the figure with one hand, with his other, kicks it with one foot and then the other, and finally, in certain versions, butts the figure with his head, in which position he is held until eventually discovered.

The story is so characteristic of West Africa, that Africanists have themselves long used Joel Chandler Harris's version of this Negro tale from the United States as a point of comparative reference. There are some who maintain that the tale, as found both in this country and in Africa, originated in India; this is a matter of specialized and somewhat acrid controversy, which is so far from settled that it is still in the realm of conjecture and need not concern us here.[29] The fact that such a complex series of incidents should have been combined into this plot sequence, both among African and among New World Negroes, brings the inescapable conclusion that, whatever its place of absolute origin, the tale as found in the

New World represents a part of the cultural luggage brought by Africans to this hemisphere.

Difficulties of folklorists in search of provenience of New World Negro tales are not dissimilar to those already discussed where the underlying unity of Old World culture must be taken into consideration. As has already been stated,[30] especially strong unity is found in animal tales over the Old World, the important place of animal stories in the repertory of Negroes in all the New World thus being a reflection of the stimuli from Europe as well as Africa. The point is best made if we again briefly summarize the distribution of such tales. The Uncle Remus, or Anansi, stories found in the United States, or Jamaica, which parallel animal tales all over the African continent, also resemble so closely as to remove the similarities from the dictates of chance the fables of Aesop, the Reynard cycle of Europe, the Panchatantra of India, and the Jataka tales of China, to name but a few of the best-known series. Stories recorded in the Philippines, in Persia, and in Tibet, wherein animals are characters, exhibit the same series of incidents combined into plots wherein similar points are made. The characters show the greatest variation, as might be expected; but whether rabbit, tortoise, or spider figures as the trickster in the New World and African Negro tales, or jackal and crow figure in the stories of India and ancient Greece, the animals do similar things in similar sequence for similar reasons.

Stories having human characters show the same tendency toward wide distribution. The "Frau Holle" motif, that takes its name from the version in the German tales collected by the Brothers Grimm, offers an example of this. The story, found over all Europe and Asia—an almost perfect parallel to the German form has been recorded from Siberia—is likewise widely spread in Africa, though this has not been pointed out until recently when, through comparisons with Dutch Guiana Negro stories, African correspondences hitherto overlooked were revealed.[31] Another tale, best entitled by the catch phase "The Magic Flight," that was long thought to be restricted to Asia and Europe and, by diffusion, to the aboriginal Indian inhabitants of the Americas, turns out to be Old World, with a considerable distribution in West Africa and many correspondences among American Negroes.[32] Students of Negro lore in the United States, who tend to refer the "John Henry" cycle to recent events in the life of a definite Negro, may be freshly stimulated by considering the implications of the "Infant Terrible"[33] cycles of Africa for their study of derivations.

The published materials in the field of folklore are so rich, indeed, that documentation here is neither possible nor necessary. An idea of the extent to which retentions are found may be obtained if the comparative notes to the collection of Dutch Guiana folklore already mentioned be consulted for the references to the motifs in these tales that are also found in West Africa, on the one hand, and in the United States and the West Indies, on the other.[34] Some indication of the nature of the acculturative accommodation can, however, be illustrated by an example taken from the recent, informally reported collection of Florida tales published by Hurston,[35] that have never been subjected to comparative analysis and are from an area where no previous collections of any appreciable size have been made. The tale explains why there are whitecaps on the water during a storm:

De wind is a woman, and de water is a woman too. They useter talk together a whole heap. Mrs. Wind useter go set down by de ocean and talk and patch and crochet. They was jus' like all lady people. They loved to talk about their chillun, and brag on 'em. Mrs. Water useter say, "Look at *my* chillun! Ah got de biggest and de littlest in de world. All kinds of chillun. Every color in de world, and every shape!" De wind lady bragged louder than de water woman: "Oh, but Ah got mo' different chilluns than anybody in de world. They flies, they walks, they swims, they sings, they talks, they cries. They got all de colors from de sun. Lawd, my chillun sho is a pleasure. 'Taint nobody got no babies like mine." Mrs. Water got tired of hearin' 'bout Mrs. Wind's chillun so she got so she hated 'em. One day a whole passle of her chillun come to Mrs. Wind and says: "Mama, wese thirsty. Kin we go git us a cool drink of water?" She says, "Yeah, chillun. Run on over to Mrs. Water and hurry right back soon." When them chillun went to squinch they thirst Mrs. Water grabbed 'em all and drowned 'em. When her chillun didn't come home, de wind woman got worried. So she went on down to de water and ast for her babies. "Good evenin', Mis' Water, you see my chillun today?" De water woman tole her, "No-oo-oo." Mrs. Wind knew her chillun had come down to Mrs. Water's house, so she passed over de ocean callin' her chillun, and every time she call de white feathers would come up on top of de water. And dat's how come we got white caps on waves. It's de feathers comin' up when de wind woman calls her lost babies. When you see a storm on de water, it's de wind and de water fightin' over dem chillun.[36]

This story, in the essential similarity of its employment of personified natural forces, its organization of plot and its explanatory point, may be compared with a tale often heard among the Yoruba and the Fon-speaking folk of Dahomey and Togoland, which in essence is as follows:

In the early days stars shone during the day as well as at night. Those seen in the daytime were the children of the sun, and those seen at night were the children of the moon. One day, however, Moon spoke to Sun and proposed that, since the children were trying to outshine them, each put his children in a sack and throw them into the sea. Sun agreed to be first, but when the turn of Moon came she did not carry out her part of the bargain. This is why, when one looks at the sea in the day-time and sees colored fish, he is looking at the sun's children, no longer in the heavens. Sun is constantly seeking vengeance from Moon, and when they meet he swallows her; so people come out when there is an eclipse and beat the drums and shout to frighten the sun and make him disgorge the moon.[37]

The malice of Moon's trickery has been lost in the transmutation of this tale in the United States, yet the other correspondences between the two leave no doubt of their historic affiliation. Other stories concerning God and Devil, or human or animal characters, which have similar explanatory bent, likewise have many parallels in West Africa, notable examples of this being in the "Bible tales" from the Sea Islands, where the process of reinterpretation stands out in stark relief.[38]

That such counterparts as these are found for explanatory tales and myths, as well as for the better-known African animal tales, would seem to indicate that the body of African mythology and folk tales has been carried over in even less disturbed fashion than has hitherto been considered the case. The changes that have occurred understandably reflect the flora, fauna, and other elements in the everyday experience of the Negroes in their new habitat. The stories also are changed, in that the supernatural figures among the char-acters are no longer vested with the power and forms of gods, as they are in African mythologies. Yet in their humbler forms, they have persisted to testify, here as in other aspects of New World Negro life, to the vitality of the African cultural endowment brought by the Negroes to this side of the Atlantic.

4

Just as folk tales are made up of the quasi-independent constitu-ents of plot, incident and character, so language consists of separate variables termed "phonetics," "vocabulary," and "grammar." In this field, the approach to the study of African survivals in the New World is to be made along two lines—an attack on phonetic and semantic carry-overs, on the one hand, and on grammar and idiom, on the other. Most of the work concerning speech survivals has dealt with the first of these, effort being directed to discover those sounds of African or European origin to be discerned in Negro dialect and the related speech of white Southerners. Grammatical structure has been given almost no attention at all, since the approach to this aspect of the problem has been dominated by the almost axiomatic principle that "pidgin" dialects reflect the lack of ability of inferior folk to take over the more complex speech habits of "higher" cultures. That this assumption is psychologically as well as linguistically untenable will be demonstrated later; for the present, it may merely be pointed out that what any individual does in learning a new language is to mobilize his new vocabulary resources in accordance with the speech patterns to which he has been conditioned, as is apparent when the phrasing of English by Frenchmen or Germans is considered.

The most recent work on Negro speech in the United States is that of Dr. Lorenzo D. Turner, of Fisk University. This research has been confined mainly to an analysis of the dialectic peculiarities of the Sea Island Negroes. This is the most distinctive form of Negro diction in this country, while study of these speech conventions has the added attraction of building on earlier linguistic research in these islands. Dr. Turner's greatest advantage over others who have studied the same problem in the same area, however, is that he alone has a background of firsthand study of African tongues, which makes it possible for him to discern survivals that would be incomprehensible to those without such training. Since his materials have not been published, abstracts from his preliminary reports and communications will be cited *in extenso*. The results of his own work may first be outlined in terms of a statement furnished by him:

Up to the present time I have found in the vocabulary of the Negroes in coastal South Carolina and Georgia approximately four thousand West African words, besides many survivals in syntax, inflections, sounds, and intonation . . . I have recorded in Georgia a few songs the words of which are entirely African. In some songs both African and English words appear. This is true also of many folk-tales. There are many compound words one part of which is African and the other English. Sometimes whole African phrases appear in Gullah without change either of meaning or of pronunciation. Frequently African phrases have been translated into English. African given names are numerous.[39]

The preliminary list to which reference has been made[40] gives but a portion of the materials this scholar now has available, since it does not include findings of the two years that have elapsed since its preparation. Turner, at the outset, assesses certain handicaps in the study of African survivals in New World culture which the linguist, like the ethnologist, has to overcome. In the first place, the conventionalized assurance of many authorities that there are no African survivals among New World Negroes figures in this linguistic field no less than in the study of other aspects of culture. This point is documented by reference to his own materials, which are called on to refute the position taken by most students:

Ambrose E. Gonzales, who edited several volumes of Gullah stories and whose interpretations and reproductions of Gullah have been generally accepted as accurate, says that, "the African brought over or retained only a few words of his jungle tongue, and even these few are by no means authenticated as part of the original scant baggage of the Negro slaves. . . . As the small vocabulary of the jungle atrophied through disuse and was soon forgotten, the contribution to the language made by the Gullah Negroes is insignificant, except through the transformation wrought upon a large body of borrowed English words." (*The Black Border*, pp. 17 f.)

Then Gonzales published what was taken to be a complete glossary of Gullah. This contains about 1700 words, most of which are English words misspelled to indicate the Negro's mispronunciation. The other words in the glossary that are in reality African have been interpreted as English words which the Negro was unable to pronounce. For instance, the English phrase *done for fat* is given as being used by the Gullahs to mean *excessively fat* (the assumption being that in the judgment of the Gullah Negro when a person is very fat he is done for). But if Gonzales had had enough training in phonetics to reproduce the word accurately, it would have been dāfa, which is the Gullah word for fat, and if he had looked into a dictionary of the Vai language, spoken in Liberia, or consulted a Vai informant, he would have found that the Vai word for *fat* is dāfa (‾ _) lit., *mouth full*.[41]

Many other words in Gonzales's glossary which, because of his lack of acquaintance with the vocabulary of certain African languages, he interprets as English, are in reality African words. Among other Gullah words which he or other American writers have interpreted as English, but which are African, are the Mende suwaŋɔ (‾ ‾ _), *to be proud* (explained by Gonzales as a corruption of the English *swagger*); the Wolof lir, *small* (taken by Gonzales to be an abbreviated form of the English *little*, in spite of the fact that the Gullah also uses *little* when he wishes to); the Wolof bɛnj (banj, bɔnj), *tooth* (explained by the Americans as a corruption of *bone*); the Twi *fa*, *to take* (explained by the Americans as a corruption of the English *for*); the Wolof fut, *to be nude* (assumed by the English to be the English *foot*); the Wolof dʒogal, *to rise*—used in Gullah in the term *dʒogal board, rise-up board, seesaw* (explained by the Americans as *juggling board*); the Mende *loni* (_ _), *stands, is standing* (explained by the Americans as a

corruption of the English *alone*, said of a child who is beginning to walk—
Mende taloni (ꓤ _ _), *he is standing*; Gullah iloni, *he* is standing, in Mende
iloni (_ ‾ _) means *he is not standing*); etc.

Apparently influenced by Gonzales's interpretation of Gullah, the late
Professor Krapp of Columbia University, author of many publications on
the American language and considered an authority in this field, without
going to the trouble to acquaint himself either with Gullah or with any of
the African languages spoken in those sections of the West Coast from
which the Negroes were brought to the United States as slaves, writes in
this fashion regarding Gullah: "The Gullah dialect," he says, "is a very
much simplified form of English with cases, numbers, genders, tenses
reduced almost to the vanishing point. . . . Very little of the dialect, however, perhaps none of it, is derived from sources other than English. In
vocabulary, in syntax, and pronunciation, practically all of the forms of
Gullah can be explained on the basis of English, and probably only a little
deeper delving would be necessary to account for those characteristics
that still seem strange and mysterious." "Generalizations are always dangerous," he continues, "but it is reasonably safe to say that not a single
detail of Negro pronunciation or Negro syntax can be proved to have other
than an English origin." ("The English of the Negro," *American Mercury*,
June, 1924)

Mr. H. L. Mencken, in the 1937 edition of *The American Language*, says
that the Negroes have inherited no given-names from their African ancestors and that the native languages of the Negro slaves seem to have left
few marks upon the American language. (pp. 112, 523) On one Georgia
island alone, St. Simons, near Brunswick, I have collected more than 3000
African words that are used as given-names. Mr. Mencken very probably
never made any inquiries of the Gullahs concerning their given-names.
Dr. Reed Smith, of the University of South Carolina, says: "What the
Gullahs seem to have done was to take a sizeable part of the English vocabulary as spoken on the coast by the white inhabitants from about 1700 on,
wrap their tongues around it, and reproduce it with changes in tonality,
pronunciation, cadence, and grammar to suit their native phonetic tendencies, and their existing needs of expression and communication. The
result has been called by one writer, 'the worst English in the world.' It
would certainly seem to have a fair claim to that distinction." "There are
curiously," he continues, "few survivals of native African words in Gullah,
a fact that has struck most students of the language"; and he lists about
twenty words which he thinks may be African in origin, but he cites no
parallels in the African languages. (Gullah, pp. 22, 23)

Dr. Guy B. Johnson, contributing to one of the chapters in T. J. Woofter's
Black Yeomanry, is of practically the same opinion as Dr. Reed Smith.
He says: "There are older Negroes in the Sea Islands who speak in such
a way that a stranger would have to stay around them several weeks before
he could understand them and converse with them to his satisfaction.
But this strange dialect turns out to be little more than the peasant English

of two centuries ago, modified to suit the needs of the slaves. From Midland and Southern England came planters, artisans, shopkeepers, indentured servants, all of whom had more or less contact with the slaves, and the speech of these poorer white folk was so rustic that their more cultured countrymen had difficulty in understanding them. From this peasant speech and from the 'baby talk' used by masters in addressing them, the Negroes developed that dialect, sometimes known as Gullah, which remains the characteristic feature of the culture of the Negroes of coastal South Carolina and Georgia. . . . The grammar of the dialect is the simplified English grammar taken over from the speech of the poorer whites. . . . The use of many archaic English words no doubt contributed to the belief held in some quarters that the Sea Island Negroes use many African words." (Pp. 49, 51)

The reaction of Dr. Turner to a methodology which is content to study a problem of provenience without taking all possible sources into account is familiar to the reader of these pages. For the assurance with which those quite innocent of any knowledge of African speech habits tend to draw sweeping conclusions regarding the presence or absence of African words in this dialect, or in American Negro speech in general, parallels a similar tendency of the students treating other aspects of Negro culture. Dr. Turner's position in this matter is, however, reinforced by a further methodological consideration:

. . . the Gullah Negro when talking to strangers is likely to use speech that is for the most part English in vocabulary, but when he talks to his associates and to the members of his family, his speech is different. My first phonograph recordings of the speech of the Gullah Negroes contain fewer African words by far than those made when I was no longer a stranger to them. One has to live among them to know their speech well.[42]

The point is well taken. Linguists are not customarily trained in the techniques of the social sciences, and any white linguist must be prepared to surmount many barriers before he can attain the confidence of the proud, free folk of the Sea Islands.

That the cautions which enlightened considerations of scholarly method dictate have not been observed by students whose concern has been with tracing African survivals in the vocabulary and phonetics of Negro speech is thus apparent; that work based on closer acquaintance with African tongues as well as with various dialectal manifestations of English is needed before adequate analyses of the linguistic acculturation of the Negro are to be made is the only conclusion that can be drawn at this time. Pending this future work, however, it would seem that far more African elements are to be looked for at least in Gullah vocabulary and pronunciation than has hitherto been realized.

The assumptions underlying the approach to the study of syntax and idiom in New World Negro speech to be given below developed out of an intensive analysis of texts recorded in Dutch Guiana in 1929,[43] and may be recapitulated as follows: The Sudanic languages of West Africa, despite their mutual unintelligibility and apparent variety of form, are fundamentally similar in those traits which linguists employ in classifying dialects, as is to be discerned when the not inconsiderable number of published grammars of native languages, spoken throughout the area from which the slaves were taken, are compared.[44] This being the case, and since grammar and idiom are the last aspects of a new language to be learned, the Negroes who reached the New World acquired as much of the vocabulary of their masters as they initially needed or was later taught to them, pronounced these words as best they were able, but organized them into their aboriginal speech patterns. Thus arose the various forms of Negro-English, Negro-French, Negro-Spanish and Negro-Portuguese spoken in the New World, their "peculiarities" being due to the fact that they comprise European words cast into an African grammatical mold. But this emphatically does not imply that these dialects are without grammar, or that they represent an inability to master the foreign tongue, as is so often claimed.

If this hypothesis is true, certain results should follow when these modes of speech are analyzed. In the first place, Negro linguistic expression should everywhere manifest greater resemblances in structure and idiom than could be accounted for by chance. Deviations from the usage of the European languages, furthermore, should all take the same direction, though the amount of deviation from accepted usage must be expected to vary with the degree of acculturation experienced by a given group. Finally, not only should these deviations be in the same direction, but they should be in accord with the conventions that mark the underlying patterns of West African languages.

Though this analysis was made some years ago, and therefore does not include reference to some of the more recent works on African languages, nor the studies of Haitian Creole that have appeared since that time, it may be cited at length, since these fresh data merely confirm its findings. Since this analysis was made from the point of view of *taki-taki*, comparisons were made to modes of Negro-English speech found elsewhere, but not to dialects deriving their vocabulary from other European languages. In quoting this analysis, references in the original discussion to the sentences in the texts have been deleted, since they can be easily checked in the original. Similarly, transpositions into *taki-taki* of the phrases taken from other Negro dialects are also omitted, being unnecessary in this present context where they would perhaps be confusing. The phonetic symbols employed are those customarily used in linguistic studies.[45]

The discussion opens with those *taki-taki* idioms which do not appear in European languages, indicating their occurrence elsewhere in the New World. The citations, it is pointed out, include texts taken from Gullah as well as from various West Indian Islands where Negro-English is spoken.

We may name some of the characteristics that stand out as forms foreign to the idiom of European languages, but which occur with a consistency that characterises grammatical forms. Among these may be noted the absence of sex-gender in pronouns, and the failure to utilise any methods of indicating sex except by employing as prefix the word for "man" or "woman," or the use of relationship terms, like "father," "mother," "brother," "sister"; the manner of indicating the possessive; of expressing comparison; of employing nouns for prepositions of place. The use of a series of verbs to express a single action, or the use of verbs to indicate habitual and completed action also characterises this speech, as does the employment of the verb "to give" as a preposition, the use of "to say" to introduce objective clauses, making the only English translation possible the word "that," the use of "make" in the sense of "let," of "back" to mean "again," "behind," "in back," and "after." Repetition of words for emphasis is a regularly employed mechanism, and this form is also used to indicate a more intense degree of the action, or to change a verb into a noun, while the verb "to go" often carries the significance of "will."[46]

Stylistic traits that appear regularly are the opening of many sentences with the word "then," the change to the future tense to mark an explanatory interval between two actions which are separated from each other in time, and the use of the adverb *te* to express emphatic distance, or effort, or emotion, or degree. Phonetically, also, deviations from the pronunciation of European words are quite regular, as, for example, the interchange of "r" and "l"; the degree of nasalisation, about which we have already commented; or the insertion of a "y" after "c" in such words as "car" and "carry" and "can't"; or the tendency to end all words with a vowel, so that "call" becomes *kari* or *kali*, "look" becomes *luku*, "must" changes to *musu*; the use of elision and the dropping of final syllables.

It soon became apparent that the characteristics which could be singled out in the Negro-English of Paramaribo were also manifested in other regions of the New World where Negroes speak English. Our first com-

parison was made with the speech of Jamaica, and in the following list
we give some of the correspondences to Suriname speech we found.[47]

one great hungry time (p. 1)	wan bɩgi pina tɛm
take out the fishes, one one (p. 1)	puru na fɩsi wan-wan
mak I bu'n you (p. 4)	mek' mi brɔn yu
belly full (p. 4)	bɛre furu
I will carry you go (p. 5)	mi sa tyari yu go
eat done (p. 11)	nyam kaba
Tiger study fe him (p. 11)	Tigri prakseri fō hɛm
knockey han' (p. 15)	naki hanu
mak me wring de neck t'row 'way in de bag (p. 16)	mek' wi broko na nɛki trowɛ na ɩni na saka
hungry tak him (p. 22)	hɔngri tek' hɛm
so-so dog-head (p. 22)	soso dagu-hɛde
carry the cow come (p. 27)	tyari na kau kɔm
it spoil (p. 33)	a pɔri
he wanted to eat him one (p. 37)	a wani nyam hɛm wɑwɑn
tell him mus' tak out piece of meat gi' him (p. 38)	taki a mus' tek' wɑn pis' meti gi 'ɛm
but me have one cock a yard fe me wife (p. 29)	ma mi habi wɑn kaka na dyari fō mi weifi
when dem ketch a pass (p. 44)	te den kɩsi 'a pasi
see one little stone a river-side deh (p. 51)	si wɑ' pikin sɩtɔ a libasei dɛ
me nyam-nyam taya (p. 54)	mi nyɑm-nyɑm taia
run go (p. 55)	lɔ' go
roll in filth today-today (p. 56)	lolo na dɔti tide-tide
so after de eat an' drink done (p. 57)	so tɛ den nyɑm ɛn drɩngi kaba
at door-mout' (p. 75)	na dɔro mɔfo
an' went away to ground (p. 93)	ɛn gowe na grɔ
kyar' me go sell (p. 153)	tya' mi go sɛri
catch half-way (p. 169)	kɩsi 'af-pasi
night catch him on de way (p. 180)	neiti kɩsi hɛm na pasi

In addition we found correspondences in such pronunciations as "bwoy"
(p. 2), for "boy," of "kyan't" (p. 2), for "can't," "kyan-crow" for "carrion-
crow" (p. 80, Suriname yankoro) of "busha" for "overseer" (p. 80, Suriname
basha or bassia), while the words "nyam" for "eat," "Buckra" for Bakra,
"white person" (p. 22), "oonoo" for "you" (p. 40, Suriname un, or unu),
as well as the exclamation "Cho!" which is often heard in Suriname, were
further indications of linguistic similarity between the two regions.

However, these correspondences in speech were true not alone of the
idiom and pronunciation of Jamaica where resemblances could be explained
on definite historical grounds, for in our next comparison with the speech

recorded by Parsons of the Andros Islanders in the Bahamas,[48] we found the following correspondences:

> says to Boukee, says (p. 1)
> day clear (p. 3)
> dat sweet (p. 6)
> but b'o' Boukee was beeg eye (p. 9)
> Vwhen he reach in de half way (p. 19)
> an' he went, an' he meet no rabbit yet (p. 11)
> de han' fasten (p. 13)
> gal, you love me so till (p. 14)
> brer, loose me (p. 16)
> next day evening (p. 19)
> finish eat (p. 24)
> Two-Yeye (p. 28)
> I sick *bad* (p. 30)
> bathed his skin (p. 37)
> . . . an killed two thousand men dead one time (p. 38)
> eat her bellyful (p. 39)
> time he hear dat, he get up an' call Lizabet, say . . . (p. 44)
> they fry fowl egg, many cake, give him (p. 53)
> yer only goin' meet poppaone . . . (p. 60)
> torectly Rabby cry . . . (p. 85)
> . . . va you dere gwine? (p. 114)
> show you macasee (p. 141)

As in Jamaica, there were also correspondences to Suriname pronunciation. Many of these have been given above, but others are "kyarry" (p. 3, Suriname *tyari*) for "carry," "kyarridge" for "carriage" (p. 28), "ooman" (p. 115) for "woman," or "kyamp" for "camp" (p. 148).

Yet another comparison was had when we analyzed the language of the tales recorded by Parsons[49] in the Sea Islands. Some of the correspondences to Suriname Negro speech we found in this collection are as follows:

> Rabbit tell Fox, said (p. 9)
> an' dat make Brer Rabbit have short tail . . . (p. 18)
> . . . an' de tail come fo' white 'til to-day (p. 19)
> she was too happy now[50] (p. 24)
> tell de gyirl fo' love him (p. 25)
> . . . de han' fasten (p. 26)
> day clean (p. 28)
> man, don't you see all dis fresh meat[51] standin' in dis lot? (p. 32)
> Rabbit lie in de sun on his so' skin (p. 44)
> an' all her people died out an' leave her one[52] (p. 46)
> so he study . . . (p. 78)
> . . . your rice too much better (p. 104)
> . . . people tell, say . . . (p. 140)

Some of the phonetic correspondences are "yeddy" (p. 1, Suriname *yere*) for "hear," "kyart" for "cart," "kyarry" for "carry," "kyan't" for "can't" (p. 1, and "shum" for "see him (or them)" (p. 18). Similar phrases and phonetic shifts[53] are to be found in the speech of the islands as reported by Peterkin, Gonzales, Stoney and Shelby, and Johnson.

The common character of the idioms in Negro speech throughout the English-speaking New World thus demonstrated, the next step was to make comparisons between the pidgin dialects of West Africa, where natives have inherited the English of their forebears, who "picked up" a knowledge of the language in earlier times in much the same way as did New World Negroes:

As only few data on pidgin are available,[54] it was necessary to go into the field to obtain the requisite material for such an investigation, and a field-trip to West Africa made this possible. During a short stay in Nigeria a small collection of tales in pidgin was made,[55] and though these numbered but seven, the following significant phrases occurred in them:

> chop no de' (p. 448)
> my neck is pain me too much (p. 448)
> I be good man, true (p. 449)
> . . . all de white man, dey fit to make men by demself . . .
>   (p. 455)
> w'en Adjapa reach inside de bird . . . (p. 451)
> an' her mother took one give to her pikin (p. 456)
> he run come from inside de hole (p. 458)
> . . . took de man fo' de house . . . (p. 458)
> . . . if I salute you two more time . . . (p. 461)

In Africa, as in the New World, we found the phonetics of Negro speech producing such changes in English pronunciation as "cyap" for "cap," "dyah" for "jar," "hyar" for "hear."

The tales told us in Nigeria were given by informants who had some degree of schooling, and whose pidgin English was therefore modified by what teaching they had received. The extracts from historical tales of Dahomey which follow were told us, however, by an informant who had learned his English entirely "by ear." This man, a son of former King Behanzin, had left Dahomey and had lived in the coastal and interior regions of Nigeria for more than ten years, where, in the course of his everyday life, he had learned what English he knew.

'Dis princess, she palaver too much. If he marry dis man today, tomorrow he go way leave 'um. He suffer everybody. He vex he fadder too much, so he sell um go 'way. He no can kill he own daughter, so he sell go 'way. When he never see he daughter no mo', he sorry now. He say, "Who find daughter, I give dash plenty," say, "I give everyt'ing." Now dey bring him come. Now he start make lau again. He fadder say, "You be my proper blood," say, "I like you too much when you be quiet." But he make too much trouble. Sell 'em again to Portuguese. White man take him go. Dey de' fo' Whydah. Dey no go fo' sea yet. . . . Dis princess he was ploud. He was fine too much. He fine pass all woman. Dere was hole in Allada, nobody mus' go. Princess he steal he fadder sandal at night. Nex' day ol' woman see someone was in hole, come tell king. Everybody go for look, see king foot. King vex, say, he no go. Princess he laugh, say, "Who go? Look, you foot." . . . He (Hwegbadja) give dem order again say, if be somebody go put faiah to anode' man house fo' burn anode' man house, if sometime he no like 'em, he burn house, if he see, kill 'um, bring him head come, show, say, "Dat man burn house." I see, I kill 'um. Den if he tell dem so, den man have enemy, take man who do not'ing, cut head and bring, den if he fin' man lie, he go kill 'um de same. Den he say, if take small small gyal (girl) no be big 'nough, if somebody spoil 'um dey go kill 'um. Make nobody see people dey pass wit' load, go sell 'um. If somebody do so, he go find out, he kill 'um . . . Den de people who de' fo' odde' king country de' lon com' fo' Hwegbadja, say, "If my fadde' die, you go bury fo' me. To put fo' stick no good." . . . So people like it too much.'

Many of the idioms and phonetic shifts of Suriname speech, the West Indies, and the United States appear in these excerpts: "too much" for "very much," "sell go 'way" for "sell and send away," "bring him come" for "bring him (her)," "take him go" for "take away," "dey de' fo' Whydah" for "they are at Whydah," "dey no go fo' sea yet," literal translation of the Suriname *den no go fō si yete*, "ploud" for "proud," "he fine pass all women," the African comparative that finds its Suriname equivalent in *a mọi mọro ala uma*, "gyal" for "girl," "if somebody spoil 'em," the Suriname equivalent of *pɔri* in the significance of "deflower," "make nobody see people dey pass . . . ," *mek' nowạ si suma den pasa*, "lon com" for *lɔ kɔm*, and, finally, the use of the term "stick" to mean "tree," a usage which has its equivalent in the Saramacca use of the term *pạụ*, also "stick," for "tree."

In Dahomey, a possession of France, this was the only English we heard. French has little pidgin, yet occasionally, in contact with a native who had not been educated in the schools, we would hear *une fois*, the French equivalent of the Suriname *wạ trọ*, used exactly as the people of the Sea Islands employ "one time." We would hear a native telling another to go *doucement, doucement,—safri, safri*, as the Suriname Negro has it,—while phonetic shifts which cause the White man to eat "flied potatoes" in Nigeria and in Suriname, make him eat *pommes flites* in the French territory of Dahomey, or cause a native to point out a young woman walking along the road with the remark *"C'est mon flere, là. C'est femme, eh?"*[56]

Still pursuing the subject of correspondences between New World and West African Negro English, we collected more tales in pidgin among the Ashanti of the British territory of the Gold Coast,—among some of these very people to whom the Suriname Negroes, in their folk-lore, owe their

trickster-hero, Anansi. We give here some of the correspondences in phraseology which are to be found in these stories:[57]

> if Kwaku Anansi chop dat co'n he go die . . .
> hunger go kill me
> hungry kill him too much
> w'y you big man sabi war, you no wan' go war, sen' pikin go?
> go kill 'em one time
> in de mawnin' time . . .
> w'en you go, don' go small, small like t'ief
> he run-go and cut it
> sasabonsam fin' dat he tail no de
> you must call my sheep come
> W'en Kwaku Anansi he come de, he no sabi, say, "Tiger sleep for
>   de sheep place."
> He tell he husband, say, "I finish."
> den he fear too much
> den he sen' all him pikin one, one
> Den himse'f say, make he go see 'em
> dey laugh, laugh, laugh te . . . make small dey all two . . .
> dey run long te . . . he no catch Aduwa
> he go cover hi'self someplace[58]
> two weeks catch
> Some small, small man say, wan' go bush. W'en he go, he meeti
>   some big wate' in bush de.
> Den he sta't to heah talk fo everyt'ing in de worl'.

While with the Ashanti, we were also able to obtain some characteristic
expressions from a member of the Mossi people from the Northern Territories of the Gold Coast, whose pidgin was as untutored and as rich in flow
as any we heard in West Africa.

> de chief hask dem say . . .
> So you be chief pikin. Make you sing, make me see. W'en you be
>   chief pikin, me go know.
> he cover he sikin all[59]
> w'en dey get up fo' dance, now dance go' 'bout six ya'ds
> he run go bush wit' pikin
> dis firs' time he de' fo' town
> rabbits den chop all bush meat[60]
> so he cali a house again, say . . .
> rabbit he pass all sense for play trick

Still other examples are to be found in Cronise and Ward's Temne tales.
These are rendered in pidgin, and beside the idiomatic expressions and constructions cited by the authors in their "Introduction,"[61] the following may
also be found:

One ooman get girl-pickin (pickaninny), (p. 49)
He go inside one big forest whey all de beef duh pass (p. 41)
Spider take de hammer soffle (softly), he hit Lion one tem . . .
   (p. 43)
De ooman ax de man: "Nar true?" (p. 47)
Spider go nah puttah-puttah, he look sotay (until) . . . (p. 48)
"Na play I duh play" (p. 48)
One day me bin say Bowman long pass dis tick . . . (p. 48)
One net big rain fa' down (p. 55)
Dat make tay (until) today . . . (p. 63)
. . . en I mus' kare dis fiah go home (p. 64)
Dey all tow, dey duh sleep (p. 66)
. . . all run go (p. 70)
. . . 'tan' up nah de do'-(door) mout' (p. 70)
Make I tie um 'roun yo' mout', make I hole um, so w'en I duh
   shake, shake, make I no fa' down (p. 72)
I done bring Trorkey come (p. 75)
Dem beef all come, dey try, dey no able (p. 83)
. . . he no bin 'tan' lek today . . . (p. 93)
Hungri tem (famine) done ketch dis Africa (p. 117)
De two beef no' know say . . . (p. 120)
. . . he drag dem nah sho' . . . (p. 121)
. . . make we come go; ef no so, ef he meet yo', he go kill yo' (p. 185)
Spider he smart man fo' true, true (p. 213)
W'en 'fraid ketch Lepped . . . (p. 225)
. . . I go mi one (alone) (p. 234)
Make yo' kare mi nah yard (p. 247)
De grabe 'plit mo' (p. 272)
Story done (p. 278)
He see white clo'es, no mo' (p. 294)
Well, de debble pull one sing (p. 182)

Among the Ashanti of the Gold Coast, an interpreter of unusual
qualifications made it possible to push the investigation a step further, by assessing the literal meaning of African idiomatic expressions. As always, working from the Guiana dialect, lists of expressions which took especially striking form were analyzed:

The following list gives some of the resultant Twi idioms, with their
literal meaning expressed in English words:

Bring	*fá bra*[62] (take come)
Take (away)	*fá kɔ'* (take go)
Run away	*djuane kɔ'* (run go)
I am hungry	*ɔkɔm di mi* (hunger eat me) or *ɔkɔm okú mĩ* (hunger kill me)
Give birth to a child	*wa nyá abɔfrá* (he catch child)

Let us go	*ma yɛ uŋkɔ'* (make we go)
I traveled for a long time	*m(i) anan ti, anan ti, anan ti* (I walked, walked, walked)
I went to look for something, I did not find it	*mi kɔ' hwi hwê, m(i) ɑn hún* (I go look for find, I no see)
Early in the morning	*anɔpa tútú* (morning early)
All of you	*mố nyìna* (you all)
She is very nice	*no hón yɛ fɛ' dodo'* (he skin is nice much-much)
Do it at once	*yɛ no prêko* (do it time one)
That is why	*asɛm nútî* (case head)
He told me	*ɔkán tchiré mísê* (he tell show me say)
A thing done at a time	*nkoró (n)koro* (one, one)
Little by little	*kakra kàkra* (small, small)
Bigger	*esún sɪne no'* (big pass it)
Edge of the road	*kwán hɔ* (road-skin)
I am angry	*m(i) akúmâ' ehuru'* (my heart burns)
In the road	*mi wɔ kwanemu* (I am road inside)
He came to a stream	*ɔba túwô esúyó bí* (he-came met river some)
Add one to it	*fa kóró toso* (take one put top)
After this	*yɛi ɛchíri* (this back)
To calm a person	*djõdjõ n(a) akómâ máno* (cool he heart give him)
"Meat" (for animal)	*o kokúm nám* (he go kill meat)
Wild animal	*wirɛm nam* (forest meat, = bush meat)
He is very foolish	*ɔyɛ kwasi á dodo'* (he is fool too much)
He is very strong	*éyɛ dɛŋ dodo'* (it is strong too much)
The tale is very nice	*asɛm' no yɛ dɛ dodo'* (story it is sweet too much)
I am afraid	*súró kâ mè* (fear touched me) or *súró chíre mî* (fear catch me)
He walked a short way	*ɔnantî kakrá* (he walked small)
Do you understand English?	*wó tê brɔfɔ̀* (you hear English)
He brought it to me to see	*ɔfá brá mi hwé* (he took come me see)

The above list shows that many of the idioms peculiar to Paramaribo, Jamaica, Andros Islands, and the Sea Islands are literal translations of Twi. The presence of similar idiomatic expressions in Yoruba, Fɔ, Ewe and Hausa speech, and as reported by Cronise and Ward and others, leads to the further hypothesis that these idioms are basic to many, if not all, of the West African tongues.

The discussion of grammatical constructions of non-English character gave results equally enlightening:

Parsons[63] makes some cogent observations on prevalent grammatical forms, and offers as a possibility that these may derive from African usage. Available grammars of West African languages throw considerable light on these perplexing constructions, and, though it is not possible here to give a complete discussion, a few examples will make the point that in this, as in the instance of many of the idioms whose literal translation we have given, the peculiarities of Negro speech are primarily due to the fact that the Negroes have been using words from European languages to render literally the underlying morphological patterns of West African tongues.

Let us consider first the tendency of New World Negroes to use the verb "to give" for the English preposition "for." In Ewe[64] *na*, "to give" is used in just this manner, and we read that ". . . what one does to another is done for him and is, as it were, given to him, e.g., . . . *he said a word (and) gave (it) to the person*, i.e., he said a word to the person; *he bought a horse (and) gave (it) to me*, i.e., he bought me a horse."

In rendering Ashanti tales, it is explained that *ma*, which is translated by the preposition "for" is really the verb "to give."[65]

In Gą, *ha*, "to give," is used as we would use "for" in English, when employed with persons.[66] The Fante-Akan language utilizes the verb *ma*, "to give" as an equivalent of the English preposition "for";[67] while, turning to a Yoruba text we find a phrase which, literally translated, reads *"Ils prennent vont donnent au roi,"* and has the meaning of "They bring to the king."[68]

In the matter of gender, we find in grammars of West African languages the explanation of the seeming lack of differentiation of sex in the use of pronouns. We have noted how "he" and "she" are interchanged in West Africa and Suriname; how, in the West Indies and the Gullah Islands, "he" is employed to indicate both a man and a woman. Ewe, we find, "has no grammatical gender."[69] Do the Ewe, then, fail to distinguish persons who differ in sex? Not at all; they must, however, employ nouns, such as "man," "woman," "youth," "maiden," "father," "mother," or they must add either *-su*, "male," or *-nɔ*, "female," to a given word as a suffix. Yet this latter method is that of New World Negro English, as, for example, when the Suriname Negro speaks of a *man-pikin*,—a boy,—as against an *umą-pikin*, a girl. In Gą,[70] as in Ewe, gender is designated by the prefixing or suffixing of an element, in this case, *yo* for a woman and *nu* for a man, though there are a few differentiating words such as "husband," "wife," "father," "mother," and the like. Similarly, in the related Fante-Akan speech,[71] it is by affixing particles or utilizing different words, that the difference of sex is indicated. Of Yoruba we read that "The Yoruba language being non-inflective, genders cannot be distinguished by their terminal syllables, but by prefixing the words *ako*, male, and *abo*, female, to the common term; . . ."[72]

Perhaps no other element in *taki-taki* proved more difficult to translate than those expressions containing what Westermann terms "substantives of place." While *taki-taki* does not have all the connotations given for each of the words listed in his Ewe grammar,[73] all the words he cites in this connection have their *taki-taki* equivalents, and many of these equivalents have retained several of their meanings in Ewe. Thus, in *taki-taki*, as in Ewe, *na mąndri* (the Ewe *dome*), not only means "a place between," but

is also used with the meaning of "between," "among," "in the midst of."
Tapu, (Ewe *dzi*), means not only "top" but also "the sky," and "over,"
"on," and "above." *Inisɛi* in Suriname (Ewe *me*), as in Africa, carries the
significance not only of "inside" but also of "the context of a word of
speech." *Na baka* is difficult to translate into English until its equivalence
to the Ewe *megbe* is perceived, when it becomes clear that it not only signi-
fies "the back" but also "behind" and "after" and "again." A last example
(though this does not exhaust the list) shows the derivation of the numerous
curious uses of the taki-taki word *hedɛ*, "head." The Ewe equivalent, *ta*,
besides its initial significance, means "point" or "peak," "on account of,"
"because," "therefore," and "for that reason," the last being the exact
translation of the Suriname word in such a phrase as *fō dati ɛdɛ*. For Gą
we find similar constructions reported.[74] Thus, the Gą people say, "he looked
at his face" for "he looked in front of him"; "my garden is at the house's
back" for "my garden is behind the house"; "he went to their middle" for
"he went among them"; "walk my back" for "walk behind me." In Fante
the same construction is found.[75]

If one wishes to know the grammatical bases of such usage as the reflexive
pronoun, *den fɔm den s'rɛfi*; the order in which those in a compound subject
involving the speaker are named, *mi nąŋga yu;* the cohortive form, which
expresses an invitation, as *mɛk' wi go* for "let us go"; forms like *mi dɛ go*,
mi bɛn go; the use of a separate term (like the taki-taki *kaba*) to denote
completed action; the use of the phrase *a taki*, "to say," to introduce objective phrases; the use of the term "more" ("surpass")[76] to make the comparative form of the adjective, he will find all these discussed in grammars
of West African languages. Let us here only indicate, from Westermann,
some other rules of Ewe that, as for other West African tongues, still are
operative for *taki-taki*. When one says, "he is four years old,"[77] he says "he
has received four years"—the Suriname *a kısi fo yari kaba*; if one wishes
to say "I know something," he says "I have come to know something,"[78]
—taki-taki *mi dɛ kɔm sabi wą sani*. In Ewe, for "tell the Governor," one
says, "say it give Governor say," our *taki gi Gramą taki;*[79] the Ewe use
of the double verb occurs also in *taki-taki* as *krɔipi a krɔipi*.[80]

On the basis of this analysis, the following statement was drawn:

It may be well to restate the conclusions arrived at on the basis of comparing *taki-taki* with Negro English in the New World, pidgin English in
Africa, Ashanti idioms, and West African grammatical forms as illustrated
in Yoruba, Ewe, Fɔ, Gą, Twi, Mende, Hausa and other West African
languages.

1. Parallels to *taki-taki* were found in Jamaican speech, in the Bahamas,
and in the Sea Islands of the United States.

2. Similar parallels were also found in pidgin English as spoken in Nigeria
and on the Gold Coast, as well as in such specimens of Negro-French spoken
by natives with no schooling as were available.

3. Phonetic peculiarities which Negro speech exhibits in the New World

were met with in African pidgin, and it was possible to trace them to African speech.[81]

Therefore, it must be concluded that not only *taki-taki*, but the speech of the other regions of the New World we have cited, and the West African pidgin dialects, are all languages exhibiting, in varying degrees of intensity, similar African constructions and idioms, though employing vocabulary that is predominantly European.

Such matters as the fate in the New World of the tonal elements in West African speech, where, as has been indicated, tone has semantic as well as phonemic significance, remain to be studied. It is a most difficult problem requiring a long-term and highly technical analysis of Negro speech in various parts of the New World. That the peculiarly "musical" quality of Negro-English as spoken in the United States and the same trait found in the speech of white Southerners[82] represent a nonfunctioning survival of this characteristic of African languages is entirely possible, especially since this same "musical" quality is prominent in Negro-English and Negro-French everywhere. One Negro who was faced with the practical task of distinguishing the registers in the tonal system of a West African language has stated[83] that he was greatly aided in this task by reference to the cadences of Negro speech he knew from Harlem. When he was confronted with the need of mastering the especially difficult combinations of tones in Ifek, the registers of such a phrase as "Yeah, boss," (⌐—) greatly simplified his task. That such an experience may offer a methodological hint for future research on the survival of tone in the speech of New World Negroes, and especially those of the United States, is not out of the range of possibility.

The materials adduced above as regards vocabulary, phonetics, grammar, and idiom in Negro speech in this country are thus to be regarded as a mere beginning of a systematic research program. They are, however, more desirable and acceptable, if only from the point of view of method, than are the many arbitrary statements concerning Negro speech based on no knowledge of even the published materials on African languages. That it is necessary to mention this point again shows the state of methodological darkness in which the scholarship of Negro studies has groped its way. If only because of this deficiency, the burden of proof rests on those who claim the descent of Negro speech from Elizabethan English[84] or from Norman French.[85] In so far as the myth of the Negro past has been accepted in the study of this aspect of culture, the stamina of the African heritage goes unrealized here as elsewhere.

Chapter IX

CONCLUSIONS

The conclusions to be drawn from the discussion in the preceding chapters may be summarized on broad lines as follows:

1. The myth of the Negro past has been outlined and the unfortunate consequences for scholarship made apparent where scholars rely on assumption rather than fact. It has been seen how student after student has been content to repeat propositions concerning Negro endowment and the Negro past without critical analysis. Those who have taken the African background into account at all have failed in the methodological task of assessing the literature to ascertain whether earlier statements retain validity in the light of modern findings. Where concern has been to explain the divergence of Negro institutions from those of the white majority, it has been uncritically held that nothing of Africa could have remained as a functioning reality in the life of Negroes in this country. This historical blind spot has resulted in a geographical provincialism, so that students have never pressed into effective research such recognition as they have shown of the importance of comparative studies among Negroes living in other parts of the New World.

2. The acceptance of this mythology has been shown to be as serious for the practical man as for scholars. Its function as a justification for prejudice has operated to aggravate the situation of the Negro, providing deep-lying sanctions for surface irritations that have their roots in convictions regarding the quality of African culture. The existence of a popular belief in the African character of certain phases of Negro custom has been seen not to vitiate the conceptual reality of the mythological system, since it is the aspects of Negro life customarily deemed least desirable that are held to be African, and are thus regarded as vestiges of a "savage" past. That the existence of survivals has been denied rather than investigated shows that the implications of this point of view have not been missed by men of good will, and this fact but emphasizes the failure of scholars to face the question of Africanisms and apply to their study all the resources of their disciplines.

If the component parts of the system are taken one by one, the specific findings applicable to each may be reviewed in these terms:

1. Negroes are naturally of childlike character, and adjust easily to the most unsatisfactory social situations, which they accept readily and even happily, in contrast to the American Indians, who preferred extinction to slavery.

The sophistication of the Negroes in Africa and the New World as exemplified by the intricacy of aboriginal world view expressed in African religious beliefs, regard for reality shown by a refusal to interpret life in terms of a dichotomy between good and evil, and the pliability with which they adapt to everyday situations of all sorts, indicates the invalidity of any ascription of childlike qualities to the Negro. This means that such maladjustments to the American scene as characterize Negro life are to be ascribed largely to the social and economic handicaps these folk have suffered, rather than to any inability to cope with the realities of life. This also means that the customary interpretation of pliancy in terms of subservience ignores pre-American traditions which, because of their consistency in all the New World as well as in Africa itself, cannot be exclusively explained in terms of adjustment to slavery and post-slavery conditions.

That Negroes refused to accept slavery, and carried on unremitting protest both individually and in groups, has been amply proved by preliminary studies of Negro discontent. In its personal manifestations this ranged from malingering to suicide, while Negro slavery was also accompanied by revolts, so endemic that fear of slave uprisings was an outstanding phase of white thought of the period, giving evidence that the Negroes could implement resentment with action. The real reasons for the success of the Negro in adapting to his New World life, a point which is brought into high relief in the mythology by the stress laid on the corresponding failure of the Indian to adjust to slavery, are found in two causes. The Negro made a satisfactory slave because he came from a social order whose economy was sufficiently complex to permit him to meet the disciplinary demands of the plantation system without any great violation of earlier habit patterns, something not true of the Indian. In the second place, the Negro's powers of physical resistance were such that various diseases which killed off the Indians in large numbers subsequent to contact with the whites, such as measles, did not affect the Negroes.

2. Only the poorer stock of Africa was enslaved, the more intel-

ligent members of the African communities raided having been clever enough to elude the slavers' nets.

It has been shown that the history of slavery gives little evidence of any kind of selectivity in the capture of Negroes. The two most important methods of procuring slaves, kidnaping and capture in war, were clearly not such as to handicap those of lesser intelligence or to give those of higher ability any advantage in escaping the slavers. This is especially true because kidnapers were more likely to make off with young Negroes than others, while the fact that in West African warfare there was no category of noncombatant operated effectively against enslavement of selected elements in a particular population. African tradition of the slave trade, though heretofore given but slight attention, indicates rather that certain categories in the upper classes of African society, especially the priests and rulers, were in some instances particularly liable to be sold to the New World. That this tradition has validity is indicated in the first place by the fact that there was no lack of leaders in the New World who could organize successful revolts and successfully administer the communities subsequently established. It is likewise supported by the need to posit the presence of many priests and other specialists in manipulating the supernatural, if the further fact that recognizable Africanisms in the New World are more numerous in the field of religion than in any other aspect of culture is to be accounted for. There is, in fact, substantiating historical evidence in the firsthand accounts of New World slavery that these upper classes were represented among the slaves, where descriptions are given of the deference paid by some slaves to others who, for them, represented their rulers in Africa.

3. Since the Negroes were brought from all parts of the African continent, spoke diverse languages, represented greatly differing bodies of custom, and, as a matter of policy, were distributed in the New World so as to lose tribal identity, no least common denominator of understanding or behavior could have possibly been worked out by them.

No element in this system has been more completely accepted than the assumptions that the Negroes of this country were derived from the most diverse ethnic stocks and linguistic units over all of Africa; that, as it is phrased, the slaves were brought to the trading centers along the African coast after a "thousand-mile long" trek across the wastes of the continent. The facts have been seen to indicate that this is far from the truth. In the light of population distribution in Africa itself, with respect to the location of European slaving factories, as evidenced in the documents of the period, and as proved by the survivals of African personal names, place names, names of deities, and specific traits of culture where these survived in the New World, the region where slaving took its greatest toll was a relatively small part of Africa; while, of these slaves, the major portion was drawn from certain fairly restricted areas lying in the coastal belt of West Africa and the Congo.

When the proposition concerning the diversity of tongues and differences in customs among the tribes providing slaves is analyzed, a result similarly different from the accepted version is found. In classifying African languages, linguists are seen to designate the dialects of most peoples of the slaving area as Sudanese, which means that, whatever the differences in vocabulary that rendered these modes of speech mutually unintelligible, they had substantial elements of similarity in basic structure. The Bantu tongues, spoken in the Congo, are generally recognized as having a high degree of homogeneity; when these are compared with the Sudanese languages in contrast to Indo-European tongues, many resemblances between the two types appear.

As in language, so with culture in general. The civilizations of the forested coastal belt of West Africa and the Congo are to be regarded as forming one of the major cultural areas of the continent; which means that they resemble each other to a far greater degree than is recognized if local differences alone are taken into account. Again, in contrast to European custom, the resemblance of these coastal cultures to those of Senegal and the prairie belt lying north of the forested region of the west coast, or in the interior of the Congo, is appreciable. In many respects the entire area of slaving may thus be thought of as presenting a far greater degree of unity than is ordinarily conceived in the face of New World contact, in the United States as elsewhere, with the language and customs of the slave-owners.

A reexamination of the facts concerning separation in the New World of slaves coming from the same tribe, in the light of modern knowledge of African culture distributions, shows that this was no barrier to the retention of African customs in generalized form, or of their underlying sanctions and values. At most, it seems to have operated to blur the edges of distinctions sharply made in the homeland; that is, to stress and thus cause the retention of linguistic and cultural least common denominators such as are called on when classifying languages or grouping civilizations. An adequate basis for communication came into existence when the slaves learned words from the language of their masters and poured these into African speech molds, thus creating linguistic forms that in structure not only resemble the aboriginal tongues, but are also similar to one another no matter what the European vehicle—English or French or Spanish or Portuguese. The same device is seen to have occurred in culture; which would mean that in the light of findings under the ethno-historical method employed in this analysis, the reasons most often advanced to account for the suppression of Africanisms in the New World turn out to be factors that encouraged their retention.

4. Even granting enough Negroes of a given tribe had the opportunity to live together, and that they had the will and ability to continue their customary modes of behavior, the cultures of Africa were so savage and relatively so low in the scale of human civilization that the apparent superiority of European custom as observed in the behavior of their masters would have caused and actually did cause them to give up such aboriginal traditions as they may otherwise have desired to preserve.

This has been seen to be poor ethnology and poorer psychology. The evaluation of one culture in terms of another has been given over by modern ethnologists for many years, since it has become increasingly apparent that, lacking adequate criteria, customs can only be subjectively compared in terms of better or worse, higher or lower. This means that scholars, drawing comparisons of this nature, have merely reacted to their own conditioning, which has given them a predisposition to bring in verdicts which favor their own customs and to place differing cultures on levels that are deemed less advanced.

This recognized, it follows that many of the terms applied to African societies, such as "simple" or "naïve," are to be discarded. Examining the cultures of West Africa, Senegal, and the Congo it has been shown how they manifest a degree of complexity that on this ground alone places them high in the ranks of the nonliterate, nonmachine societies over the world, and makes them comparable in many respects to Europe of the Middle Ages. Some of the traits of these West African civilizations are: well-organized, intricate economic systems, which in many areas include the use of money to facilitate exchange; political systems which, though founded on the local group, were adequate to administer widespread kingdoms; a complex social organization, regularized through devices such as the sanctions of the ancestral cult in its kinship aspects, and including societies of all kinds, secret and nonsecret, performing functions of insurance, police, and other character; involved systems of religious belief and practice, which comprise philosophically conceived world views and sustained cult rituals; and a high development of the arts, whether in folk literature, the graphic and plastic forms, or music and the dance.

Coming, then, from relatively complex and sophisticated cultures, the Negroes, it has been seen, met the acculturative situation in its various manifestations over the New World far differently than is customarily envisaged. Instead of representing isolated cultures, their endowments, however different in detail, possessed least common denominators that permitted a consensus of experience to be drawn on in fashioning new, though still Africanlike, customs. The presence of members of native ruling houses and priests and diviners among the slaves made it possible for the cultural lifeblood to coagulate through reinterpretation instead of ebbing away into the pool of European culture. In some parts of the New World full-blown African civilizations resulted from successful revolts which permitted the establishment of independent or quasi-independent Negro communities. Elsewhere the process of acculturation resulted in varied degrees of reinterpretation of African custom in the light of the new situations.

The force of aboriginal sanctions has been seen to be such, however, that even where reinterpretation was most thoroughgoing, as in the United States, African points of view and African fundamental drives were not entirely lost. Slaves in different parts of the New World were exposed to European custom in differing degrees of intensity. In the same region, slaves assigned to various types of work had different kinds of contact with their masters. Yet even where acculturation along specialized lines was greatest, as in the case of mechanics and those trained in other crafts, acceptance of European modes of life by no means always followed; in Brazil, for example, this is seen to have resulted in the use of the unsupervised leisure of such specialists to preserve and further the retention of African traditions.

A factor of importance, consistently unrecognized in evaluating the acculturative processes operative among the Negroes, has been found to be the African traditional attitude toward what is new, what is foreign. Aboriginally manifested most strongly in the field of religion, where both conquered and conquerors often took over the gods of their opponents, it has operated to endow the African with a psychological resilience in facing new situations that has proved of good stead in his New World experience. To term an old deity by a new name is but one manifestation of a device which, in the field of social organization, has made for disregard of European sanctions underlying family structure while accepting European terminology relating to the family; for the adaptation of African patterns of mutual self-help in matters pertaining to death to outward Euro-American conventions of lodges and funerals; for the reworking of song and dance in accordance with the demands of the new setting. In instance after instance that has been cited from the literature bearing on the highly acculturated Negroes of the United States, it has been demonstrated how a proper assessing of these vestigial forms of African practice has led to the recognition of slightly modified African sanctions supporting forms of a given institution that are almost entirely European. This principle of disregard for outer form while retaining inner values, characteristic of Africans everywhere, is thus revealed as the most important single factor making for an understanding of the acculturative situation. That it reveals intellectual sophistication rather than naïveté negates the proposition in the mythology which holds that the force of superior European custom was so overwhelming that nothing of Africa could stand in the face of it.

5. *The Negro is thus a man without a past.*

The implications of this final culminating belief concerning the Negro have been seen to be of the greatest importance in shaping attitudes toward Negroes on the part of whites and attitudes of Negroes toward themselves. It has been indicated how, in the patterned values of this country, the past characteristically operates as a psychological support for the present; that it is held as explanation and justification of any cultural peculiarities a group may manifest. Where the ancestral endowment is a matter for pride, these special cultural traits are regarded with pride; where the past is "savage" and to be forgotten, specialized aspects of custom require apology where they cannot be concealed. To recognize that the past of the Negro in slavery and the physical differences that mark off this group from the American majority have aggravated attitudes toward the presumed savagery of their pre-American past, is merely to favor an explanation in terms of multiple causation rather than to employ the simplistic approach of conventionally minded students. To neglect any of these elements in the situation must distort perspective—which means that it is as necessary to realize the force of

African tradition making for the special cultural traits that mark off the Negro as it is to bear in mind the slave past or high pigmentation. To rephrase the matter, it is seen that the African past is no more to be thought of as having been thrown away by those of African descent than it is to assume that the traits that distinguish Italians or Germans or Old Americans or Jews or Irish or Mexicans or Swedes from the entire population of which they form a part can be understood in their present forms without a reference to a preceding cultural heritage.

In the evaluative processes of this country, then, the past counts more heavily than is realized, from which it follows that the extent to which the past of a people is regarded as praiseworthy, their own self-esteem will be high and the opinion of others will be favorable. The tendency to deny the Negro any such past as all other minority groups of this country own to is thus unfortunate, especially since the truth concerning the nature of Negro aboriginal endowment, and its tenaciousness in contact with other cultures, is not such as to make it suffer under comparison. The recognition by the majority of the population of certain values in Negro song and Negro dance has already heightened Negro self-pride and has affected white attitudes toward the Negro. For the Negro to be similarly proud of his entire past as manifested in his present customs should carry further these tendencies.

It has been seen that for the contribution to scientific knowledge to be gained from the study of Negro cultures, it is equally important to consider the operational significance of the Negro past. Only in this way can the laboratory that history has set up for the scientist be best utilized. Comparative studies, which recognize the historical affiliations of Negroes in West Africa and all the New World, must, especially for students in the United States, supplement the provincialism which refuses to look beyond the borders of a single country. The principle of differing degrees of acculturation to be discerned in the behavior of Negroes in various parts of the New World, and in various socio-economic strata of Negro populations everywhere, can thus be called on to implement the analysis of changes that have taken place during the historic adventure of the Negro people. Conventions which defy explanation except in terms of devious manipulation of logical possibilities become straightened into plain historical sequences when this principle is used, so that as concerns both scientific and practical considerations, it is possible to reason with greater cogency and act with greater assurance.

References

CHAPTER I

[1] "The Conflict and Fusion of Cultures with Special Reference to the Negro," *Journal of Negro History,* 4:116, 1919.

[2] *The Negro Family in the United States,* Chicago, 1939, pp. 21 f.

[3] "The Negro Family in the United States" (book review), *American Journal of Sociology,* 14:799, 1940.

[4] *Shadow of the Plantation,* Chicago, 1934, p. 3.

[5] *Brown America, the Story of a New Race,* New York, 1931, p. 11.

[6] *Ibid.,* pp. 10 f.

[7] *The Relation of the Alabama-Georgia Dialect to the Provincial Dialects of Great Britain,* Baton Rouge, 1935, p. 64.

[8] *Folk Culture on St. Helena Island, South Carolina,* Chapel Hill, N. C., 1930, p. 6.

[9] *The Etiquette of Race Relations in the South,* Chicago, 1937, *passim.*

[10] Thus, in a letter to an inquirer after African survivals in the behavior of Negroes of the United States (Mr. Joseph Ralph of Long Beach, Calif.), written early in 1925, the following statement was made:

As to the preservation of African cultural elements, I do not believe that such are to be observed in any of the modes of behavior of the American Negro.

In writing of the Negroes of Harlem, New York City, at about this time ("The Negro's Americanism," in Alain Locke (ed.), *The New Negro,* New York, 1925, pp. 359 f.), the same position was emphasized:

What there is today in Harlem distinct from the white culture that surrounds it is, as far as I am able to see, merely a remnant from the peasant days in the South. Of the African culture, not a trace. Even the spirituals are an expression of the emotion of the Negro playing through the typical religious patterns of white America. . . . As we turn to Harlem we see . . . it represents, as do all American communities which it resembles, a case of complete acculturation. And so, I return again to my reaction on first seeing this center of Negro activity, as the complete description of it: "Why, it's the same pattern, only a different shade!"

Two years later the identical point of view was stressed ("Acculturation and the American Negro," *Southwestern Political and Social Science Quarterly,* 8:216, 224, 1927):

Perhaps the best instance which may be given of this fashion in which one people may accept and validate for themselves the culture of another folk is contained in the Negroes of this country, particularly in the Negroes who have migrated to the northern cities and settled there in large communities . . . The African Negro may be of the same racial stock as some of his American brothers. But culturally, they are as widely separated as the Bostonian whose ancestry came to this country in the Mayflower, and the descendant of the King of Ashanti who lives today in West Africa.

In this latter paper, the relationship between physical type and ability to handle one culture as against another was primarily the subject under discussion, and there is no reason to assume that the conclusion reached in the argument, that such a relationship cannot be shown, is invalid. Yet the sentences quoted, when considered solely in the light of the principal concern of our discussion here, show that Negro behavior was believed to be "the same pattern, only a different shade" from that of the general white population in every aspect of activity.

[11] The methodological challenge this research presents, one which has by no means been adequately met, is in itself of real moment. For since no adequate attack on it is limited to any one discipline, or to any single geographic region, it demands a constant attention to techniques of utilizing cross-disciplinary resources as the analyses move ever more widely over the areas and the circumstances of Negro-white contact. By virtue of this fact, the problem in its largest aspects may be thought of as a significant lead toward achieving an integration in the sciences such as is becoming increasingly recognized as the next essential step in the development of knowledge.

[12] *American Negro Slavery,* New York, 1918, p. 46.

[13] R. Redfield, R. Linton, and M. J. Herskovits, "Memorandum for the Study of Acculturation," *American Anthropologist,* 38:149-152, 1936, *passim.* See also M. J. Herskovits, "The Significance of the Study of Acculturation for Anthropology," *American Anthropologist,* 39:250-264, 1937, *passim.*

[14] M. J. Herskovits, *Acculturation, the Study of Culture Contact,* New York, 1938. See also R. Linton, *Acculturation in Seven American Indian Tribes,* New York, 1940.

[15] *The Negro in the New World,* London, 1910.

[16] *Warning from the West Indies,* London, 1938 (rev. ed.).

[17] "The West Indies as a Sociological Laboratory," *American Journal of Sociology,* 29:290-291, 304, 1923-1924.

[18] "The Conflict and Fusion of Cultures with Special Reference to the Negro," *Journal of Negro History,* 4:115, 1919.

[19] "Magic, Mentality, and City Life," in: R. E. Park, *The City,* Chicago, 1925.

[20] *The American Race Problem, A Study of the Negro,* New York, 1938 (2nd ed.), pp. 15-16.

[21] Weatherly, *op. cit.,* p. 292.

[22] "The Conflict and Fusion of Cultures with Special Reference to the Negro," *Journal of Negro History,* 4:129, 1919.

[23] M. J. Herskovits, "The Negro in the New World: The Statement of a Problem," *American Anthropologist,* 32:145-156, 1930.

[24] M. J. and F. S. Herskovits, *Rebel Destiny, Among the Bush Negroes of Dutch Guiana,* New York, 1934, pp. viii-xii; *Suriname Folk-Lore,* New York, 1937, pp. 1-135.

[25] M. J. Herskovits, "The Negro in the New World . . . ," *American Anthropologist,* 32:149 f., 1930.

[26] Cf., for example, Arthur Ramos, *As Culturas Negras no Novo Mundo,* Rio de Janeiro, 1937; M. J. Herskovits, "The Social History of the Negro," in: C. Murchison, *Handbook of Social Psychology,* Worcester, Mass., 1935.

[27] E.g., the numerous works of Arthur Ramos, among which may be cited *O Negro Brasileiro,* Rio de Janeiro, 1934, *O Folk-Lore Negro do Brasil,* Rio de Janeiro, 1935, and *The Negro in Brazil,* trans. Richard Pattee, Washington, D. C., 1939; of Gilberto Freyre, especially his *Casa-Grande & Senzala,* Rio de Janeiro, 1st ed., 1934, 2nd ed., 1936, 3rd ed., 1938; of Edison Carneiro, *Religiões Negras,* Rio de Janeiro, 1936; of Jacques Raimundo, *O Elemento Afro-Negro na Lingua Portuguesa,* Rio de Janeiro, 1933, and *O Negro Brasileiro,* Rio de Janeiro, 1936; of João Dornas Filho, *A Escravidao no Brasil,* Rio de Janeiro, 1939; and the proceedings of the two Afro-Brazilian Congresses, *Estudos Afro-Brasileiros,* Rio

de Janeiro, 1935, *Novos Estudos Afro-Brasileiros,* Rio de Janeiro, 1937, and *O Negro no Brasil,* Rio de Janeiro, 1940; likewise Rüdiger Bilden, "Brazil, Laboratory of Civilization," *The Nation,* 128:71-74, 1929, and Donald Pierson, "The Negro in Bahia, Brazil," *American Sociological Review,* 4:524-533, 1939.

[28] Dr. Price-Mars, *Ainsi Parla l'Oncle,* Port-au-Prince, 1928; J. C. Dorsainvil, *Vodun et Névrose,* Port-au-Prince, 1931; M. J. Herskovits, *Life in a Haitian Valley,* New York, 1937; Harold Courlander, *Haiti Singing,* Chapel Hill, 1940.

[29] M. J. Herskovits, "African Gods and Catholic Saints in New World Negro Belief," *American Anthropologist,* 39:635-643, 1937.

[30] A. Ramos, *O Folk-Lore Negro do Brasil;* Fernando Ortiz, *Los Negros Brujos,* Madrid, 1917.

[31] Martha Beckwith, *Black Roadways, a Study in Jamaican Folk Life,* Chapel Hill, 1929.

[32] E.g., R. S. Rattray, *Ashanti,* Oxford, 1923; *Religion and Art in Ashanti,* Oxford, 1927; *Ashanti Law and Constitution,* Oxford, 1929; *Akan-Ashanti Folk Tales,* Oxford, 1930; H. Labouret, "Les Tribus du Rameau Lobi," *Tr. et Mèm. de l'Institut d'Ethnologie,* No. XV, Paris, 1931; M. J. Herskovits, *Dahomey,* New York, 1938; C. K. Meek, *A Sudanese Kingdom,* Oxford, 1931, and *Law and Authority in a Nigerian Tribe,* 1937; and the volumes of the journal *Africa.* Unpublished results of field work done in West Africa under fellowship grants of the Social Science Research Council by W. R. Bascom (among the Yoruba, 1937-1938), Joseph Greenberg (among the Hausa and Maguzawa, 1938-1939), and by J. S. Harris (among the Ibo, 1939-1940), are also of considerable importance in filling out our knowledge of the range of West African custom. The wealth of materials available on Gold Coast tribes alone is strikingly indicated by the number of titles listed in A. W. Cardinall, *A Bibliography of the Gold Coast,* Accra (Gold Coast), not dated, esp. Sections I-IX.

[33] Carried on by J. C. Trevor in 1936, under the auspices of Northwestern and Columbia universities, and A. A. Campbell, in 1939-1940, as Fellow of the Social Science Research Council.

[34] Carried on by M. J. and F. S. Herskovits in 1939, under a grant made by the Carnegie Corporation of New York.

[35] *Indigenous Races of the Earth,* Philadelphia, 1854; J. C. Nott, "The Diversity of the Human Race," *Du Bow's Review,* 10:113-132, 1851.

[36] "The Negro as a Contrast Conception," in: E. T. Thompson, *Race Relations and the Race Problem,* Durham, N. C., 1939, p. 174.

[37] *Ibid.,* p. 171. The reference is to B. T. Washington, *The Story of the Negro,* New York, 1909, Vol. 1, pp. 8 f.; the author adds, "A similar contrast was made by William McDougall in *Is America Safe for Democracy?* by juxtaposing a picture of Lincoln and an African savage."

[38] Pp. 5 f.

[39] *Ibid.,* p. 407. The citation is from J. M. Mecklin, *Democracy and Race Friction, a Study in Social Ethics,* New York, 1914, p. 43.

[40] Dowd, *op. cit.,* pp. 401 f. The first citation is from William H. Thomas, *The American Negro: What He Was, What He Is, and What He May Become,* New York, 1901, p. 134; the second is from H. W. Odum, *Social and Mental Traits of the Negro,* New York, 1910, p. 224.

[41] *Ibid.,* p. 39.

[42] *Ibid.,* pp. 48 f.

[43] *In Freedom's Birthplace, a Study of the Boston Negroes,* New York, 1914, pp. 399 f.

[44] *Caste and Class in a Southern Town,* New Haven, 1937, p. 370. Citation taken from J. E. Lind, "Phylogenetic Elements in the Psychoses of the Negro," *Psychoanalytic Review,* 4:303 f., 1917.

[45] *Race Traits and Tendencies of the American Negro,* New York, 1896, p. 312.

[46] J. A. Tillinghast, *The Negro in Africa and America,* New York, 1902, pp. 29 f.

[47] *American Negro Slavery,* pp. 4 and 8. The recommendation of Tillinghast's volume already cited as a "convenient sketch of the primitive African regime" and of Dowd's *The Negro Races,* New York, 1907, as "a fuller survey" causes one to speculate regarding the carry-over of historical method, in so far as the use of source materials is concerned, into a nonhistorical field. For both of these are what historians would call secondary or, better, tertiary sources!

[48] *Folk Beliefs of the Southern Negro,* Chapel Hill, 1926, pp. 8 ff., *passim.* References are to Tillinghast, to early travelers such as Bosman, and to later travelers such as Cruickshank or Miss Kingsley.

[49] Pp. 20 f.

[50] *Ibid.,* p. 24.

[51] *Ibid.,* p. 42.

[52] W. D. Weatherford and C. S. Johnson, *Race Relations,* New York, 1934, pp. 27 f.; the footnote reference appended to the passage is to Franz Boas, *The Mind of Primitive Man,* New York, 1910, Chap. I.

[53] See Chap. III.

[54] "Voodoo Worship and Child Sacrifice in Hayti," *Journal of American Folk-Lore,* 1:17 f., 1888.

[55] *After Freedom,* New York, 1939, p. xi.

[56] *The English Language in America,* New York, 1925 (2 vols.), and "The English of the Negro," *American Mercury,* 2:190 ff., 1924.

[57] *The English Language in America,* Vol. I, pp. 60 f. and 155. For an independent analysis of Krapp's position, see p. 278.

[58] "The English of the Negro," p. 190.

[59] M. J. Herskovits, "The Ancestry of the American Negro," *The American Scholar,* 8:93 f., 1938-1939.

[60] *Journal of Negro History,* 22:367, 1937.

[61] E. F. Frazier, "Traditions and Patterns of Negro Family Life in the United States," in: E. B. Reuter, *Race and Culture Contacts,* New York, 1934, p. 194.

CHAPTER II

[1] *The History, Civil and Commercial of the British Colonies in the West Indies* . . . London, 1801 (3rd ed.), Vol. II, pp. 126 f.

[2] *Travels in the Interior Districts of Africa Performed Under the Direction and Patronage of the African Association in the Years 1795, 1796, and 1797,* London, 1799 (2nd ed.).

[3] Cf. M. J. and F. S. Herskovits, "A Footnote to the History of Negro Slaving," *Opportunity,* 11:178 f., 1933.

[4] M. L. E. Moreau de St. Méry, *Description* . . . *de la partie française de l'Isle Saint-Domingue,* Philadelphia, 1797-98, Vol. I, pp. 237 f.

[5] Personal communication.

[6] David D. Wallace, *The Life of Henry Laurens* . . . , New York, 1915, pp. 76 f.

[7] Phillips, *American Negro Slavery,* p. 43. Just how Phillips reached his conclusion regarding the pygmoid character of these Negroes cannot be said, but his comment bespeaks slight knowledge of the geography and ethnic types of the region.

[8] Cf. Ramos, *O Folk-Lore Negro do Brasil, passim.*

[9] Personal communication.

[10] *La Traite et l'Esclavage des Congolais par les Européens,* Wettern, Belgium, 1929, pp. 88 f. The reference to Grandpré, a slave trader whose experience covered more than thirty years along the African coast is contained in a volume by this dealer entitled, *Voyage à la Côte Occidentale d'Afrique fait dans les années 1786 et 1787,* Paris, 1801, Vol. I, pp. 223 f. For further discussion of the sources of Congo slaves by Rinchon see his *Le Trafic Négrier, d'après les livres de commerce du capitaine gantois Pierre-Ignace-Lévin Van Alstein,* Brussels, 1938, pp. 89 ff.

[11] J. Maes and O. Boon, "Les Peuplades du Congo Belge, Nom et Situation Géographique," *Monographies Idéologiques,* Publications de Bureau de Documentation Ethnographique, Musée du Congo Belge, Tervueren, Belgium, 1935, Vol. I, sér. 2.

[12] M. J. Herskovits, "The Significance of West Africa for Negro Research," *Journal of Negro History,* 21:15-30, 1936.

[13] *Ibid.,* pp. 21-22.

[14] *American Negro Slavery,* p. 31.

[15] *Ibid.,* pp. 30 f.

[16] *Ibid.,* p. 44; the data were gathered from the file of the *Royal Gazette* of Kingston, Jamaica, for 1803.

[17] *Folk Beliefs of the Southern Negro,* p. 3. The number of slaves imported is derived from the *Negro Year Book* for 1918-1919, p. 151. As so often in discussing Africa, Puckett took his statement of locale from Tillinghast, *The Negro from Africa to America,* in this case pp. 7 ff.

[18] *Ibid.,* pp. 3 f. The extraordinary statement concerning the docile coastal tribes —the warlike Ashanti and Dahomeans, for example!—is taken from Tillinghast's excogitations, to be found on page 10 of his work.

[19] *The American Race Problem,* p. 133.

[20] "The Conflict and Fusion of Cultures . . . ," *Journal of Negro History*, p. 117.

[21] Anthony Benezet, *Some Historical Account of Guinea*, Philadelphia, 1771.

[22] *Race Relations*, p. 124. Where these authors obtained the spellings of tribal names they use cannot be said, but the errors are striking—Wydyas for Whydahs, Fulis for Fulas, etc.

[23] *The Negro in the New World*, pp. 82, 133, 275 f., 314.

[24] *Ibid.*, p. 470.

[25] *Nantes au XVIIIᵉ siècle; l'ère des Négriers (1714-1774), d'après des documents inédits*, Paris, 1931.

[26] E.g., Du Bois, *Black Folk, Then and Now*, p. 143.

[27] *Le Trafic Négrier . . .* , pp. 304 f., based on preceding tables.

[28] *Geschichte des Missionen der evangelischen Brüder auf den Inseln S. Thomas, S. Croix und S. Jan*, Barby, 1777, pp. 270 ff.

[29] Some of the relevant passages from Oldendorp are to be found in M. J. Herskovits, "On the Provenience of New World Negroes," *Social Forces*, 12:250 f., 1933.

[30] J. J. Hartsinck, *Beschryving van Guiana, of de Wilde Kust in Suid-America . . .* Amsterdam, 1770; Capt. J. G. Stedman, *Narrative of a five years' expedition against the Revolted Negroes of Suriname*, London, 1776.

[31] Moreau de St. Méry, *op. cit.*; F. X. Charlevoix, *Historie de l'Isle Espagnole ou de S. Domingue*, Paris, 1730-1731; Père J. B. Labat, *Nouveau Voyage aux Isles d'Amérique*, The Hague, 1724.

[32] Lewis, *Journal of a West India Proprietor*, London, 1834; Edwards, *op. cit.*

[33] Wm. Bosman, *A New and Accurate Description of the Coast of Guinea . . .* (English trans.), London, 1721 (2nd ed.) ; Capt. Wm. Snelgrave, *A New Account of Some Parts of Guinea, and the Slave-Trade . . .* , London, 1734.

[34] *Captain Canot; or Twenty Years of an African Slaver*, New York, 1854.

[35] "The Slave Trade in South Carolina Before the Revolution," *American Historical Review*, 33:809-828, 1928; *Documents Illustrative of the Slave Trade to America*, Carnegie Institution Publication No. 409, Vols. I-IV, 1930-1935.

[36] *American Negro Slavery*, p. 113.

[37] *Documents Illustrative of the Slave Trade to America*, Vol. III, pp. 462 ff.

[38] *Ibid.*, p. 318.

[39] *Ibid.*, pp. 43, 45.

[40] M. J. Herskovits, "The Significance of West Africa for Negro Research," *loc. cit.*; computations from Donnan, *op. cit.*, Vol. IV, pp. 175 ff., *passim*.

[41] J. D. Wheeler, *A Practical Treatise on the Law of Slavery . . .* , New York, 1837; see also Jeffrey R. Brackett, *The Negro in Maryland . . .* , Baltimore, 1889.

[42] *Op. cit.*, Vol. IV, pp. 278 ff., *passim*. The points of origin in this table are equated as closely as possible with those in the preceding one; most notable is the fact that only 1,168 slaves were brought in ships sailing from "Benin," "Bonny," "New Calabar," and "Old Calabar."

[43] *Le Trafic Négrier . . .* , pp. 247 ff.

[44] Personal communication.

[45] These figures are to be found in M. J. Herskovits, "The Significance of West Africa for Negro Research," *loc. cit.*, p. 27.

[46] Stephen Fuller, *Two Reports . . . on the Slave-Trade*, London, 1798, pp. 20 ff.

[47] Cf. also L. E. Bouet-Willaumez, *Commerce et Traite des Noirs aux Côtes Occidentales d'Afrique*, Paris, 1848; particularly Part II and maps.

[48] Rinchon, *Le Trafic Négrier . . .* , pp. 274 ff.; the author's sources are indicated on pp. 243 ff. of his book.

[49] Herskovits, "The Significance of West Africa for Negro Research," *loc. cit.*, pp. 27 f.

CHAPTER III

::

[1] *The Negro in Africa and America.*

[2] *The Negro Races.*

[3] *The Negro from Africa to America.*

[4] *The American Race Problem,* p. 199.

[5] See p. 302, n. 32.

[6] *The Tshi-Speaking Peoples of the Gold Coast,* London, 1887; *The Ewe-Speaking Peoples of the Slave Coast of West Africa,* London, 1890; *The Yoruba-Speaking Peoples of the Slave Coast of West Africa,* London, 1894.

[7] Bosman, *A New and Accurate Description of the Coast of Guinea . . .* ; John Barbot, "A Description of the Coast of North and South Guinea . . . ," *Churchill's Voyages,* Vol. VI, London, 1732; Robert Norris, *Memoirs of the Reign of Bossa Ahadee . . . ,* London, 1789; Abbé Proyart, *Histoire de Loango, Kakonga et autres Royaumes d'Afrique . . . ,* Paris, 1776; Snelgrave, *A New Account of Some Parts of Guinea, and the Slave Trade. . . ,* London, 1734.

[8] T. E. Bowdich, *Mission from Cape Coast Castle to Ashantee,* London, 1819; R. J. Burton, "A Mission to Gelele, King of Dahome . . . ," *Memorial Edition of Burton's Works,* Vols. III and IV, London, 1893.

[9] Mary A. Kingsley, *Travels in West Africa,* London, 1897; *West African Studies,* London, 1899; Robert H. Nassau, *Fetichism in West Africa; Forty Years' Observation of Native Customs and Superstitions,* New York, 1904.

[10] *The Negro in Africa and America,* p. 28.

[11] Collected during field work in Eastern Nigeria, 1938-1939.

[12] "Land and Labour in a Cross River Village, Southern Nigeria," *Geographical Journal,* 90:24-51, 1937.

[13] *Op. cit.,* p. 29.

[14] *Ibid.,* p. 31.

[15] *Ibid.,* p. 33.

[16] *Ibid.,* pp. 31 f.

[17] *Ibid.,* p. 72.

[18] *Ibid.,* p. 80.

[19] *Ibid.,* p. 86.

[20] *Democracy and Race Friction,* pp. 82 f. A footnote reference after the first sentence of the quotation is to an article by Reinsch, "The Negro Race and European Civilization," *American Journal of Sociology,* 11:155, 1905. Reinsch's paper, one of the most extreme examples of the position being considered here, is not cited because, except for Mecklin, references to it are practically never encountered.

[21] *The American Race Problem,* New York, 1927 (1st ed.), pp. 197 ff.

[22] *Ibid.* (2nd ed.), pp. 310 f.

[23] " 'Secret Societies,' Religious Cult-Groups, and Kinship Units among the West African Yoruba," Unpublished Doctor's Thesis, Northwestern University, 1939.

[24] *Ashanti,* pp. 90 f.

[25] *Nights with Uncle Remus,* Boston, 1911; *Uncle Remus Returns,* Boston, 1918; *Uncle Remus, His Songs and Sayings,* New York, 1929.

[26] By M. J. Herskovits, principally in Dahomey and among the Ashanti.

[27] See pp. 269 ff.

[28] *Religion and Art in Ashanti, passim.*

[29] Cf. numerous articles in *Africa*, London; *Journal de la Soc. des Africanistes*, Paris; *Anthropos*, Mödlingbei-Wien; *American Anthropologist*, Menasha, Wis.; *Journal of the Royal Anthropological Institute*, London; *Congo*, Brussels; *Zeitschrift für Ethnologie*, Berlin; and other reviews.

[30] "Nupe State and Community," *Africa*, 8:257-303, 1935; "Witchcraft and Anti-Witchcraft in Nupe Society," *Africa*, 8:423-447, 1935.

[31] "Ritual Festivals and Social Cohesion in the Hinterland of the Gold Coast," *American Anthropologist*, 38:590-604; *Marriage Law Among the Tallensi*, Accra (Gold Coast), 1937; "Communal Fishing and Fishing Magic in the Northern Territories of the Gold Coast," *Jour. Royal Anth. Inst.*, 67:131-142, 1937; "Social and Psychological Aspects of Education in Taleland," Supplement to *Africa*, xi, No. 4, London, 1938; M. and S. L. Fortes, "Food in the Domestic Economy of the Tallensi," *Africa*, 9:237-276, 1936.

[32] Deborah Lifszyc and Denise Paulme, "Les Animaux dans le Folklore Dogon," *Rev. de Folklore Français et de Folklore Colonial*, 6:282-292, 1936; "La Fête des Semailles en 1935 chez les Dogon de Sanga," *Jour. de la Soc. des Africanistes*, 6:95-110, 1936; Michel Leiris and André Schaeffner, "Les rites de circoncision chez les Dogon de Sanga," *Jour. de la Soc. des Africanistes*, 6:141-162, 1936; Marcel Griaule, "Blason totémiques des Dogon," *Jour. de la Soc. des Africanistes*, 7:69-78, 1937; Denise Paulme, "La Divination par les chaculs chez les Dogon de Sanga," *Jour. de la Soc. des Africanistes*, 7:1-14, 1937; Deborah Lifszyc, "Les formules propitiatoires chez les Dogon," *Jour. de la Soc. des Africanistes*, 7:33-56, 1937.

[33] C. D. Forde, "Land and Labour in a Cross River Village"; "Fission and Accretion in the Patrilineal Clans of a Semi-Bantu Community in Southern Nigeria," *Jour. Royal Anth. Inst.*, 68:311-338, 1938; "Government in Umor," *Africa*, 12:129-162, 1939; J. S. Harris, "The Position of Women in a Nigerian Society," *Trans. New York Academy of Science*, Ser. II, 2:141-148, 1940.

[34] N. de Cleene, "Les Chefs Indigènes au Mayombe," *Africa*, 8:63-75, 1935; "La Famille dans l'Organization Social du Mayombe," *Africa*, 10:1-15, 1937; C. Estermann, "La Tribu Kwangama en Face de la Civilisation Européenne," *Africa*, 7:431-443, 1934; "Les Forgerous Kwangama," *Bull. de la Soc. Neuchâteloise de Géographie*, 44:109-116, 1936; "Coutumes des Mbali du Sud d'Angola," *Africa*, 12:74-76, 1939.

[35] "Les Tribus du rameau Lobi, Volta Noire moyenne," *Tr. et Mèm. de l'Inst. d'Ethnologie*, Vol. XV, Paris, 1931; Louis Tauxier, *Le Noir du Soudan; Pays Mossi et Gourounsi*, Paris, 1912; *Le Noir du Yatenga*, Paris, 1917; Charles Monteil, *Les Khassonké, Monographie d'une peuplade du Soudan français*, Paris, 1915; *Les Bambaras de Segon et du Kaarta*, Paris, 1924; Maurice Delafosse, *Haut Senegal-Niger* (Soudan français), 3 vols., Paris, 1912; Louis Desplagnes, *Le Plateau Central Nigérien*, Paris, 1907; Marcel Griaule, "Jeux Dogons," *Tr. et Mèm. de l'Inst. d'Ethnologie*, Vol. XXXII, Paris, 1938; "Masques Dogons," *Tr. et Mèm. de l'Inst. d'Ethnologie*, Vol. XXXIII, Paris, 1938.

[36] N. W. Thomas, *Anthropological Report on Sierra Leone*, London, 1916; D. Westermann, *Die Kpelle, ein Negerstamm in Liberia*, Göttingen, 1921.

[37] S. W. Koelle, *African Native Literature . . . in the Kanuri or Bornu Language . . .*, London, 1854.

[38] *Nègres Gouro et Gagou (centre de la Côte d'Ivoire)*, Paris, 1924; *Religion, Mœurs et Coutumes des Agnis de la Côte d'Ivoire*, Paris, 1932.

[39] *The Northern Tribes of Nigeria*, London, 1925 (2 vols.); *A Sudanese Kingdom*, London, 1931; *Tribal Studies in Northern Nigeria*, London, 1931 (2 vols.); "The Kulu in Northern Nigeria," *Africa*, 7:257-269, 1934; *Law and Authority in a Nigerian Tribe*, Oxford, 1937.

[40] *In the Shadow of the Bush*, London, 1916; *Life in Southern Nigeria*, London,

1923; *The Peoples of Southern Nigeria,* London, 1926 (4 vols.) ; *Some Nigerian Fertility Cults,* London, 1927.

[41] *Anthropological report on the Edo-speaking peoples in Nigeria,* London, 1910; *Anthropological report on the Ibo-speaking peoples of Nigeria,* London, 1913-1914 (6 vols.).

[42] *Urwald-documente. Vier jahre unter den crossflussnegern Kameruns,* Berlin, 1908.

[43] *Die Pangwe,* Berlin, 1913 (2 vols.) ; *Die Baja, ein Negerstamm in Mittleren Sudan,* Stuttgart, 1934.

[44] "Notes on the ethnography of the BaMbala," *Jour. Royal Anth. Inst.,* 35: 398-426, 1905; "Notes ethnographiques sur les peuples communément appelés Bakuba, ainsi que sur les peuplades apparentées. Les Bushongo," *Annales de la Musée du Congo Belge. Ethnographie, Anthropologie,* sér. 3, t. 2, Brussels, 1910; "Notes ethnographiques sur les populations habitant les bassins du Kasai et du Congo belge," *Annales de la Musée du Congo Belge. Ethnographie, Anthropologie,* sér. 3, t. 2, Brussels, 1910.

[45] "The Ovimbundu of Angola," *Field Museum of Natural History, Anth. Ser.,* 21 :90-362, Chicago, 1934.

[46] *Collection de Monographies ethnographiques,* Brussels, 1907-1911, Vols. I-VIII.

[47] *Among Congo Cannibals,* Philadelphia, 1913; *Among the Primitive BaKongo,* London, 1914.

[48] Ad. Cureau, *Les Sociétés Primitives de l'Afrique Équatoriale,* Paris, 1921; E. Verhulpen, *Baluba et Balubaisés du Katanga,* Antwerp (not dated).

[49] Alice Werner, *Structure and Relationship of African Languages,* London, 1930.

[50] *Ibid.,* pp. 13 ff.

[51] *Ibid.,* pp. 32 f.

[52] *Ibid.,* pp. 47 f.

[53] See pp. 275 ff.

[54] M. J. Herskovits, "The Culture Areas of Africa," *Africa,* 3 :59-77, 1930.

CHAPTER IV

▪▪

[1] Redfield, Linton and Herskovits, "Memorandum for the Study of Acculturation," *American Anthropologist, loc. cit.,* p. 152, IV, C.

[2] *An Abstract of the Evidence delivered before a Select Committee of the House of Commons in the Years 1790, and 1791 . . .* London, 1791, pp. 38 ff.

[3] *An Account of the Slave Trade on the Coast of Africa,* London, 1788, p. 30.

[4] *A New Account of some parts of Guinea, and the Slave-Trade . . . ,* pp. 162 ff.

[5] *American Negro Slavery,* p. 35.

[6] "American Slave Insurrections before 1861," *Journal of Negro History,* 22: 303 ff., 1937.

[7] *Ibid.,* pp. 302 f. The citation to the quotation and the captain's statement are Donnan, *Documents Illustrative of the Slave Trade to America,* Vol. III, pp. 293, 325; that to the lawsuit is Catterall, *Judicial Cases Concerning American Slavery and the Negro,* Vol. I, pp. 19 f., where a full description of the revolt on which action was based is to be read. Other instances of revolt insurance are cited by Wish as these are found in the same works, Donnan, Vol. III, p. 217, and Catterall, Vol. III, p. 568.

[8] Cf. Ramos, *The Negro in Brazil,* pp. 24 f., for a discussion of this same point as concerns Brazil.

[9] Herskovits, *Life in a Haitian Valley,* pp. 59 f.; references to data cited will be found in the notes to this passage.

[10] This account is abstracted from Ramos, *The Negro in Brazil,* pp. 42 ff., and C. E. Chapman, "Palmares; The Negro Numantia," *Journal of Negro History,* 3:29-32, 1918; see also Ramos, *ibid.,* pp. 24 ff., and Johnston, *The Negro in the New World,* pp. 95 f., for a long series of later Brazilian slave revolts.

[11] Cf. Stedman, *Narrative of a five years' expedition against the Revolted Negroes of Suriname, passim*; M. J. and F. S. Herskovits, *Rebel Destiny, passim.* This most recent incident has not been published, as far as is known.

[12] L. A. Pendleton, "Our New Possessions—the Danish West Indies," *Journal of Negro History,* 2:267-288, 1917. The data concerning the revolt are from C. E. Taylor, *Leaflets from the Danish West Indies,* London, 1888.

[13] W. Westergaard, "Account of the Negro Rebellion on St. Croix, Danish West Indies, 1759," *Journal of Negro History,* 11:50-61, 1926.

[14] Pendleton, *op. cit.,* pp. 277 ff.

[15] See p. 90; see also Phillips, *American Negro Slavery,* p. 464.

[16] Herskovits, *Life in a Haitian Valley,* pp. 60 ff.; citations to sources are appended to the quotations.

[17] The only published data on the Black Caribs are in a paper by Eduard Conzemius, "Ethnographical Notes on the Black Carib (Garif)," *American Anthropologist,* 30:183-205, 1928.

[18] Cf. Johnston, *The Negro in the New World,* p. 314.

[19] *Ibid.,* pp. 217 ff.; Phillips, *American Negro Slavery,* pp. 464 f., dates the first revolt at 1675, and gives slightly differing versions of subsequent events from those of Johnston.

[20] The best source for the Maroon uprising and deportation is Bryan Edwards, *The History, Civil and Commercial, of the British Colonies in the West Indies,* Vol. I, Appendix No. 2, pp. 522 ff.

[21] Cf. Phillips, *American Negro Slavery,* p. 466.

[22] *The Homes of the New World,* New York, 1868, Vol. II, p. 346.

[23] *Ibid.,* pp. 331 f. The Luccomees, as far as can be discovered, are the counterpart of the people termed Yoruba or Nago by the French and British writers.

[24] *The Rise of American Civilization,* New York, 1930 (1-vol. ed.).

[25] Fred A. Shannon, *Economic History of the United States,* New York, 1934, p. 324.

[26] *The Rise of the Common Man, 1830-1850,* New York, 1935, pp. 282 f.

[27] Curtis P. Nettels, *The Roots of American Civilization,* New York, 1938, p. 468; Guion G. Johnson, *Ante-Bellum North Carolina,* Chapel Hill, 1937, pp. 510 ff.

[28] Wish, "American Slave Insurrections before 1861," pp. 306 ff.; Aptheker, "American Negro Slave Revolts," *Science and Society,* 1:512-538, 1937, and *Negro Slave Revolts in the United States,* New York, 1939.

[29] *Negro Slave Revolts in the United States,* pp. 16 f.

[30] *Ibid.,* pp. 71 f.

[31] "The Slave Insurrection Panic of 1856," *Journal of Southern History,* 5:206, 1939.

[32] *Ibid.,* p. 222.

[33] This revolt has been the inspiration of a powerful novel, almost alone in its exploitation of this type of situation—Arna Bontemps, *Black Thunder,* New York, 1936.

[34] J. C. Ballagh, *A History of Slavery in Virginia,* Baltimore, 1902, p. 89.

[35] See Herbert Aptheker, "Maroons within the Present Limits of the United States," *Journal of Negro History,* 24:167-184, 1939; and Joshua R. Giddings, *The Exiles of Florida,* Columbus, Ohio, 1858.

[36] B. Schrieke, *Alien Americans, a Study of Race Relations,* New York, 1936, pp. 123 ff.

[37] Frederick L. Olmsted, *A Journey in the Back Country,* New York, 1863, p. 228.

[38] *Ibid.,* pp. 65 f.

[39] Olmsted, *A Journey in the Seaboard Slave States,* New York, 1856, pp. 481 f.

[40] *Ibid.,* pp. 480 f.

[41] *Ibid.,* p. 100.

[42] *Ibid.,* p. 91.

[43] *Ibid.,* p. 388.

[44] *Economic History of Virginia in the Seventeenth Century,* New York, 1907, Vol. II, p. 108.

[45] *Slavery in the United States,* New York, 1837, pp. 69 f.

[46] W. S. Drewry, *Slave Insurrections in Virginia, 1830–1865,* Washington, 1900, p. 27.

[47] *The Plantation Overseer, as Revealed in his Letters,* Northampton (Mass.), 1925, pp. 20 f.

[48] *The Negro in Maryland, a Study of the Institution of Slavery,* Baltimore, 1889, pp. 132 f.

[49] *A Journey in the Back Country,* p. 476.

[50] William Still, *The Underground Railroad,* Philadelphia, 1872.

[51] *Ibid.,* p. 57.

[52] *Ibid.,* pp. 58 f.

[53] Harvey Wish, "Slave Disloyalty under the Confederacy," *Journal of Negro History,* 23:435-450, 1938; Herbert Aptheker, *The Negro in the Civil War,* New York, 1938.

[54] Cf. Elizabeth Hyde Botume, *First Days amongst the Contrabands,* Boston, 1893, for a vivid picture of the reaction of the Negroes in this situation.

[55] *The Homes of the New World,* Vol. II, p. 338.

[56] *Description . . . de la partie française de l'Isle Saint Domingue,* Vol. I, pp. 29 f.

[57] Cf. M. J. and F. S. Herskovits, "A Footnote to the History of Negro Slaving," *loc. cit.,* and M. J. Herskovits, "The Social History of the Negro," *loc. cit.,* pp. 239 ff.

[58] *Nouveau Voyage aux Isles d'Amérique,* Vol. II, p. 39.

[59] *A New Account of some Parts of Guinea, and the Slave-Trade . . . ,* pp. 158 f.

[60] *An Account of the Slave Trade on the Coast of Africa,* p. 18.

[61] M. J. and F. S. Herskovits, "A Footnote to the History of Negro Slaving," *loc. cit.,* p. 178.

CHAPTER V

[1] *The Negro Family in the United States,* pp. 5 ff.

[2] See pp. 11-12.

[3] Guion G. Johnson, *A Social History of the Sea Islands, with Special Reference to St. Helena Island, South Carolina,* Chapel Hill, 1930, p. 31.

[4] Ramos, *The Negro in Brazil,* pp. 17 ff.

[5] *The Plantation Overseer, as Revealed in his Letters,* p. 3.

[6] *The Southern Plantation, A Study in the Development and the Accuracy of a Tradition,* New York, 1925, p. 148.

[7] Brackett, *The Negro in Maryland . . . ,* pp. 38 f.

[8] Johnson, *A Social History of the Sea Islands,* p. 127.

[9] *Ibid.,* p. 131.

[10] Johnson, *Ante-Bellum North Carolina,* p. 526.

[11] *Ibid.,* p. 469.

[12] *American Negro Slavery,* p. 75.

[13] *Ibid.,* pp. 83 f.

[14] *Ibid.,* pp. 232 f.

[15] *Ibid.,* p. 84.

[16] *Ibid.,* pp. 95 f.

[17] *A Second Visit to the United States of North America,* New York, 1849, Vol. I, pp. 268 f.

[18] C. S. Johnson, *Shadow of the Plantation,* p. 8.

[19] "Plantations with Slave Labor and Free," *American Historical Review,* 30: 743 f., 1924-1925.

[20] Herskovits, *Life in a Haitian Valley,* pp. 39 f.; translated from Pierre de Vassière, *Saint-Domingue (1629-1789), la société et la vie créole sous l'ancien régime,* Paris, 1909, pp. 280 f.

[21] *Folk Beliefs of the Southern Negro,* pp. 10 f.

[22] M. J. Herskovits, *The American Negro, A Study in Racial Crossing,* New York, 1928, and "Social Selection and the Formation of Human Types," *Human Biology,* 1:250-262, 1929.

[23] Herskovits, *op. cit.;* see also Dollard, *Caste and Class in a Southern Town,* p. 70.

[24] Olmsted's commentary is germane here: "In the French, Dutch, Danish, German, Spanish, and Portuguese colonies, the white fathers of colored children have always been accustomed to educate and emancipate them and endow them with property. In Virginia, and the English colonies generally, the white fathers of mulatto children have always been accustomed to use them in a way that most completely destroys the oft complacently-asserted claim, that the Anglo-Saxon race is possessed of deeper natural affection than the more demonstrative sort of mankind." *A Journey in the Seaboard Slave States,* New York, 1856, p. 232. For data indicating the relative numbers of mulattoes among the free Negroes of pre-Civil War times see E. F. Frazier, *The Free Negro Family, a Study of Family Origins before the Civil War,* Nashville, 1932, pp. 12 f., and "Traditions and Patterns of

Negro Family Life in the United States," in: E. B. Reuter, *Race and Culture Contacts*, New York, 1934, pp. 204 ff.

[25] *American Negro Slavery*, p. 75.

[26] *Ibid.*, p. 291.

[27] *A Second Visit to the United States of North America*, Vol. I, p. 263.

[28] *Memorials of a Southern Planter*, Baltimore, 1887, p. 192.

[29] Frederick Douglass, *My Bondage and My Freedom*, New York and Auburn, 1855, p. 59.

[30] Johnson, *Ante-Bellum North Carolina*, p. 83.

[31] Bremer, *The Homes of the New World*, Vol. II, p. 449.

[32] R. Bickell, *The West Indies as They Are; or a Real Picture of Slavery . . . in the Island of Jamaica*, London, 1825, pp. 54 f.

[33] *Op. cit.*, Vol. II, p. 20.

[34] *Ibid.*, pp. 24 f.

[35] Bruce, *Economic History of Virginia in the Seventeenth Century*, Vol. II, pp. 105 f.

[36] Olmsted, *A Journey in the Seaboard Slave States*, pp. 47 f.

[37] *Ibid.*, p. 17.

[38] Johnson, *A Social History of the Sea Islands . . .* , pp. 77 f.

[39] The effect of such procedures in the way of obtaining a foothold for African cooking traditions in the South has, incidentally, been consistently overlooked; yet it is not unlikely that the slaves exerted an appreciable influence in shaping the cuisine regarded at present as characterizing various regions of the South.

[40] Johnson, *A Social History of the Sea Islands . . .* , p. 130.

[41] Cf. F. G. Speck, "The Negroes and the Creek Nation," *Southern Workman*, 37:106-110, 1908, and K. W. Porter, "Relations between Negroes and Indians within the Present Limits of the United States," *Journal of Negro History*, 17: 287-367, 1932.

[42] Reuter, *The Mulatto in the United States*, Boston, 1918, pp. 378 f.

[43] E. F. Frazier, "The Negro Slave Family," *Journal of Negro History*, 15:215, 1930. The quotation is from R. E. Park, "The Conflict and Fusion of Cultures . . ." *loc. cit.*, p. 119.

[44] Frazier, *op. cit.*, p. 258.

[45] Puckett, *Folk Beliefs of the Southern Negro*, p. 10.

[46] *Ibid.*, p. 284; see also the explanation given by this author on p. 167 for the retention of beliefs in magic, or on p. 31 for folk tales.

CHAPTER VI

[1] Cf. D. Young, *American Minority Peoples*, New York, 1932, and "Research Memorandum on Minority Peoples in the Depression," *Bull. 31*, Soc. Sci. Research Council, New York, 1937, for the setting of the Negro in the larger minority group situation in this country.

[2] W. R. Bascom, "Acculturation among the Gullah Negroes," *Amer. Anth.*, 43: 43-50, 1941.

[3] Herskovits, *Dahomey*, Vol. I, pp. 32 f., Plate 3, and *Life in a Haitian Valley*, p. 254, plate opposite p. 100.

[4] Botume, *First Days amongst the Contrabands*, p. 53.

[5] Caroline Couper Lovell, *The Golden Isles of Georgia*, Boston, 1932, pp. 187 f.

[6] The former collected by W. R. Bascom; the latter by M. J. Herskovits.

[7] *Folk Beliefs of the Southern Negro*, p. 27.

[8] *Folk-Lore of the Sea Islands, South Carolina*, Cambridge, 1923, p. 204.

[9] Mary A. Owen, *Old Rabbit the Voodoo and Other Sorcerers*, London, 1893, pp. 10 f.

[10] Doyle, *The Etiquette of Race Relations in the South*, p. 76, quoting William Ferguson, *America by River and Rail*, London, 1856, p. 149.

[11] M. J. and F. S. Herskovits, *Suriname Folk-Lore*, pp. 4 ff.

[12] *First Days amongst the Contrabands*, p. 59.

[13] *The Etiquette of Race Relations in the South*, p. 76.

[14] *My Bondage and My Freedom*, pp. 69 f.

[15] *Folk Beliefs of the Southern Negro*, p. 393.

[16] *Ibid.*, p. 23.

[17] *Ibid.*, p. 394.

[18] For illustrations of this and other instances of how elaborate the rules of etiquette can be in a Negro tribe, see M. J. and F. S. Herskovits, *Rebel Destiny*, various passages indicated under "Etiquette" in the index.

[19] *Caste and Class in a Southern Town*, p. 6.

[20] *The Etiquette of Race Relations in the South*, p. 161.

[21] *Ibid.*, pp. 79 f. The first illustration is given as from Olmsted, *The Cotton Kingdom: a Traveler's Observations on Cotton and Slavery in the American Southern States*, New York, 1861, Vol. II, pp. 1 f.; the second from Douglass, *My Bondage and My Freedom*, pp. 252 f.

[22] *Folk Beliefs of the Southern Negro*, p. 50.

[23] Charles C. Jones, *The Religious Instruction of the Negroes*, Savannah, 1842, pp. 130 f.

[24] M. J. Herskovits, "Adjiboto, an African Game of the Bush-Negroes of Dutch Guiana," *Man*, 29:122-127, 1929, and "Wari in the New World," *Jour. of Royal Anth. Inst.*, 62:23-37, 1932.

[25] Rattray, *Ashanti Law and Constitution*, pp. 107 ff., and Herskovits, *Dahomey*, Vol. I, pp. 106 ff., Vol. II, pp. 72 ff.

[26] Powdermaker, *After Freedom*, p. 126.

[27] "African Institutions in America," *Journal of American Folk-Lore*, 18:15-32, 1905; see also Bernard C. Steiner, "History of Slavery in Connecticut," *Johns Hopkins Univ. Stud. in Hist. and Pol. Sci.*, 11th Ser., September-October, 1893.

[28] Aimes, *ibid.*, p. 16; Steiner, *ibid.*, p. 78.

[29] Steiner, *op. cit.*, pp. 78 f.; see also Aimes, *op. cit.*, p. 16.

[30] *Ibid.*, p. 19.

[31] Paul Lewinson, *Race, Class, and Party; a History of Negro Suffrage and White Politics in the South*, New York, 1932, *passim*.

[32] Cf. among others Forde, "Land and Labour in a Cross River Village"; René Maunier, "La Construction Collective de la Maison en Kabylie," *Tr. et Mèm., Inst. d'Eth.*, Vol. III, Paris, 1926; Herskovits, *Dahomey*, Vol. I, pp. 75 f.

[33] W. R. Bascom, "Acculturation among the Gullah Negroes," *Amer. Anth.*, 43: 44-46, 1941.

[34] H. W. Odum, *Social and Mental Traits of the American Negro*, New York, 1910, pp. 98 f.

[35] *Ibid.*, pp. 104 f.

[36] *Ibid.*, p. 249.

[37] *After Freedom*, p. 122.

[38] *Shadow of the Plantation*, p. 183, n. 4.

[39] W. E. B. Du Bois (Ed.), "Economic Co-operation among Negro Americans," *Atlanta University Publications*, No. 12, Atlanta, 1907, p. 92.

[40] *Ibid.*, p. 96.

[41] "The Beginnings of Insurance Enterprise among Negroes," *Journal of Negro History*, 22:417-432, 1937.

[42] *Ibid.*, p. 417. Cf. the comments of Cornelius King, an official of the Farm Credit Administration, on Negro cooperation, in an article entitled, "Cooperation— Nothing New," *Opportunity*, 18:328, 1940.

[43] Cf. Gist, Noel, "Secret Societies: A Cultural Study of Fraternalism in the United States," *Univ. Missouri Studies*, 15:1-184, 1940.

[44] Herskovits, *Life in a Haitian Valley*, pp. 107 ff., 258 ff.

[45] *Shadow of the Plantation*, p. 49.

[46] *The Negro Family in the United States*, pp. 109 f., 343 ff., 620 ff.

[47] *Negro Illegitimacy in New York City*, New York, 1926, *passim*.

[48] As indicated, for example, by Parsons, *Folk-Lore of the Sea Islands, South Carolina*, p. 206.

[49] *Shadow of the Plantation*, pp. 66 f.

[50] M. J. Herskovits, "A Note on 'Woman Marriage' in Dahomey," *Africa*, 10: 335-341, 1937.

[51] *After Freedom*, p. 149.

[52] *Ibid.*, pp. 156 ff.

[53] Frazier, *The Negro Family in the United States*, pp. 126 f.

[54] *Ibid.*, p. 326.

[55] *Ibid.*, pp. 461 f.

[56] *Shadow of the Plantation*, pp. 29, 32 f., 39 f.

[57] *Ibid.*, p. 37.

[58] *After Freedom*, pp. 146 f.

[59] *The Negro Family in the United States*, p. 153.

[60] *Ibid.*, p. 158.

[61] *After Freedom*, p. 147.

[62] *Folk Beliefs of the Southern Negro*, p. 23.

[63] Tradition and Patterns of Negro Family Life in the United States," *loc. cit.*. p. 198.

[64] *Shadow of the Plantation*, p. 29.

[65] *The Negro Family in the United States*, pp. 57 f.

[66] *Ibid.*, p. 55.

[67] "The Negro Slave Family," *loc. cit.*, p. 234.

[68] *Ibid.*

[69] *Shadow of the Plantation*, pp. 48 f.

[70] Powdermaker, *After Freedom*, p. 146.

[71] *Ibid.*, p. 127.

[72] M. J. Herskovits, V. K. Cameron, and Harriet Smith, "The Physical Form

of Mississippi Negroes," *American Journal of Physical Anthropology,* 16:193-201, 1931.

[73] Frazier, *The Negro Family in the United States,* pp. 258 f.

[74] Herskovits, *Dahomey,* Vol. I, pp. 139 ff.

[75] Frazier, *op. cit.,* pp. 257 f.

[76] *First Days amongst the Contrabands,* p. 48.

[77] *Folk Beliefs of the Southern Negro,* p. 559; the reference made is to Fanny D. Bergen, "Animal and Plant Lore," *Mem. Amer. Folk-Lore Society,* Vol. VII, 1899, p. 84.

[78] Frazier, *op. cit.,* p. 259.

[79] Johnson, *Shadow of the Plantation,* pp. 57 f.

[80] *Ibid.,* pp. 64 f.

[81] *Ibid.,* p. 71.

[82] Powdermaker, *After Freedom,* pp. 201 ff.

[83] The most complete collection of data of this sort is in V. K. Cameron, "Folk Beliefs Pertaining to Health of the Southern Negro," unpublished Master's Thesis, Northwestern University, 1930, pp. 18 ff.

[84] The fullest materials on these points are to be found in Parsons, *"Folk-Lore of the Sea Islands, South Carolina," passim;* and Puckett, *Folk Beliefs of the Southern Negro,* pp. 332 ff.

[85] Cf. Herskovits, *Dahomey,* Vol. I, pp. 262 f., 270 ff., for instances of this.

[86] "Record of Negro Folk-Lore," *Journal of American Folk-Lore,* 19:76 f., 1906.

[87] *Loc. cit.,* as from M. N. Work, "Some Geechee Folk-Lore," *Southern Workman,* 35:633-635, 1905.

[88] *Folk Beliefs of the Southern Negro,* p. 100.

[89] *Op. cit.,* p. 197.

[90] "Braziel Robinson Possessed of Two Spirits," *Journal of American Folk-Lore,* 13:226-228, 1900.

[91] *Op. cit.,* pp. 335 f.

[92] *Folk-Lore of the Sea Islands, South Carolina,* p. 198.

[93] *Op. cit.,* pp. 334 f.

[94] Rattray, *Religion and Art in Ashanti,* pp. 51 ff.; and Herskovits, *Dahomey,* Vol. I, pp. 259 ff. For lists of Dahomean names of this sort, see Herskovits, especially pp. 263 ff.

[95] Turner's materials are not as yet available in published form; for a preliminary report on Puckett's elaborate project in the study of Negro names and their derivation see his paper "Names of American Negro Slaves," in: G. P. Murdock, *Studies in the Science of Society Presented to Albert Galloway Keller,* New Haven, 1937, pp. 471-494.

[96] *Ibid.,* pp. 474 f.; the references are to J. C. Cobb, *Mississippi Scenes,* Philadelphia, 1815, p. 173, and to Carter G. Woodson, *Free Negro Heads of Families in the United States* in 1830, Washington, 1925.

[97] Puckett, *op. cit.,* p. 475.

[98] Personal communication.

[99] Given in his mimeographed report.

[100] *First Days amongst the Contrabands,* pp. 48 f.

[101] *Folk Beliefs of the Southern Negro,* pp. 340 ff.

[102] *Ibid.,* p. 340.

[103] *Folk-Lore of the Sea Islands, South Carolina,* p. 199; the importance of the crossroads, like the calling of the child's soul, comes directly from Africa, despite the footnoted comment by the author on the resemblance of the calling practice having been observed among the Zuñi Indians, "who have taken it, no doubt, from their Mexican neighbors."

[104] M. J. and F. S. Herskovits, *Suriname Folk-Lore,* pp. 49 ff.

[105] *Folk Beliefs Pertaining to Health of the Southern Negro,* p. 50.

[106] *Op. cit.,* p. 199.

[107] *Folk Beliefs of the Southern Negro,* p. 347.

[108] Powdermaker, *After Freedom*, p. 208.

[109] *Ibid.*, pp. 208 f.

[110] Frazier, *The Negro Family in the United States*, pp. 255 f.; see also a passage pp. 495 f.

[111] M. J. and F. S. Herskovits, *Rebel Destiny*, chap. I.

[112] Beckwith, *Black Roadways, a Study in Jamaican Folk Life*, pp. 71 ff.; Ramos, *O Negro Brasileiro*, pp. 140 ff.; Herskovits, *Life in a Haitian Valley*, pp. 205 ff.

[113] *Social and Mental Traits of the American Negro*, pp. 133 f.

[114] *After Freedom*, p. 122.

[115] *Ibid.*, p. 133.

[116] *Shadow of the Plantation*, p. 183; cf. also Puckett, *Folk Beliefs of the Southern Negro*, p. 87.

[117] Herskovits, *Dahomey*, Vol. I, pp. 353 ff.; cf. also the reference in R. S. Rattray, *Religion and Art in Ashanti*, p. 178, which indicates that this custom was also known in the Gold Coast.

[118] Frazier, "The Negro Slave Family," *loc. cit.*, p. 216. The citation is from William E. Hatcher, *John Jasper, The Unmatched Negro Philosopher and Preacher*, New York, 1908, pp. 36 ff.

[119] Frazier, *op. cit.*; this quotation is from Bishop L. J. Coppin, *Unwritten History*, Philadelphia, 1919, p. 55.

[120] *Folk Beliefs of the Southern Negro*, pp. 93 f.

[121] For an account of an African rite performed in connection with sending the soul of a dead infant back to Africa during the days of slavery, see Ball, *Slavery in the United States*, pp. 264 f.; this may be compared with the procedure among the Guiana Negroes in returning an African spirit to its home as reported in M. J. and F. S. Herskovits, *Suriname Folk-Lore*, p. 86.

[122] Johnson has described the funeral of a man belonging to the group studied by him (*Shadow of the Plantation*, pp. 162 ff.), and Powdermaker has given a less complete account of a "middle-class" funeral (*After Freedom*, pp. 249 ff.). Neither of these students, however, "carries through" his description by describing premortuary and immediate postmortuary and postfuneral rites outside the church.

[123] Puckett, *Folk Beliefs of the Southern Negro*, p. 90.

[124] *Ibid.*, p. 92.

[125] *Ibid.*, pp. 92 f.

[126] *Ibid.*, p. 88.

[127] *Folk-Lore of the Sea Islands, South Carolina*, p. 215.

[128] Herskovits, *Dahomey*, Vol. II, pp. 195 ff.

[129] Herskovits, *Life in a Haitian Valley*, pp. 213 f.

[130] *Shadow of the Plantation*, p. 165.

[131] *Folk Beliefs of the Southern Negro*, p. 80.

[132] *Shadow of the Plantation*, p. 22.

[133] *Folk Beliefs of the Southern Negro*, p. 85.

[134] M. J. and F. S. Herskovits, *Rebel Destiny*, p. 18.

[135] *Folk-Lore of the Sea Islands, South Carolina*, p. 213.

[136] *Op. cit.*, p. 99; for a parallel to the rite given in the last sentence cf. Herskovits, *Dahomey*, Vol. I, pp. 374 f., where an account is given of the manner in which the young men run with the corpse through the village.

[137] Puckett, *op. cit.*, p. 82.

[138] *Ibid.*, p. 87.

[139] *Ibid.*, p. 84.

[140] *Ibid.*, p. 128.

[141] *Ibid.*, pp. 107 ff., *passim;* Zora Hurston, *Mules and Men*, Philadelphia, 1935, pp. 283 ff.

[142] Puckett, *op. cit.*, pp. 102 ff.

[143] *Ibid.*, pp. 104 f.; Parsons, *Folk-Lore of the Sea Islands, South Carolina*, pp. 213 f.

CHAPTER VII

[1] Bertram W. Doyle, "Racial Traits of the Negro as Negroes Assign Them to Themselves," unpublished Master's Thesis, University of Chicago, 1924, p. 90, citing W. J. Gaines, *The Negro and the White Man*, Philadelphia, 1910, p. 185.

[2] *Caste and Class in a Southern Town*, pp. 224 f.

[3] L. P. Jackson, "Religious Development of the Negro in Virginia from 1760 to 1860," *Journal of Negro History*, 16:198, 1931.

[4] *Ibid.*, p. 170.

[5] *Ibid.*, p. 211 (footnote 115).

[6] *Ibid.*, p. 198.

[7] *Ibid.*, p. 199.

[8] Doyle, *Etiquette of Race Relations*, p. 45, quoting Mary Roykin Chestnut, *A Diary from Dixie*, New York, 1905, p. 354.

[9] *Shadow of the Plantation*, pp. 151 f.

[10] *The Religious Instruction of the Negroes*, pp. 49 ff., *passim*.

[11] *Op. cit.*, in numerous passages, e.g., pp. 232 f.

[12] *Op. cit.*, pp. 125 f.

[13] John B. Cade, "Out of the Mouths of Ex-Slaves," *Journal of Negro History*, 20:331, 1935.

[14] Powdermaker, *After Freedom*, p. 270; Frazier, *The Negro Family in the United States*, pp. 30 f.; Doyle, *op. cit.*, pp. 43 f.

[15] Doyle, *op. cit.*, p. 32.

[16] *Op. cit.*, pp. 223, 239.

[17] Raymond J. Jones, "A Comparative Study of Religious Cult Behavior Among Negroes with Special Reference to Emotional Group Conditioning Factors," *Howard University Studies in the Social Sciences*, Vol. II, no. 2, Washington, 1939, p. 2 ff.

[18] *Ibid.*, p. 5; the "Classified Table of Religious Cults in the United States" given by this author on pp. 124 f. of his work will be found useful.

[19] *The Homes of the New World*, Vol. I, p. 311.

[20] *Op. cit.*, pp. 71 ff. It is somewhat difficult to understand why this student was content to report only services of secondary importance. Meetings on Monday and Friday nights and Saturday mornings are hardly those at which gatherings large enough to be typical would be found. That no services taking place on Sundays or on religious holidays are analyzed by him is unfortunate, since at these times larger congregations heighten tensions and enhance an emotional tone sufficiently deep even on lesser occasions.

[21] *In Freedom's Birthplace . . .*, pp. 244 ff.

[22] *Social and Mental Traits of the Negro*, pp. 74 f., 83 ff.

[23] *Folk Beliefs of the Southern Negro*, pp. 532 ff.

[24] *Caste and Class in a Southern Town*, pp. 344 ff.

[25] *After Freedom*, pp. 232 ff.

[26] W. E. Barton, *Old Plantation Hymns*, Boston, 1899, pp. 41-42.

[27] *Shadow of the Plantation*, pp. 159 f.

[28] M. J. and F. S. Herskovits, *Rebel Destiny*, pp. 228 ff.; 307 ff.

[29] M. J. and F. S. Herskovits, *Suriname Folk-Lore*, p. 92.

[30] See p. 16 f., notes 27 and 30.

[31] M. J. Herskovits, "African Gods and Catholic Saints in New World Negro Belief," *Amer. Anth.*, 39:635-647, 1937. The point is made the more striking by the recent discovery that the identical mechanism is operative among the Mohammedanized tribes of West Africa itself, where the *jinn* are identified with the pagan *iska* by the Hausa. Cf. J. H. Greenberg, *The Religion of a Sudanese Culture as Influenced by Islam*, unpublished Doctor's Thesis, Northwestern University, 1940, and *idem.*, "Some Aspects of Negro-Mohammedan Culture-Contact among the Hausa," *Amer. Anth.*, 43:51-61, 1941.

[32] *O Negro Brasileiro*, Figs. 26, 27, 31, 32.

[33] This field work was carried out in accordance with the systematic program of study of New World Negro cultures described on pp. 6 ff., 15 ff.

[34] Cf., for example, his *Folk Beliefs of the Southern Negro*, pp. 27 ff., 114, 532, 548 ff., 567, and 574, for some very cogent references to African aspects of Negro religion.

[35] *Ibid.*, pp. 545 f.

[36] *Op. cit.*, p. 56.

[37] See pp. 12-14.

[38] Jones, *op. cit.*, pp. 45 f.

[39] E.g., Johnson, *Shadow of the Plantation*, p. 151.

[40] Jones, *op. cit.*, p. 49.

[41] *After Freedom*, p. 232.

[42] *Ibid.*, pp. 259 f.

[43] *Ibid.*, p. 273.

[44] F. M. Davenport, *Primitive Traits in Religious Revivals*, New York, 1905, pp. 94, 125 f., 133, and 142.

[45] Cf. James Mooney, "The Ghost-Dance Religion with a Sketch of the Sioux Outbreak of 1890," *14th Ann. Rep., Bureau of Amer. Ethnology*, Part II, Washington, 1897; Leslie Spier, "The Prophet Dance of the Northwest and Its Derivatives: the Source of the Ghost Dance," *Gen. Ser. in Anthropology*, No. 1, Menasha (Wis.), 1935.

[46] Davenport, *op. cit.*, p. 73.

[47] *Ibid.*, p. 77.

[48] *Folk Beliefs of the Southern Negro*, pp. 539 f.

[49] "Religious Folk-Beliefs of Whites and Negroes," *Journal of Negro History*, 16:9-35, 1931.

[50] *Op. cit.*, p. 92.

[51] "Religious Folk-Beliefs of Whites and Negroes," *loc. cit.*, pp. 26 f.

[52] *Ibid.*, pp. 20 f.

[53] Cf. the description in Dollard, *Caste and Class in a Southern Town*, p. 236.

[54] James Mooney, "The Cherokee River Cult," *Journal of American Folk-Lore*, 13:1 ff., 1900.

[55] *Op. cit.*, p. 262.

[56] *Ibid.*

[57] See p. 18.

[58] *Folk-Beliefs of the Southern Negro*, p. 219, from G. W. Cable, *The Grandissimes*, New York, 1898.

[59] *Ibid.* (Puckett), p. 221; *ibid.* (Cable), pp. 91 f.

[60] *Green Thursday*, New York, p. 28.

[61] Herskovits, *Dahomey*, Vol. I, p. 35.

[62] Puckett, *Folk-Beliefs of the Southern Negro*, p. 421.

[63] *Ibid.*, pp. 257, 319, 381, 424.

[64] *Ibid.*, p. 553.

[65] *Folk Beliefs Pertaining to Health of the Southern Negro*, p. 37.

[66] Henry C. Davis, "Negro Folklore in South Carolina," *Journal of American Folk-Lore*, 37:245, 1914; Puckett, *op. cit.*, p. 290.

[67] Puckett, *op. cit.*, p. 399.

[68] B. A. Botkin, "'Folk-Say' and Folk-Lore," in W. T. Couch, *Culture in the South*, Chapel Hill, 1934, p. 590.

[69] *Folk-Beliefs of the Southern Negro*, p. 311.

[70] *Folk Beliefs Pertaining to Health of the Southern Negro*, pp. 36 f.

[71] *Ibid.*, pp. 40 ff.

[72] *Ibid.*, pp. 46 f.

[73] *Judicial Cases Concerning American Slavery and the Negro*, Vol. II, pp. 520 f.

[74] *My Bondage and My Freedom*, pp. 238 f.

[75] Herskovits, *Dahomey*, Vol. II, pp. 256 ff.

[76] *Folk-Beliefs of the Southern Negro*, p. 287.

[77] *Ibid.*, p. 296; the quotation is from A. M. Bacon, "Conjuring and Conjure-Doctors," *Southern Workman*, 24:211, 1895, and the footnoted references state that "this paper gives several illustrative cases."

[78] Herskovits, *Dahomey*, Vol. II, pp. 285 ff.

[79] *Folk-Beliefs of the Southern Negro*, p. 229; citing Mary A. Owen, "Among the Voodoos," *Proc. of the Int. Folk-Lore Congress*, 1891, pp. 232 ff.

[80] Herskovits, *op. cit.*, Vol. II, pp. 263 ff.

[81] *Old Rabbit the Voodoo and Other Sorcerers*, London, 1893.

[82] *Mules and Men*, Philadelphia, 1935.

[83] E.g., Louis Pendleton, "Negro Folk-Lore and Witchcraft in the South," *Journal of American Folk-Lore*, 3:201-207, 1890; (Miss) Herron and A. M. Bacon, "Conjuring and Conjure-Doctors in the Southern United States," *ibid.*, 9: 143-147, 224-226, 1896; Ruby Adams Moore, "Superstitions of Georgia," *ibid.*, 5: 230-231, 1892, and 9:227-228, 1896; Julian A. Hall, "Negro Conjuring and Tricking," *ibid.*, 10:241-243, 1897.

[84] E.g., 3:281 ff., 1890; 12:288 ff., 1899; and 19:76 f., 1906, among others.

[85] "Hoodoo in America," *Journal of Amercian Folk-Lore*, 44:318 ff., 1931.

[86] *Mules and Men*, pp. 239 ff.

[87] *Fabulous New Orleans*, New York, 1928, pp. 309 ff.

[88] *The Grandissimes*, pp. 85, 167.

[89] *Ibid.*, pp. 281, 380.

[90] *Ibid.*, pp. 91 f.

[91] See p. 236.

[92] Cable, *op. cit.*, pp. 167, 281.

[93] *Ibid.*, p. 296.

[94] *Ibid.*, p. 229.

[95] *Ibid.*, pp. 135 f.

[96] *Journal of American Folk-Lore*, 10:76, 1897.

[97] Article "Vaudou," *Grand Dictionnaire Universel du XIXe Siècle*, ed. Larousse, Paris, 1866-1890, Vol. XV, p. 812.

[98] *Mules and Men*, p. 242.

[99] *Ibid.*, p. 253.

[100] *Ibid.*, p. 248.

[101] *Ibid.*, p. 299 f.

[102] *Ibid.*, p. 300.

[103] Herskovits, *Life in a Haitian Valley*, pp. 169 ff.

[104] Herskovits, *Dahomey*, Vol. II, pp. 189, 200.

[105] Cf. especially his work, *O Negro Brasileiro*, Chap. V, "O syncretismo religioso," pp. 75 ff., and photographs such as Fig. 19, of an altar in one of the Bahian *candomblés*.

[106] *Los Negros Brujos*, pp. 53 ff.

[107] Dr. Price-Mars, *Ainsi Parla l'Oncle*, Port-au-Prince, 1928, esp., pp. 180 f., and Herskovits, *Life in a Haitian Valley*, pp. 277 ff.

[108] Puckett, *Folk Beliefs of the Southern Negro*, p. 193. This passage is abstracted from a novel, Mrs. Helen Pitkin's *An Angel by Brevet*. Puckett states

that these happenings, "although written in the form of fiction, are scientifically accurate, being an exact reproduction of what she herself has seen or obtained from her servants and absolutely free from imagination" (p. 192). In the light of the internal evidence, there is no reason to doubt the validity of the performances described; certainly the names of African gods, which check with our scientific knowledge of the region and with such other works dealing with New Orleans as are available indicate that the paragraph reproduced here, and the entire section of which it is a part, are factually valid. It is worth noting that João do Rio (Paulo Barreto), writing at the turn of the century, noted the identification by the Negroes in Rio de Janeiro of S. Antonio (Saint Anthony) with Verequete. *As Religiões no Rio,* Paris and Rio de Janeiro, n.d., p. 16.

[109] *Ibid.,* pp. 195, 362, 563 ff.

[110] *Ibid.,* pp. 562 f.; the reference is to M. A. Owen, "Among the Voodoos," *Proc. of the Int. Folk-Lore Congress,* 1891, pp. 232 f.

[111] *After Freedom,* p. 290.

[112] *Op. cit.,* pp. 196 f.

[113] Puckett, *Folk Beliefs of the Southern Negro,* p. 548; the quotation is from "Race Problems of the South," *Report of the Proceedings of the First Annual Conference of the Southern Society for the Promotion of the Study of Race Conditions and Problems in the South,* Montgomery, Ala., 1900, p. 143.

[114] Barton, *Old Plantation Hymns,* p. 11.

[115] *Mules and Men,* p. 306.

[116] Herskovits, *Dahomey,* Vol. II, p. 223.

[117] Louis Pendleton, "Negro Folk-Lore and Witchcraft in the South," *loc. cit.,* pp. 201 f.

[118] M. J. and F. S. Herskovits, *Suriname Folk-Lore,* pp. 105 f.

[119] *Folk Beliefs of the Southern Negro,* pp. 202 f.; the quotation is from A. E. Gonzales, *The Black Border,* Columbia, S. C., 1922, p. 107.

[120] *Ibid.,* p. 203.

[121] *Ibid.,* p. 541.

[122] *Ibid.,* p. 51.

[123] *Religion and Art in Ashanti,* p. 25.

[124] *Ibid.,* fig. 19.

[125] *Ibid.,* p. 26.

[126] London, 1899, p. 117.

[127] *Ibid.,* p. 28.

[128] Samuel Johnson, *The History of the Yoruba,* London, 1921, p. 29; Stephen S. Farrow, *Faith, Fancies and Fetich, or Yoruba Paganism,* London, 1926, p. 19; and Herskovits, *Dahomey,* Vol. II, pp. 259-262.

[129] *Folk-Lore of the Sea Islands, South Carolina,* p. 213.

[130] Owen, *Old Rabbit the Voodoo and other Sorcerers,* p. 11.

[131] Puckett, *Folk Beliefs of the Southern Negro,* p. 153.

[132] *Ibid.,* pp. 143 f.

[133] *Ibid.,* pp. 154 f.; the latter two citations are from E. Dayrell, *Folk-Stories from Southern Nigeria,* London, 1910, pp. 11 ff., and R. H. Milligan, *The Fetish Folk of West Africa,* London, 1912, p. 240; and from G. E. Ellis, *Negro Culture in West Africa,* New York, 1914, p. 63.

[134] *Religion and Art in Ashanti,* pp. 29 f.

CHAPTER VIII

¹ This section gives in essence the findings of a report written for the Committee on Research in Comparative Musicology, American Council of Learned Societies.

² Cf. W. F. Allen, C. P. Ware, and Lucy Garrison, *Slave Songs of the United States*, New York, 1868 (reprinted, 1929).

³ *Afro-American Folksongs*, New York, 1914.

⁴ "American Negro Songs," *Int. Rev. of Missions*, 15:748-753, 1926.

⁵ *Ibid.*

⁶ *American Negro Folk-Songs*, Cambridge, 1928.

⁷ "The Genesis of the Negro Spiritual," *American Mercury*, 26:243-248, 1932; *White Spirituals of the Southern Uplands*, Chapel Hill, 1933; *Spiritual Folk-Songs of Early America*, New York, 1937.

⁸ *Folk Culture on St. Helena Island*, Chapel Hill, 1930; "The Negro Spiritual, a Problem in Anthropology," *American Anthropologist*, 33:151-171, 1931; "Negro Folk Songs in the South," in: W. T. Couch, *Culture in the South*, Chapel Hill, 1934, pp. 547-569.

⁹ E.g., G. B. Johnson, *American Anthropologist*, 33:170, 1931.

¹⁰ Cf. the report of his paper read before the American Musicological Society as given in the *Christian Science Monitor* for Sept. 15, 1939, p. 1.

¹¹ *Journal of American Folk-Lore*, 48:394-397.

¹² Cf. George Herzog, *Research in Primitive and Folk Music in the United States*, Bulletin No. 24, 1936, American Council of Learned Societies, where both publications and record collections are listed.

¹³ For Haiti we have Harold Courlander's *Haiti Singing*, New York, 1939; for Jamaica the melodies transcribed by Helen H. Roberts, "A Study of Folk Song Variants Based on Field Work in Jamaica," *Journal of American Folk-Lore*, 38: 149-216, 1925, and "Possible Survivals of African Songs in Jamaica," *Musical Quarterly*, 11:340-358, 1926, among other titles by this student. A contribution by Fernando Ortiz on Cuban Negro Music, especially drum rhythms ("La Música Sagrada de los Negros Yorubá en Cuba," *Estudios Afro-Cubanos*, 2:89-104, 1938) will be found of great value. Musicological analyses and transcriptions of songs from Dutch Guiana by Dr. M. Kolinski will be found in M. J. and F. S. Herskovits, *Suriname Folk-Lore*, New York, 1936, pp. 491 ff.

¹⁴ Some Negro melodies are to be found in the work by Mme. Elsie Houston-Péret, *Chants populaires du Brésil*, Paris, 1930.

¹⁵ Attention may be called to the new Argentinian musical review *Pauta*, the first number of which includes a paper entitled "Folklore de la Costa Zamba; la Marinera," pp. 5, 32, by the Peruvian student of Negro life, Fernando Romero.

¹⁶ A publication by Douglas H. Varley, "African Native Music, an Annotated Bibliography," *Royal Empire Society Bibliographies*, No. 8, London, 1936, will be found as useful for its New World entries (pp. 86 ff.), as for its African citations.

¹⁷ An approach such as that suggested by M. Metfessel (in his *Phonophotography in Folk Music*, Chapel Hill, 1928) may be of help in studying problems of this order.

[18] See pp. 15-18.

[19] M. J. and F. S. Herskovits, *Suriname Folk-Lore, loc. cit.*

[20] Electrical recordings, subsequently made in Haiti by Allan Lomax in 1936 for the Library of Congress and by Harold Courlander in 1939 for the Department of Anthropology of Columbia University, should be noted.

[21] Those drawn on were R. N. Dett (ed.), *Religious Folk-Songs of the Negro as Sung at Hampton Institute,* Hampton, Va., 1927; N. Ballanta, *Saint Helena Island Spirituals,* New York, 1925; J. W. Johnson, *Second Book of Negro Spirituals,* New York, 1926; and Allen, Ware, and Garrison, *Slave Songs of the United States* (reprint of 1929).

[22] M. Griaule, "Masques Dogons," *Tr. et Mèm. de l'Inst. d'Ethnologie,* Vol. XXXIII, Paris, 1938, pp. 716 ff.

[23] *Ibid.,* Fig. 251, p. 736, especially column four, figures three from top to end, and the foot movements noted by the small arrows.

[24] Carried out with the support of a Fellowship grant of the Rosenwald Fund, under the sponsorship of the Department of Anthropology, Northwestern University.

[25] "The Dance in Place Congo," *Century Magazine,* 31:517-532, 1885-1886.

[26] *Mules and Men,* pp. 299 ff.

[27] *Fabulous New Orleans, passim.*

[28] The appropriate titles will be found in the Bibliography to M. J. and F. S. Herskovits, *Suriname Folk-Lore,* pp. 762 ff. Since this volume has appeared, certain other collections have been published: E. C. Parsons, "Folk-Lore of the Antilles, French and English," *Mem. American Folk-Lore Society,* Vol. XXVI, Pt. 2, New York, 1936; Suzanne Comhaire-Sylvain, "Creole Tales from Haiti," *Journal of American Folk-Lore,* 50:207-295, 1937, and 51:219-346, 1938; Hurston, *Mules and Men,* 1935; Samuel G. Stoney and Gertrude Mathews Shelby, *Black Genesis,* New York, 1930; M. J. and F. S. Herskovits, "Tales in Pidgin English from Ashanti," *Journal of American Folk-Lore,* 51:52-101, 1937.

[29] E. C. Parsons, "The Provenience of Certain Folk Tales. III, Tar Baby," *Journal of American Folk-Lore,* 30:227-234, 1919; W. N. Brown, "The Tar-Baby Story at Home," *Scientific Monthly,* 15:228-234, 1922; Aurelio M. Espinosa, "Notes on the Origin and History of the Tar-Baby Story," *Journal of American Folk-Lore,* 43:129-209, 1930.

[30] See pp. 18 f.

[31] M. J. and F. S. Herskovits, *Suriname Folk-Lore,* pp. 316 f., n. 2.

[32] *Ibid.,* pp. 326 ff.

[33] *Ibid.,* pp. 324 f.

[34] *Ibid.,* pp. 151 ff., *passim.*

[35] *Mules and Men,* pp. 25 ff.

[36] *Ibid.,* pp. 166 f.

[37] For the Togoland version, see Jakob Spieth, *Die Ewe-Stamme,* Berlin, 1906, p. 557.

[38] Stoney and Shelby, *Black Genesis.*

[39] Personal communication.

[40] Lorenzo D. Turner, *West African Survivals in the Vocabulary of Gullah,* presented before the Modern Language Association, New York meeting, December, 1938.

[41] The marks after these African words indicate their tonal registers, (⌐) being high tone, (_) low tone, (⌐) a glide of high to low. The system is an adaptation of that worked out by Miss Ida Ward in her *Introduction to the Ibo Language* (Cambridge, 1936) and her work, *The Phonetic and Tonal Structure of Efik* (Cambridge, 1933). The chief advantage of this system is its freedom from the diacritical marks that otherwise must be used to denote the all-important element of significant tone in African words.

[42] These citations are from a paper, *Some Problems Involved in the Study of the Negroes in the New World with Special Reference to African Survivals,* de-

livered before a Conference on Negro Studies held in April, 1940, in Washington, D. C., under the auspices of the American Council of Learned Societies.

[43] M. J. and F. S. Herskovits, *Suriname Folk-Lore,* pp. 116 ff.

[44] See pp. 78-81.

[45] Cf. M. J. and F. S. Herskovits, *op. cit.,* "Table of Phonetic Symbols," p. xi.

[46] G. Merrick in "Notes on Hausa and Pidgin English" (*Journal African Society,* Vol. VIII, 1908, pp. 304 f.), says: "Intention is expressed by the idea of motion. Example 'I will do' by 'I go do.' . . . The above remarks though probably applicable to other African languages, have been written with speech reference to Hausa."

[47] We take our examples from Martha Beckwith, "Jamaica Anansi Stories" (*Mem. American Folk-Lore Society,* Vol. XVII, New York, 1924), and the page numbers in parenthesis after each quoted phrase refer to this work. In this, as in the lists that follow, only the first occurrence of a given idiom is referred to, though all those we cite are quite common. Following the example, we give the corresponding *taki-taki* equivalent.

[48] "Folk-Tales of Andros Island, Bahamas," *Mem. American Folk-Lore Society,* Vol. XVIII, New York, 1918.

[49] "Folk-Lore of the Sea Islands, South Carolina," *ibid.,* Vol. XVI, New York, 1923.

[50] The footnoted explanation of "too" as "a characteristic use for 'very' " exactly corresponds to the way Suriname Negroes employ *tumusi.*

[51] Note the use of the word "meat" with the meaning of "live animal."

[52] Once again the use of "one" for "alone" is to be remarked.

[53] Tribute must be paid to the insight with which Hugo Schuchardt ("Die Sprache der Saramakkaneger in Surinam," *Verh. der K. Akad. van Wetenschappen te Amsterdam, Afd. Letterkunde* [n.s.], Vol. XIV, No. 6, 1914, pp. ix-xiv), discerned the resemblances between the speech of various groups of Negroes in the New World and *taki-taki,* on the basis of a vastly smaller amount of data than is available today.

[54] There are the tales of Cronise and Ward (*Cunnie Rabbit, Mr. Spider and the Other Beef: West African Folk Tales,* New York, 1903), and it is worthy of remark that several students of New World Negro dialect have noticed correspondences between the speech recorded in these tales and that of the Negroes which those students have investigated. The paper by Merrick is perhaps the only study extant of West African pidgin as such.

[55] M. J. and F. S. Herskovits, "Tales in Pidgin English from Nigeria," *Journal of American Folk-Lore,* Vol. XLIV, 1931.

[56] Not enough data in Negro-French were available when this section was written to make the sort of comparisons we make here between Negro-English in Africa and in the New World. Our experience with Negro-French in Dahomey, however, compared with the few examples of Haitian French we were able to find in the literature, and with the sketch of (Louisiana) Creole grammar by Fortier (*Louisiana Studies,* New Orleans, 1894, pp. 125 ff.), convinced us that study would show a unity of Negro-French wherever spoken that would be akin to that of Negro-English, and more, that a basic similarity in idiom between Negro-French and Negro-English would also be found to exist. These assumptions have been more than validated by the texts published by Parsons ("Folk-Lore of the Antilles, French and English," *Mem. American Folk-Lore Society,* Vol. XXV, Pt. 1, New York, 1933), which appeared while the work from which the above section is quoted was in press, by her unpublished manuscripts of Haitian tales, which we have been privileged to examine, and by the findings of our own field work in Haiti during the summer of 1934.

[57] M. J. and F. S. Herskovits, "Tales in Pidgin English from Ashanti," *Journal of American Folk-Lore,* 50:52-101, 1937 (issued 1938).

[58] The use of the word "cover" having the sense of "hide" is to be remarked.

[59] Again one finds the use of "skin" for "body." One morning our steward-boy, after receiving a message from the chief of Asokore for us, translated as follows: "De chief he sen' hask how you sikin be tiday." It was a formal inquiry about our health.

[60] "Bush-meat," i.e., wild animals.

[61] *Cunnie Rabbit, Mr. Spider and the Other Beef: West African Folk Tales,* pp. 32 ff.

[62] In transcribing Twi, the same phonetic system employed for *taki-taki* has been used except that an apostrophe here stands for a glottal stop, and that tonal marks (á = high, a = middle, à = low, â = middle to low, ă = middle to high, å = high to low) are employed.

[63] *Folk Lore of the Sea Islands, South Carolina,* p. xx. On p. xvii, n. 5, similarities in idiom between the Sea Island speech and that of Sierra Leone, as recorded by Cronise and Ward, are cited.

[64] D. Westermann, *A Study of the Ewe Language,* London, 1930, p. 50. The examples cited for Ewe also apply in the case of Fɔ, the related language of Dahomey, as can be seen by referring to Maurice Delafosse, *Manuel Dahoméen,* Paris, 1894, *passim.*

[66] M. B. Wilkie, *Ga Grammar, Notes, and Exercises,* London, 1930, p. 30.

[67] W. T. Balmer and F. C. F. Grant, *A Grammar of the Fanti-Akan Language,* London, 1921, p. 24.

[68] Abbé Pierre Bouche, *Contes Nagos,* Mélusine, Vol. II, 1884-1885, cols. 129-130. Other Yoruba examples may be found in J. A. de Gaye and W. S. Beecroft, *Yoruba Grammar* (2nd ed.), London, 1923, *passim.*

[69] Westermann, *op. cit.,* p. 43.

[70] Wilkie, *op. cit.,* p. 7.

[71] Balmer and Grant, *op. cit.,* pp. 62 ff.

[72] S. Johnson, *The History of the Yorubas, from the Earliest Times to the Beginning of the Protectorate,* London, 1921, p. xxxvi; Gaye and Beecroft, *op. cit.,* p. 8.

[73] Westermann, *op. cit.,* pp. 52 ff.

[74] Wilkie, *op. cit.,* p. 29.

[75] Balmer and Grant, *op. cit.,* chap xi.

[76] Thus Philip V. King ("Some Hausa Idioms," *Journal African Society,* Vol. VIII, 1908, p. 196) states of Hausa, "The absence of any proper comparative is one of the weakest spots in the language. The English . . . 'too many,' 'too good,' etc., can only be rendered by the use of the verb *fi*—'to pass or excell' . . . 'He is cleverer than you'—*Ya fika nankali* (lit., 'He surpasses you as to sense'). . . ."

[77] Westermann, *op. cit.,* p. 116.

[78] *Ibid.,* p. 119.

[79] *Ibid.,* pp. 126 f.

[80] *Ibid.,* p. 129.

[81] Cf., for example, Balmer and Grant, *op. cit.,* p. 14, sections 12 and 13, for their remarks on the "glide" in Fanti.

[82] M. J. Herskovits, "What Has Africa Given America?" *The New Republic,* 84:92-94, 1935.

[83] Personal communication.

[84] G. B. Johnson, *Folk Culture on St. Helena Island, South Carolina;* and C. Brooks, *The Relation of the Alabama-Georgia Dialect to the Provincial Dialects of Great Britain.*

[85] Jules Faine, *Philologie créole,* Port-au-Prince, 1936.

BIBLIOGRAPHY

AIMES, HUBERT S., "African Institutions in America." *Jour. Amer. Folk-Lore*, 18:15-32, 1905.

ALLEN, W. F., WARE, C. P., and GARRISON, LUCY, *Slave Songs of the United States*. New York, 1868 (reprinted, 1929).

An Abstract of the Evidence Delivered before a Select Committee of the House of Commons in the Years 1790, and 1791 . . . London, 1791.

ANON., "Record of Negro Folk-Lore." *Jour. Amer. Folk-Lore*, 19: 75-77, 1906.

ANON., "Vaudou." *Grand Dictionnaire Universel du XIXᵉ Siècle Français*, ed. Pierre Larousse, Vol. XV, 1866-1890, p. 812.

APTHEKER, HERBERT, "American Negro Slave Revolts." *Science and Society*, 1:512-538, 1937.

——, *The Negro in the Civil War*. New York, 1938.

——, "Maroons within the Present Limits of the United States." *Jour. Negro Hist.*, 24:167-184, 1939.

——, *Negro Slave Revolts in the United States*. New York, 1939.

BALL, CHARLES, *Slavery in the United States*. New York, 1837.

BALLAGH, J. C., *A History of Slavery in Virginia*. Baltimore, 1902.

BALLANTA, N., *Saint Helena Island Spirituals*. New York, 1925.

BALMER, W. T., and GRANT, F. C. F., *A Grammar of the Fanti-Akan Language*. London, 1921.

BARBOT, JEAN, "A Description of the Coast of North and South Guinea . . ." *Churchill's Voyages*, Vol. V. London, 1732.

BARTON, W. E., *Old Plantation Hymns*. Boston, 1899.

BASCOM, W. R., " 'Secret Societies,' Religious Cult-Groups, and Kinship Units among the West African Yoruba." Unpublished Doctor's Thesis, Northwestern University, 1939.

——, "Acculturation among the Gullah Negroes," *Amer. Anth.*, 43: 43-50, 1941.

BASSETT, JOHN S., *The Plantation Overseer, as Revealed in his Letters*. Northampton (Mass.), 1925.

BEARD, CHARLES and MARY, *The Rise of American Civilization*. New York, 1930. (1-vol. ed.)

BECKWITH, MARTHA, "Jamaica Anansi Stories." *Mem. Amer. Folk-Lore Soc.*, Vol. XVII, New York, 1924.

BECKWITH, MARTHA, *Black Roadways, a Study in Jamaican Folk Life.* Chapel Hill, 1929.

BENEZET, ANTHONY, *Some Historical Account of Guinea, and of the Slave Trade.* Philadelphia, 1771.

BERGEN, FANNY D., "Animal and Plant Lore." *Mem. Amer. Folk-Lore Soc.,* Vol. VII, Boston and New York, 1899.

BICKELL, R., *The West Indies as They Are; or a Real Picture of Slavery . . . in the Island of Jamaica.* London, 1825.

BILDEN, RÜDIGER, "Brazil, Laboratory of Civilization." *The Nation,* 128:71-74, 1929.

BOAS, FRANZ, *The Mind of Primitive Man.* New York, 1910.

BONTEMPS, ARNA, *Black Thunder.* New York, 1936.

BOSMAN, WM., *A New and Accurate Description of the Coast of Guinea . . .* (Engl. trans., 2nd ed.), London, 1721.

BOTKIN, BENJAMN A., " 'Folk-Say' and Folk-Lore," in W. T. Couch (ed.), *Culture in the South,* pp. 570-593. Chapel Hill, 1934.

BOTUME, ELIZABETH HYDE, *First Days Amongst the Contrabands.* Boston, 1893.

BOUCHE, L'ABBÉ PIERRE, "Contes Nagos." *Mélusine,* Vol. II, 1884-1885.

BOUET-WILLAUMEZ, LOUIS ÉDOUARD, *Commerce et Traite des Noirs aux Côtes Occidentales d'Afrique.* Paris, 1848.

BOWDICH, T. EDWARD, *Mission from Cape Coast Castle to Ashantee.* London, 1819.

BRACKETT, JEFFREY R., *The Negro in Maryland, a Study of the Institution of Slavery.* Baltimore, 1889.

BREMER, FREDERIKA, *The Homes of the New World.* New York, 1868. (2 vols.)

BRINTON, D. G., *Races and Peoples.* New York, 1890.

BROOKS, CLEANTH, JR., *The Relation of the Alabama-Georgia Dialect to the Provincial Dialects of Great Britain.* Baton Rouge, 1935.

BROWN, W. NORMAN, "The Tar-Baby Story at Home." *Scientific Monthly,* 15:228-234, 1922.

BROWNING, JAMES B., "The Beginnings of Insurance Enterprise among Negroes." *Jour. of Negro Hist.,* 22:417-432, 1937.

BRUCE, P. A., *Economic History of Virginia in the Seventeenth Century.* New York, 1907. 2 vols.

BURTON, RICHARD F., "A Mission to Gelele, King of Dahome . . ." *Memorial Edition of Burton's Works,* Vols. III and IV. London, 1893.

CABLE, G. W., "The Dance in Place Congo." *Century Magazine,* 31:517-532, 1885-1886.

————, *The Grandissimes.* New York, 1898.

CADE, JOHN B., "Out of the Mouths of Ex-Slaves." *Jour. Negro Hist.,* 20:294-337, 1935.

CAMERON, VIVIAN K., "Folk Beliefs Pertaining to Health of the Southern Negro." Unpublished Master's Thesis, Northwestern University, 1930.

CARDINALL, A. W., *A Bibliography of the Gold Coast.* Accra (Gold Coast), not dated.

CARNEIRO, EDISON, *Religiões Negras.* Rio de Janeiro, 1936.

CATTERALL, HELEN T. (ed.), *Judicial Cases Concerning American Slavery and the Negro.* Carnegie Institute, Washington, 1926. (5 vols.)

CHAPMAN, CHARLES E., "Palmares; The Negro Numantia." *Jour. Negro Hist.,* 3:29-32, 1918.

CHARLEVOIX, F. X., *Histoire de l'Isle Espagnole ou de S. Domingue.* Paris, 1730-1731.

CHESNUT, MARY ROYKIN, *A Diary from Dixie.* New York, 1905.

CLEENE, N. DE, "Les Chefs Indigènes au Mayombe." *Africa,* 8:63-75, 1935.

———, "La Famille dans l'Organization Social du Mayombe." *Africa,* 10:1-15, 1937.

COBB, J. C., *Mississippi Scenes.* Philadelphia, 1851.

COMHAIRE-SYLVAIN, SUZANN, "Creole Tales from Haiti." *Jour. Amer. Folk-Lore,* 50:207-295, 1937, and 51:219-346, 1938.

CONZEMIUS, EDUARD, "Ethnographical Notes on the Black Carib (Garif)." *Amer. Anth.,* 30:183-205, 1928.

COPELAND, LEWIS C., "The Negro as a Contrast Conception," in: E. T. Thompson, *Race Relations and the Race Problem.* Durham, N. C., 1939.

COPPIN, BISHOP L. J., *Unwritten History.* Philadelphia, 1919.

COUCH, W. T., *Culture in the South.* Chapel Hill, 1934.

COURLANDER, HAROLD, *Haiti Singing.* Chapel Hill, 1939.

CRONISE, FLORENCE M., and WARD, HENRY W., *Cunnie Rabbit, Mr. Spider and the Other Beef: West African Folk Tales.* New York, 1903.

CUREAU, AD., *Les Sociétés Primitives de l'Afrique Equatoriale.* Paris, 1912.

DANIELS, JOHN, *In Freedom's Birthplace, a Study of the Boston Negros.* Boston and New York, 1914.

DAVENPORT, F. M., *Primitive Traits in Religious Revivals.* New York, 1905.

DAVIS, HENRY C., "Negro Folklore in South Carolina." *Jour. Amer. Folk-Lore,* 27:241-248, 1914.

DAYRELL, E., *Folk-Stories from Southern Nigeria.* London, 1910.

DELAFOSSE, MAURICE, *Manuel Dahoméen.* Paris, 1894.

———, *Haut Sénégal-Niger (Soudan français).* Paris, 1912. (3 vols.)

DESPLAGNES, LOUIS, *Le Plateau Central Nigérien.* Paris, 1907.

DETT, R. N. (ed.), *Religious Folk-Songs of the Negro as Sung at Hampton Institute.* Hampton (Va.), 1927.

DOLLARD, J., *Caste and Class in a Southern Town.* New Haven, 1937.

DONNAN, ELIZABETH, "The Slave Trade into South Carolina before the Revolution." *Am. Hist. Rev.,* 33:804-828, 1927-1928.

———, "Documents Illustrative of the Slave Trade to America." *Carnegie Institution Publication* No. 409, Vols. I-IV, 1930-1935.

DORAINVIL, J. C., *Vodun et Névrose.* Port-au-Prince, 1931.

DORNAS, JOÃO FILHO, *A Escravidão no Brasil.* Rio de Janeiro, 1939.

DOUGLASS, FREDERICK, *My Bondage and My Freedom.* New York and Auburn, 1855.

DOWD, JEROME, *The Negro Races.* New York, 1907-1914.

———, *The Negro in American Life.* New York, 1926.

DOYLE, BERTRAM W., "Racial Traits of the Negro as Negroes Assign Them to Themselves." Unpublished Master's Thesis, University of Chicago, 1924.

———, *The Etiquette of Race Relations in the South.* Chicago, 1937.

DREWRY, W. S., *Slave Insurrections in Virginia, 1830-1865.* Washington, 1900.

DU BOIS, E. B. (ed.), "Economic Cooperation among Negro Americans." *Atlanta University Publications,* Vol. 12. Atlanta, 1907.

———, *Black Folk, Then and Now.* New York, 1939.

EDWARDS, BRYAN, *The History, Civil and Commercial, of the British Colonies in the West Indies . . . ,* Vols. I, II, III. 3rd ed., London, 1801.

ELLIS, A. B., *The Tshi-Speaking Peoples of the Gold Coast.* London, 1887.

———, *The Ewe-Speaking Peoples of the Slave Coast of West Africa.* London, 1890.

———, *The Yoruba-Speaking Peoples of the Slave Coast of West Africa.* London, 1894.

ELLIS, G. E., *Negro Culture in West Africa.* New York, 1914.

EMBREE, E. R., *Brown America, the Story of a New Race.* New York, 1931.

ESPINOSA, AURELIO M., "Notes on the Origin and History of the Tar-Baby Story." *Jour. Amer. Folk-Lore,* 43:129-209, 1930.

ESTERMANN, C., "La Tribu Kwangama en Face de la Civilisation Européenne." *Africa,* 7:431-443, 1934.

———, "Les Forgerous Kwangama." *Bull. de la Soc. Neuchâteloise de Géographie,* 44:109-116, 1936.

———, "Coutumes des Mbali du Sud d'Angola." *Africa,* 12:74-86, 1939.

Estudos Afro-Brasileiros. Trabalhas apresentados ao 1° Congresso Afro-Brasileiro reunido no Recife em 1934, 1° vol. Rio de Janeiro, 1935.

FAINE, JULES, *Philologie créole.* Port-au-Prince, 1936.

FALCONBRIDGE, ALEXANDER, *An Account of the Slave Trade on the Coast of Africa.* London, 1788.

FARROW, STEPHEN S., *Faith, Fancies and Fetich, or Yoruba Paganism.* London, 1926.

FERGUSON, WILLIAM, *America by River and Rail.* London, 1856.

FISH, CARL RUSSEL, *The Rise of the Common Man, 1830-1850.* New York, 1927.

FORDE, C. D., "Land and Labour in a Cross River Village, Southern Nigeria." *Geogr. Jour.,* 90:24-51, 1937.

———, "Fission and Accretion in the Patrilineal Clans of a Semi-Bantu Community in Southern Nigeria." *Jour. Royal Anth. Inst.,* 68:311-338, 1938.

———, "Government in Umor." *Africa,* 12:129-162, 1939.

FORTES, M., "Ritual Festivals and Social Cohesion in the Hinterland of the Gold Coast." *Amer. Anth.,* 38:590-604, 1936.

———, "Communal Fishing and Fishing Magic in the Northern Territories of the Gold Coast." *Jour. Royal Anth. Inst.,* 67:131-142, 1937.

———, *Marriage Law Among the Tallensi.* Accra (Gold Coast), 1937.

———, "Social and Psychological Aspects of Education in Taleland." Supplement to *Africa,* Vol. XI, No. 4, 1938.

FORTES, M. and S. L., "Food in the Domestic Economy of the Tallensi." *Africa,* 9:237-276, 1936.

FORTIER, ALCEE, *Louisiana Studies.* New Orleans, 1894.

FRAZIER, E. F., "The Negro Slave Family." *Jour. Negro Hist.,* 15:198-259, 1930.

———, *The Free Negro Family, a Study of Family Origins before the Civil War.* Nashville, 1932.

———, "Traditions and Patterns of Negro Family Life in the United States," in: E. B. Reuter, *Race and Culture Contacts.* New York, 1934, pp. 191-207.

———, *The Negro Family in the United States.* Chicago, 1939.

FREYRE, GILBERTO, *Casa-Grande & Senzala.* Rio de Janeiro, 1934. 2nd ed., 1936; 3rd ed., 1938.

FULLER, STEPHEN, *Two Reports . . . on the Slave-Trade.* London, 1789.

GAINES, FRANCIS PENDLETON, *The Southern Plantation, A Study in the Development and the Accuracy of a Tradition.* New York, 1925.

GAINES, W. J., *The Negro and the White Man.* Philadelphia, 1910.

GASTON-MARTIN, *Nantes au XVIIIe siècle; l'ère des Négriers (1714-1744), d'après des documents inédits.* Paris, 1931.

GAYE, J. A. DE, and BEECROFT, W. S., *Yoruba Grammar.* 2nd ed., London, 1923.

GIDDINGS, JOSHUA R., *The Exiles of Florida.* Columbus (Ohio), 1858.

GIST, NOEL P., "Secret Societies; A Cultural Study of Fraternalism in the United States." *Univ. Missouri Studies,* 15:1-184, 1940.

GONZALES, A. E., *The Black Border*. Columbia (S.C.), 1922.

GRANDPRÉ, *Voyage à la Côte Occidentale d'Afrique fait dans les années 1786 et 1787*. Paris, 1801.

GREENBERG, JOSEPH, *The Religion of a Sudanese Culture as Influenced by Islam*. Unpublished Doctor's Thesis, Northwestern University, 1940.

——, "Some Aspects of Negro-Mohammedan Culture-Contact among the Hausa," *Amer. Anth.*, 43:30-42, 1941.

GRIAULE, MARCEL, "Blason totémiques des Dogon." *Jour. de la Soc. des Africanistes*, 7:69-78, 1937.

——, "Jeux Dogons." *Tr. et. Mém. de l'Inst. d'Ethnologie*, Vol. XXXII, Paris, 1938.

——, "Masques Dogons." *Tr. et Mém. de l'Inst. d'Ethnologie*, Vol. XXXIII, Paris, 1938.

HALL, JULIEN A., "Negro Conjuring and Tricking." *Jour. Amer. Folk-Lore*, 10:241-243, 1897.

HAMBLY, WILFRID D., "The Ovimbundu of Angola." Field Museum of Natural History, *Anthropological Series, Publication* 329, 21:90-362. Chicago, 1934.

HARRIS, JOEL CHANDLER, *Nights with Uncle Remus*. Boston, 1911.

——, *Uncle Remus Returns*. Boston and New York, 1918.

——, *Uncle Remus, His Songs and Sayings*. New York, 1929.

HARRIS, J. S., "The Position of Women in a Nigerian Society." *Trans. New York Acad. of Sci.*, Ser. II, 2:141-148, 1940.

HARTSINCK, JAN J., *Beschryving van Guiana, of de Wilde Kust in Suid-America . . .* Amsterdam, 1770.

HATCHER, WILLIAM E., *John Jasper, The Unmatched Negro Philosopher and Preacher*. New York, 1908.

HERRON (MISS), and BACON, A. M., "Conjuring and Conjure-Doctors in the Southern United States." *Jour. Amer. Folk-Lore*, 9:143-147, 224-226, 1896.

HERSKOVITS, MELVILLE J., "The Negro's Americanism," in: Alain Locke, *The New Negro*. New York, 1925, pp. 353-360.

——, "Acculturation and the American Negro." *Southw. Pol. and Soc. Sci. Quart.*, 8:211-225, 1927.

——, *The American Negro, A Study in Racial Crossing*. New York, 1928.

——, "Adjiboto, An African Game of the Bush-Negroes of Dutch Guiana." *Man*, 29:122-127, 1929.

——, "Social Selection and the Formation of Human Types." *Human Biology*, 1:250-262, 1929.

——, "The Culture Areas of Africa." *Africa*, 3:59-77, 1930.

——, "The Negro in the New World: The Statement of a Problem." *Amer. Anth.*, 32:145-156, 1930.

——, "Wari in the New World." *Jour. Royal Anth. Inst.*, 62:23-37, 1932.

HERSKOVITS, MELVILLE J., "On the Provenience of New World Negroes." *Social Forces,* 12:247-262, 1933.

——, "The Social History of the Negro," in: C. Murchison, *Handbook of Social Psychology.* Worcester (Mass.), 1935, pp. 207-267.

——, "What Has Africa Given America?" *The New Republic,* 84: 92-94, 1935.

——, "The Significance of West Africa for Negro Research." *Jour. Negro Hist.,* 21:15-30, 1936.

——, "African Gods and Catholic Saints in New World Negro Belief." *Amer. Anth.,* 39:635-643, 1937.

——, "A Note on 'Woman Marriage' in Dahomey." *Africa,* 10:335-341, 1937.

——, *Life in a Haitian Valley.* New York, 1937.

——, "The Significance of the Study of Acculturation for Anthropology." *Amer. Anth.,* 39:259-264, 1937.

——, *Acculturation, The Study of Culture Contact.* New York, 1938.

——, *Dahomey.* New York, 1938. (2 vols.)

——, "The Ancestry of the American Negro." *The American Scholar,* 8:84-94, 1938-1939.

HERSKOVITS, M. J. and F. S., "Tales in Pidgin English from Nigeria." *Jour. Amer. Folk-Lore,* 44:448-466, 1931.

——, "A Footnote to the History of Negro Slaving." *Opportunity,* 11:178-181, 1933.

——, *Rebel Destiny, Among the Bush Negroes of Dutch Guiana.* New York, 1934.

——, *Suriname Folk-Lore.* New York, 1937.

——, "Tales in Pidgin English from Ashanti." *Jour. Amer. Folk-Lore,* 50:52-101, 1937.

HERSKOVITS, M. J., CAMERON, VIVIAN K., and SMITH, HARRIET, "The Physical Form of Mississippi Negroes." *Amer. Jour. Phys. Anthr.,* 16:193-201, 1931.

HERZOG, GEORGE, *Research in Primitive and Folk Music in the United States.* Bulletin No. 24, 1936, American Council of Learned Societies.

——, American Musicological Society paper (*African Influence on North American Indian Music*) as reported in the *Christian Science Monitor,* Sept. 15, 1939.

——, review of Jackson, J. P., "White Spirituals of the Southern Uplands," and Johnson, G. B., "Folk Culture on St. Helena Island." *Jour. Amer. Folk-Lore,* 48:394, 397, 1935.

HOFFMAN, FREDERICK L., *Race Traits and Tendencies of the American Negro.* New York, 1896.

HORNBOSTEL, ERICH M. VON, "American Negro Songs." *Int. Rev. Missions,* 15:748-753, 1926.

HOUSTON-PÉRET, ELSIE (Mme.), *Chants Populaires du Brésil.* Paris, 1930.

HURSTON, ZORA, "Hoodoo in America." *Jour. Amer. Folk-Lore*, 44: 317-417, 1931.
———, *Mules and Men*. Philadelphia, 1935.

JACKSON, GEORGE PULLEN, "The Genesis of the Negro Spiritual." *Amer. Mercury*, 26:243-248, 1932.
———, *White Spirituals of the Southern Uplands*. Chapel Hill, 1933.
———, *Spiritual Folk-Songs of Early America*. New York, 1937.
JACKSON, L. P., "Religious Development of the Negro in Virginia from 1760 to 1860." *Jour. Negro Hist.*, 16:168-239, 1931.
JOÃO DO RIO (Paulo Barreto), *As Religiões no Rio*. Paris and Rio de Janeiro, n.d.
JOHNSON, CHARLES S., *Shadow of the Plantation*. Chicago, 1934.
JOHNSON, GUION G., *A Social History of the Sea Islands, with Special Reference to St. Helena Island, South Carolina*. Chapel Hill, 1930.
———, *Ante-Bellum North Carolina*. Chapel Hill, 1937.
JOHNSON, GUY B., *Folk Culture on St. Helena Island, South Carolina*. Chapel Hill, 1930.
———, "The Negro Spiritual, a Problem in Anthropology." *Amer. Anth.*, 33:157-171, 1931.
———, "Negro Folk Songs in the South," in: W. T. Couch, *Culture in the South*. Chapel Hill, 1934, pp. 547-569.
JOHNSON, J. W., *Second Book of Negro Spirituals*. New York, 1926.
JOHNSON, (REV.) SAMUEL, *The History of the Yorubas, from the Earliest Times to the Beginning of the Protectorate*. London, 1921.
JOHNSTON, SIR HARRY H., *The Negro in the New World*. London, 1910.
JONES, CHARLES COLCOCK, *The Religious Instruction of the Negroes*. Savannah, 1842.
JONES, RAYMOND J., "A Comparative Study of Religious Cult Behavior Among Negroes with Special Reference to Emotional Group Conditioning Factors." *Howard University Studies in the Social Sciences*, Vol. II, no. 2, Washington, 1939.

KEANE, A. H., *Man: Past and Present*. Cambridge, 1920.
KING, CORNELIUS, "Cooperation—Nothing New." *Opportunity*, 18: 331, 1940.
KING, PHILIP V., "Some Hausa Idioms." Jour. African Soc., 8:1908.
KINGSLEY, MARY A., *Travels in West Africa*. London, 1897.
———, *West African Studies*. London, 1899.
KOELLE, S. W., *African Native Literature . . . in the Kanuri or Bornu Language . . .* London, 1854.
KOLINSKI, M., "Suriname Music," in: M. J. and F. S. Herskovits, *Suriname Folk-Lore*, pp. 491-739. New York, 1936.
KRAPP, GEORGE P., "The English of the Negro." *Amer. Mercury*, 2: 190-195, 1924.

KRAPP, GEORGE P., *The English Language in America.* New York, 1925. (2 vols.)

KREHBIEL, H. E., *Afro-American Folksongs.* New York, 1914.

LABAT, JEAN BAPTISTE, *Nouveau Voyage aux Isles de l'Amerique.* The Hague, 1724. (2 vols.)

LABOURET, HENRI, "Les Tribus du rameau Lobi, Volta Noire moyenne." *Tr. et Mém. de l'Inst. d'Ethnologie,* Vol. XV, Paris, 1931.

LEIRIS, MICHEL, et SCHAEFFNER, ANDRÉ, "Les Rites de Circoncision chez les Dogon de Sanga." *Jour. de la Soc. des Africanistes,* 6:141-162, 1936.

LEWINSON, PAUL, *Race, Class, and Party; a History of Negro Suffrage and White Politics in the South.* New York, 1932.

LEWIS, M. G., *Journal of a West India Proprietor.* London, 1834.

LIFSZYC, DEBORAH, "Les formules propitiatoires chez les Dogon." *Jour. de la Soc. des Africanistes,* 8:33-56, 1938.

LIFSZYC, D., and PAULME, DENISE, "Les Animaux dans le Folklore Dogon." *Rev. de Folklore Français et de Folklore Colonial,* 6:282-292, 1936.

———, "La Fête des Semailles en 1935 chez les Dogon de Sanga." *Jour. de la Soc. des Africanistes,* 6:95-110, 1936.

LIND, J. E., "Phylogenetic Elements in the Psychoses of the Negro." *Psychoanalytic Review,* 4:303-332, 1917.

LINTON, RALPH, *Acculturation in Seven American Indian Tribes.* New York, 1940.

LOCKE, ALAIN (ed.), *The New Negro.* New York, 1925.

LOVELL, CAROLINE C., *The Golden Isles of Georgia.* Boston, 1932.

LYELL, SIR CHARLES, *A Second Visit to the United States of North America.* New York, 1849. (2 vols.)

MACDOUGALL, WILLIAM, *Is America Safe for Democracy?* New York, 1921.

MACMILLAN, W. M., *Warning from the West Indies.* London, 1938. (rev. ed.)

MAES, J., and BOON, O., "Les Peuplades du Congo Belge, Nom et Situation Géographiques." *Monographies Idéologiques,* Publications de Bureau de Documentation Ethnographique, Musée du Congo Belge, Tervueren, Belgium, Vol. I, ser. 2, 1935.

MANSFELD, ALFRED, *Urwald-documente. Vier jahre unter den cross-flussnegern Kameruns.* Berlin, 1908.

MARCH, WM., *Come in at the Door.* New York, 1934.

MAUNIER, RENÉ, "La Construction Collective de la Maison en Kabylie." *Tr. et Mém., Inst. d'Eth.,* Vol. III. Paris, 1926.

MAYER, BRANTZ, *Captain Canot; or Twenty Years on an African Slaver.* New York, 1854.

MECKLIN, J. M., *Democracy and Race Friction, a Study in Social Ethics.* New York, 1914.

MEEK, C. K., *The Northern Tribes of Nigeria*. London, 1925. (2 vols.)
———, *A Sudanese Kingdom*. Oxford, 1931.
———, *Tribal Studies in Northern Nigeria*. London, 1931. (2 vols.)
———, "The Kulu in Northern Nigeria." *Africa*, 7:257-269, 1934.
———, *Law and Authority in a Nigerian Tribe*. Oxford, 1937.
MENCKEN, H. L., *The American Language*. New York, 1936 ed.
MERRICK, G., "Notes on Hausa and Pidgin English." *Jour. African Soc.*, 8:303-307, 1908.
METFESSEL, M., *Phonophotography in Folk Music*. Chapel Hill, 1928.
MILLIGAN, R. H., *The Fetish Folk of West Africa*. New York and Chicago, 1912.
MONTEIL, CHARLES, *Les Khassonké, Monographie d'une peuplade du Soudan français*. Paris, 1915.
———, *Les Bambara de Segou et du Kaarta*. Paris, 1924.
MOONEY, JAMES, "The Ghost-Dance Religion, with a Sketch of the Sioux Outbreak of 1890." *14th Ann. Rep. Bur. of Amer. Ethnology*, Part II. Washington, 1896.
———, "The Cherokee River Cult." *Jour. Amer. Folk-Lore*, 13:1-10, 1900.
MOORE, (MISS) RUBY ANDREWS, "Superstitions of Georgia." *Jour. Amer. Folk-Lore*, 5:230-231, 1892, and 9:227-228, 1896.
MOREAU DE ST. MÉRY, MEDERIC LOUIS ELIE, *Description . . . de la partie françoise de l'Isle Saint-Dominque*. Philadelphia, 1797-1798. (2 vol.)

NADEL, S. F., "Nupe State and Community." *Africa*, 8:257-303, 1935.
———, "Witchcraft and Anti-Witchcraft in Nupe Society." *Africa*, 8:423-447, 1935.
NASSAU, ROBERT H., *Fetichism in West Africa; Forty Years' Observation of Native Customs and Superstitions*. New York, 1904.
NETTELS, CURTIS P., *The Roots of American Civilization*. New York, 1938.
NEWELL, WM. W., "Myths of Voodoo Worship and Child Sacrifice in Hayti." *Jour. Amer. Folk-Lore*, 1:16-30, 1888.
NORRIS, ROBERT, *Memoirs of the Reign of Bossa Ahádee . . .* London, 1789.
NOTT, J. C., "The Diversity of the Human Race." *Du Bow's Review*, 10:113-132, 1851.
NOTT, J. C., and GLIDDEN, C. R., *Indigenous Races of the Earth*. Philadelphia, 1857.
Novos Estudos Afro-Brasileiros (Segundo Tomo). Trabalhas apresentadas ao 1°. Congresso Afro-Brasileiro do Recife em 1934. Rio de Janeiro, 1937.

ODUM, H. W., "Social and Mental Traits of the Negro." *Col. Univ. Studies in History, Econ., and Public Law*. New York, 1910.

OLDENDORP, C. G. A., *Geschichte des Missionen der Evangelischen Brüder auf den inseln S. Thomas, S. Croix und S. Jan.* Barby, 1777.

OLMSTED, FREDERICK L., *A Journey in the Seaboard Slave States.* New York, 1863.

———, *The Cotton Kingdom: A Traveler's Observations on Cotton and Slavery in the American Southern States.* Vol. II, New York, 1862. 2d ed.

———, *A Journey in the Back Country.* New York, 1863.

O Negro no Brasil. Trabalhas apresentadas ao 2°. Congresso Afro-Brasileiro (Bahia), em 1937. Rio de Janeiro, 1940.

ORTIZ, FERNANDO, *Los Negros Brujos.* Madrid, 1917.

———, "La Música Sagrada de los Negros Yorubá en Cuba." *Estudios Afro-Cubanos,* 2:89-104, 1938.

OVERBERGH, CYR. VAN, *Collection de Monographies ethnographiques.* Vols. I-VIII, Brussels, 1907-1911.

OWEN, MARY A., "Among the Voodoos." *Proc. of the Int. Folk-Lore Congress, 1891.* London, 1892.

———, *Old Rabbit the Voodoo and Other Sorcerers.* London, 1893.

PARK, MUNGO, *Travels in the Interior Districts of Africa Performed under the Direction and Patronage of the African Association in the Years 1795, 1796, and 1797.* 2nd ed., London, 1799.

PARK, ROBERT E., "The Conflict and Fusion of Cultures with Special Reference to the Negro." *Jour. Negro Hist.,* 4:111-133, 1919.

———, "Magic, Mentality, and City Life," in: R. E. Park (ed.), *The City.* Chicago, 1925.

PARSONS, ELSIE CLEWS, "Folk-tales of Andros Island, Bahamas." *Mem. Amer. Folk-Lore Soc.,* Vol. XIII, New York, 1918.

———, "The Provenience of Certain Folk Tales. III, Tar Baby." *Folk-Lore,* 30:227-234, 1919.

———, "Folk-Lore of the Sea Islands of South Carolina." *Mem. Amer. Folk-Lore Soc.,* Vol. XVI. Cambridge (Mass.), 1923.

———, "Folk-Lore of the Antilles, French and English." *Mem. Amer. Folk-Lore Soc.,* Vol. XXVI, Pt. 1, New York, 1936.

PAULME, DENISE, "La Divination par les chacals chez les Dogon de Sanga." *Jour. de la Soc. des Africanistes,* 7:1-14, 1937.

PENDLETON, L. A., "Our New Possessions—the Danish West Indies." *Jour. Negro Hist.,* 2:267-288, 1917.

PENDLETON, LOUIS, "Negro Folk-Lore and Witchcraft in the South." *Jour. Amer. Folk-Lore,* 3:201-207, 1890.

PETERKIN, JULIA, *Green Thursday.* New York, 1924.

PHILLIPS, ULRICH BONNELL, *American Negro Slavery, A Survey of the Supply, Employment and Control of Negro Labor as Determined by the Plantation Regime.* New York, 1918.

———, "Plantations with Slave Labor and Free." *Amer. Hist. Rev.,* 30:738-753, 1924-1925.

PIERSON, DONALD, "The Negro in Bahia, Brazil." *Amer. Soc. Review,*
    4:524-533, 1939.

PITKIN, HELEN, *An Angel by Brevet.* Philadelphia, 1904.

PORTER, KENNETH W., "Relations between Negroes and Indians within
    the Present Limits of the United States." *Jour. Negro Hist.,* 17:
    287-367, 1932.

POWDERMAKER, HORTENSE, *After Freedom.* New York, 1939.

PRICE-MARS, L., *Ainsi Parla l'Oncle.* Port-au-Prince, 1928.

PROYART, ABBÉ, *Histoire de Loango, Kakonga et autres Royaumes
    d'Afrique* . . . Paris and Lyon, 1776.

PUCKETT, NEWBELL N., *Folk Beliefs of the Southern Negro.* Chapel
    Hill, 1926.

———, "Religious Folk-Beliefs of Whites and Negroes." *Jour. Negro
    Hist.,* 16:9-35, 1931.

———, "Names of American Negro Slaves," in: G. P. Murdock,
    *Studies in the Science of Society Presented to Albert Galloway
    Keller,* pp. 471-494. New Haven, 1937.

RAIMUNDO, JACQUES, *O Elemento Afro-Negro na Lingua Portuguesa.*
    Rio de Janeiro, 1933.

———, *O Negro Brasileiro.* Rio de Janeiro, 1936.

RAMOS, ARTHUR, *O Negro Brasileiro.* Rio de Janeiro, 1934.

———, *O Folk-Lore Negro do Brasil.* Rio de Janeiro, 1935.

———, *As Culturas Negras no Novo Mundo.* Rio de Janeiro, 1937.

———, *The Negro in Brazil.* Washington, 1939. (Trans. Richard Pattee)

RATTRAY, R. S., *Ashanti.* Oxford, 1923.

———, *Religion and Art in Ashanti.* Oxford, 1927.

———, *Ashanti Law and Constitution.* Oxford, 1929.

———, *Akan-Ashanti Folk Tales.* Oxford, 1930.

RATZEL, FRIEDRICH, *History of Mankind.* New York, 1896.

RECLUS, ELISEE, *Universal Geography.* Paris, 1893.

REDFIELD, R., LINTON, R., and HERSKOVITS, M. J., "Memorandum for
    the Study of Acculturation." *Amer. Anth.,* 38:149-152, 1936.

REED, RUTH, *Negro Illegitimacy in New York City.* New York, 1926.

REINSCH, PAUL S., "The Negro Race and European Civilization."
    Amer. Jour. Sociol., 11:145-167, 1905.

REUTER, E. B., *The Mulatto in the United States.* Boston, 1918.

———, *The American Race Problem, A Study of the Negro.* New
    York, 1927. 2nd ed., 1938.

———, "The Negro Family in the United States" (Book Review).
    Amer. Jour. of Sociol., 45:798-801, 1940.

RINCHON, PÈRE DIEUDONNE, *La Traite et l'Esclavage des Congolais
    par les Européens.* Wetteren, Belgium, 1929.

———, *Le Trafic Négrier, d'après les livres de commerce du capitaine
    gantois Pierre-Ignace-Liévin Van Alstein.* Brussels, 1938.

ROBERTS, HELEN H., "A Study of Folk Song Variants Based on Field
    Work in Jamaica." *Jour. Amer. Folk-Lore,* 38:149-216, 1925.

ROBERTS, HELEN H., "Possible Survivals of African Songs in Jamaica."
    Mus. Quarterly, 12:340-358, 1926.
ROMERO, FERNANDO, "Folklore de la Costa Zamba: la Marinera." *Pauta,*
    No. 1, pp. 5, 32, 1939.

SAXON, LYLE, *Fabulous New Orleans.* New York, 1928.
SCHRIEKE, B., *Alien Americans, A Study of Race Relations.* New York,
    1936.
SCHUCHARDT, HUGO, "Die Sprache der Saramakkaneger in Surinam."
    *Verh. der K. Akad. van Wetenschappen te Amsterdam, Afd. Letter-
    kunde* (n.s.), Vol. XIV, No. 6, 1914.
SHANNON, FRED A., *Economic History of the People of the United
    States.* New York, 1934.
SMEDES, SUSAN DABNEY, *Memorials of a Southern Planter.* Baltimore,
    1887.
SMITH, REED, "Gullah." *Bulletin of the University of South Carolina,*
    No. 190, 1926.
SNELGRAVE, CAPT. WM., *A New Account of Some Parts of Guinea,
    and the Slave-Trade* . . . London, 1734.
Southern Society for the Promotion of the Study of Race Conditions
    and Problems in the South. *Report of the Proceedings of the First
    Annual Conference* at Montgomery (Ala.), 1900. Richmond (Va.),
    1900.
SPECK, F. G., "The Negroes and the Creek Nation." *Southern Work-
    man,* 37:106-110, 1908.
SPIER, LESLIE, "The Prophet Dance of the Northwest and its Deriva-
    tives: The Source of the Ghost Dance." *Gen. Ser. in Anth.,* No. 1,
    pp. 1-74. Menasha (Wis.), 1935.
SPIETH, JAKOB, *Die Ewe-Stamme.* Berlin, 1906.
STEDMAN, CAPT. J. G., *Narrative of a Five Years' Expedition against
    the Revolted Negroes of Suriname.* London, 1796.
STEINER, BERNARD C., "History of Slavery in Connecticut." *Johns Hop-
    kins University Studies in History and Political Science,* 11th
    Series, Vol. XI, pp. 377-452. Sept.-Oct., 1893.
STEINER, ROLAND, "Braziel Robinson Possessed of Two Spirits." *Jour.
    Amer. Folk-Lore,* 13:226-228, 1900.
STILL, WILLIAM, *The Underground Railroad.* Philadelphia, 1872.
STONEY, SAMUEL G., and SHELBY, GERTRUDE MATHEWS, *Black Genesis.*
    New York, 1930.

TALBOT, P. A., *In the Shadow of the Bush.* London, 1912.
———, *Life in Southern Nigeria.* London, 1923.
———, *The Peoples of Southern Nigeria.* London, 1926. (4 vols.)
———, *Some Nigerian Fertility Cults.* London, 1927.
TAUXIER, LOUIS, *Le Noir du Soudan; Pays Mossi et Gourounsi.* Paris,
    1912.
———, *Le Noir du Yatenga.* Paris, 1917.

TAUXIER, LOUIS, *Négres Gouro et Gagou (center de la Côte d'Ivoire).* Paris, 1924.

———, *Religion, Moeurs et Coutumes des Agnis de la Côte d'Ivoire.* Paris, 1932.

TAYLOR, CHARLES E., *Leaflets from the Danish West Indies.* London, 1888.

TESSMANN, GÜNTER, *Die Pangwe.* Berlin, 1913 (2 vols.)

———, *Die Baja, ein Negerstamm in Mittleren Sudan.* Stuttgart, 1934-1937.

THOMAS, N. W., *Anthropological report on the Edo-speaking peoples of Nigeria.* London, 1910.

———, *Anthropological report on the Ibo-speaking peoples of Nigeria.* London, 1913-1914. (6 vols.)

———, *Anthropological report on Sierre Leone.* London, 1916.

THOMAS, WM. H., *The American Negro: What He Was, What He Is, and What He May Become.* New York, 1901.

TILLINGHAST, JOSEPH A., *The Negro in Africa and America.* New York, 1902. (Publication of the American Economic Assoc., 3rd series, Vol. III, No. 2, May, 1902. Pp. 407-637.)

TORDAY, E., and JOYCE, T. A., "Notes on the Ethnography of the BaMbala." *Jour. Royal Anth. Inst.,* 35:398-426, 1905.

———, "Documents ethnographiques concernants les populations du Congo belge." *Annales de la Musée du Congo Belge. Ethnographie, Anthropologie,* sér. 3, t. 2. Brussels, 1910.

———, "Notes ethnographiques sur les peuples communément appelés Bakuba, ainsi que sur les peuplades apparentées. Les Bushongo." *Annales de la Musée du Congo Belge. Ethnographie, Anthropologie,* sér. 3, t. 2. Brussels, 1910.

TURNER, LORENZO D., *West African Survivals in the Vocabulary of Gullah,* presented before the Modern Language Association, New York Meeting, December, 1938.

———, *Some Problems Involved in the Study of the Negroes in the New World with Special Reference to African Survivals.* A paper delivered before a Conference on Negro Studies held in April, 1940, in Washington, D.C., under the auspices of the American Council of Learned Societies.

VARLEY, DOUGLAS H., "African Native Music, an Annotated Bibliography." *Royal Empire Society Bibliographies,* No. 8. London, 1936.

VASSIÈRE, PIERRE DE, *Saint-Domingue (1629-1789), la société et la vie créole sous l'ancien régime.* Paris, 1909.

VERHULPEN, EDMOND, *Baluba et Balubaïsés du Katanga.* Antwerp, n.d.

WAITZ, FRANZ THEODOR, *Anthropologie der Naturvölker.* Leipzig, 1877.

WALLACE, DAVID D., *The Life of Henry Laurens with a Sketch of the Life of Lieutenant-Colonel John Laurens.* New York, 1915.

WARD, IDA, *The Phonetic and Tonal Structure of Efkik*. Cambridge, 1933.

———, *Introduction to the Ibo Language*. Cambridge, 1936.

WASHINGTON, B. T., *The Story of the Negro*. New York, 1909. (2 vols.)

WEATHERFORD, W. D., *The Negro from Africa to America*. New York, 1924.

WEATHERFORD, W. D., and JOHNSON, CHAS. S., *Race Relations*. New York, 1934.

WEATHERLY, U. G., "The West Indies as a Sociological Laboratory." *Amer. Jour. Sociol.*, 29:290-304, 1923.

WEEKS, JOHN H., *Among Congo Cannibals*. Philadelphia, 1913.

———, *Among the Primitive BaKongo*. London, 1914.

WERNER, ALICE, *Structure and Relationship of African Languages*. London, 1930.

WESTERGAARD, W., "Account of the Negro Rebellion on St. Croix, Danish West Indies, 1759." *Jour. Negro Hist.*, 11:50-61, 1926.

WESTERMANN, D., *Die Kpelle, ein Negerstamm in Liberia*. Göttingen, 1921.

———, *A Study of the Ewe Language*. London, 1930.

WHEELER, J. D., *A Practical Treatise on the Law of Slavery, being a Compilation of all the Decisions made on that Subject in the Several Courts of the United States and State Courts*. New York, 1837.

WHITE, NEWMAN, *American Negro Folk-Songs*. Cambridge, 1928.

WILKIE, M. B., *Ga Grammar, Notes, and Exercises*. London, 1930.

WISH, HARVEY, "American Slave Insurrections before 1861." *Jour. Negro Hist.*, 22:299-320, 1937.

———, "Slave Disloyalty under the Confederacy." *Jour. Negro Hist.*, 23:435-450, 1938.

———, "The Slave Insurrection Panic of 1856." *Jour. Southern Hist.*, 5:206-222, 1939.

WOODSON, CARTER G., *Free Negro Heads of Families in the United States in 1830*. Washington, 1925.

———, *The African Background Outlined*. Washington, 1936.

———, "Life in a Haitian Valley" (Book Review.) *Jour. Negro Hist.*, 22:366-369, 1937.

WOOFTER, T. J., *Black Yeomanry*. New York, 1930.

WORK, M. N., "Some Geechee Folk-Lore." *Southern Workman*, 35:633-635, 1905.

———, (ed.), *Negro Year Book, 1918-1919*. Tuskegee Institute, Ala.

YOUNG, DONALD, *American Minority Peoples*. New York, 1932.

———, "Research Memorandum on Minority Peoples in the Depression." *Bulletin 31*, Social Science Research Council, New York, 1937.

SUPPLEMENTARY BIBLIOGRAPHY

Note: This Bibliography includes two groups of titles. The first, and by far the most numerous, represents works that have appeared since this book was originally published. In addition, however, I have included a few historical studies, slave autobiographies and other volumes that antedate the earlier edition, but which were not accessible to me, or which I had not encountered when I was gathering my materials.

M. J. H.

ALEGRÍA, RICARDO E., "The Fiesta of Santiago Apostol (St. James the Apostle) in Loiza, Puerto Rico." *Jour. Amer. Folklore,* 69: 123-130, 1956.

AMES, DAVID, "Negro Family Types in a Cuban Solar." Phylon, 11: 159-163, 1950.

APTER, DAVID, *The Gold Coast in Transition.* Princeton (New Jersey), 1955.

ARBOLEDA, JOSE R., *The Ethnohistory of the Colombian Negro.* M.A. Thesis, Northwestern University, 1950.

BALDUS, HERBERT (ed.), Section "Estudos Afrobrasileiros." *Annais,* XXXI Cong. Int. de Americanistas, São Paulo, I: 463-554, 1954.

BALL, CHARLES, *Fifty Years in Chains; or, The Life of an American Slave.* New York, 1858.

BASCOM, W. R., "The Focus of Cuban Santeria," *Southwestern Jour. of Anth.,* 6: 64-68, 1952.

BASCOM, W. R., "The Yoruba in Cuba," *Nigeria,* 39: 15-24, 1950.

BASCOM, W. R., "The Esusu: A Credit Institution among the Yoruba," *Jour. Royal Anth. Inst.,* 62: 63-70, 1952.

BASCOM, W. R. and WATERMAN, R. A., "African and New World Negro Folklore." *Funk and Wagnalls Standard Dictionary of Folklore,* New York, 1949-50: 18-24.

BASTIDE, R., "L'Islam Noir au Brésil" *Hespéris,* 3e-4e trimestres, 1952.

BASTIDE, R., *Estudos Afro-Brasileiros,* 3ª Serie. São Paulo, 1953.

BAUER, RAYMOND and ALICE, "Day to Day Resistance to Slavery." *Jour. Negro Hist.,* 27: 388-419, 1942.

BELTRÁN, GONZALO AGUIRRE, *La Poblacion Negra de Mexico (1519-1890).* Mexico, D. F., 1946.

BELTRÁN, GONZALO AGUIRRE, "Tribal Origins of Slaves in Mexico." *Jour. Negro Hist.,* 31: 269-352, 1945.

BELTRÁN, GONZALO AGUIRRE, *Cuijla, Esbozo etnográfico de un pueblo negro.* Mexico, D. F., 1958.

BOTKIN, BENJAMIN, *Lay My Burden Down: A Folk History of Slavery.* Chicago, 1945.

BOURGUINON, ERIKA E., "Class Structure and Acculturation in Haiti." *Ohio Jour. of Sci.,* 52: 317-20, 1952.

BROWN, STERLING A., "Negro Folk Expression." *Phylon,* 11: 318-327, 1950; 14: 45-67, 1955.

BUSIA, K., *The Position of the Chief in the Modern Political System of Ashanti.* London, 1951.

CAMPBELL, A. A., "St. Thomas Negroes: A Study of Personality and Culture." *Psych. Monographs,* 55: no. 5 (whole no. 253), 1943.

CARMICHAEL, MRS. A. C., *Domestic Manners and Social Activities of the White, Coloured and Negro Population of the West Indies.* (2 vols.), London, 1833.

CARNEIRO, EDISON, "The Structure of African Cults in Bahia." *Jour. Amer. Folklore,* 53: 271-78, 1940.

CARR, ANDREW T., "A Rada Community in Trinidad." *Caribbean Quarterly*, 3 : 35-54, 1953.

CHRISTENSEN, J. B., *Double Descent among the Fanti.* New Haven (Conn.), 1954.

COELHO, RUY, *The Black Carib of Honduras.* Ph.D. Dissertation, Northwestern University, 1955.

COFFIN, J., *An Account of Some of the Principal Slave Insurrections.* New York, 1860.

CROWLEY, DANIEL J., *Tradition and Individual Creativity in Bahamian Folktales,* Ph.D. Dissertation, Northwestern University, 1956.

CURTIN, PHILIP D., *Two Jamaicas, The Role of Ideas in a Tropical Colony, 1830-1865.* Cambridge (Mass.), 1955.

DARK, PHILIP, *Bush Negro Art.* London, 1954.

DAVIE, MAURICE R., *Negroes in American Society.* New York, 1949.

DAVIS, ALLISON, GARDINER, B. B., and GARDINER, M. R., *Deep South.* Chicago, 1941.

DENUCE, J., *L'Afrique au XVIe Siècle et le Commerce Anversois.* Antwerp, 1937.

DIGGS, IRENE, "Zumbi and the Republic of Os Palmares." *Phylon,* 14 : 62-69, 1953.

DIKE, K. ONWUKA, *Trade and Politics in the Niger Delta, 1830-1885.* Oxford, 1956.

DRAKE, ST. CLAIR and CAYTON, H. R., *Black Metropolis, a Study of Negro Life in a Northern City.* New York, 1945.

EDUARDO, OCTAVIO DA COSTA, *The Negro in Northern Brazil.* New York, 1948.

ESCALANTE, AQUILAS, "Notas Sobre El Palenque de San Basilio." *Inst. de Investigación Etnológica, Divulgaciones Etnológicas, 3 :* 207-359, 1954.

FORDE, D. (ed.), *African Worlds.* London, 1954.

FORTES, M., *The Dynamics of Clanship among the Tallensi.* London, 1945.

FORTES, M., *The Web of Kinship among the Tallensi.* London, 1949.

FORTES, M. and EVANS-PRITCHARD, E. E. (eds.), *African Political Systems.* London, 1940.

FRANKLIN, JOHN HOPE, *From Slavery to Freedom.* New York, 1947.

FRAZIER, FRANKLIN, "The Negro in Bahia, Brazil." *Amer. Soc. Rev.,* 7 : 465-475, 1942.

FRAZIER, FRANKLIN, *Black Bourgeosie,* Glencoe (Ill.), 1957.

FREYRE, GILBERTO, *The Masters and the Slaves.* New York, 1946.

GREENBERG, J. H. *The Influence of Islam on a Sudanese Religion.* New York, 1946.

HALL, ROBERT A., JR., "The Linguistic Structure of Taki-Taki." *Language,* 24 : 92-116, 1948.

HALL, ROBERT A., JR., "The Genetic Relationships of Haitian Creole." *Richerche Linguistiche,* 1 : 1-10, 1950.

HENRIQUES, FERNANDO M., *Family and Colour in Jamaica.* London, 1953.

HEPWORTH, GEORGE H., *The Whip, Hoe and Sword; or, The Gulf Department in '63.* Boston, 1864.

HERSKOVITS, M. J., "Education and Cultural Dynamics." *Amer. Jour. Soc.,* 48: 737-49, 1942.

HERSKOVITS, M. J., "The Negro in Bahia, Brazil: a Problem in Method." *Amer. Soc. Rev.,* 7: 394-402, 1943.

HERSKOVITS, M. J., "The Southernmost Outposts of New World Africanisms." *Amer. Anth.,* 45: 495-510, 1943.

HERSKOVITS, M. J., "Drums and Drummers in Afro-Brazilian Cult Life." *Musical Quarterly,* 30: 477-492, 1944.

HERSKOVITS, M. J., "Trinidad Proverbs ('Old Time Saying So')." *Jour. Amer. Folklore,* 58: 195-207, 1945.

HERSKOVITS, M. J., "Problem, Method and Theory in Afroamerican Studies." *Afroamérica,* 1: 5-24, 1945; also published in *Phylon,* 7: 337-354, 1946.

HERSKOVITS, M. J., "African Literature (Negro Folklore)." *Ency. of Literature* (J. T. Shipley, ed.), I: 3-15. New York, 1946.

HERSKOVITS, M. J., "The Contribution of Afroamerican Studies to Africanist Research." *Amer. Anth.,* 50: 1-10, 1948.

HERSKOVITS, M. J., "The Present Status and Needs of Afroamerican Research." *Jour. of Negro Hist.,* 36: 123-47, 1951.

HERSKOVITS, M. J., "The Social Organization of the Candomble." *Annais,* XXXI Cong. Int. de Americanistas, São Paulo, 1955, I: 505-532.

HERSKOVITS, M. J., "On Some Modes of Ethnographic Comparison." *Bijdragen tot de Taal-, Land- en Volkenkunde,* 112:129-148, 1956.

HERSKOVITS, M. J. and F. S., "The Negroes of Brazil." *Yale Review,* 32: 263-279, 1943.

HERSKOVITS, M. J. and F. S., *Dahomoan Narrative.* Evanston, 1958.

HERSKOVITS, M. J. and F. S., *Trinidad Village.* New York, 1947.

HERSKOVITS, M. J. and WATERMAN, R. A., "Musica de Culto Afro-bahiana." *Revista de Estudios Musicales,* 1: 65-127, 1949.

HISS, PHILIP H., *Netherlands America.* New York, 1943.

JOHNSON, CHARLES S. (ed.), *Education and the Cultural Process.* Chicago, 1943.

KARDINER, A. and OVESEY, L., *The Mark of Oppression.* New York, 1951.

KATZIN, MARGARET, *Higglers of Jamaica,* Ph.D. Dissertation, Northwestern, 1958.

KEMBLE, FANNY, *Journal of a Residence on a Georgia Plantation in 1838-39.* New York, 1863.

KERR, MADELINE, *Personality and Conflict in Jamaica.* Liverpool (Eng.), 1952.

KING, JAMES F., "The Colored Castes and American Representation in the Cortez of Cádiz." *Hispanic American Historical Rev.* 33: 33-64, 1953.

KING, JAMES F., "Negro Slavery in New Grenada," *Greater America,* Berkeley (Cal.), 1945, 295-318.

LARSEN, JENS, *Virgin Islands Story.* Philadelphia, 1950.

LEYBURN, JAMES G., *The Haitian People.* New Haven, 1941.

LLOYD, CHRISTOPHER, *The Navy and the Slave Trade. The Suppression of the Slave Trade in the Nineteenth Century.* London, 1949.

LOGAN, RAYFORD W., *The Negro in American Life and Thought, 1877-1901.* New York, 1954.

MACFALL, HALDANE, *The Wooings of Jezebel Pettifer.* New York, 1925.

MAUPOIL, B., *La Géomancie à l'Ancienne Côte des Esclaves.* Paris, 1943.

MERRIAM, ALAN P., *Songs of the Afro-Bahian Cults, a Musicological Analysis,* Ph.D. Dissertation, Northwestern University, 1951.

MERRIAM, ALAN P., "An Annotated Bibliography of African and African-Derived Music since 1936." *Africa,* 21: 319-29, 1951.

MÉTRAUX, A., "Reactions psychologiques à la christianisation de la Vallée de Marbial (Haïti)." *Rev. de Psychologie des Peuples,* 8: 250-267, 1950.

MÉTRAUX, A., "Croyances et Pratiques Magiques dans la Vallée de Marbial, Haïti." *Jour. de la Société des Americanistes,* n. s., 42: 135-198, 1953.

MÉTRAUX, A., "Divinités et cultes vodou dans la vallée de Marbial (Haïti)." *Zaire,* 7:675-707, 1954.

MONOD, THEODORE (ed.), "Le Monde Noire." *Présence Africaine,* nos. 8-9, 1950.

MONOD, THEODORE (ed.), "Les Afro-Americaines." *Mem. de l'Institut Français d'Afrique Noire,* no. 27, 1953.

MOORE, GEORGE H., *Notes on the History of Slavery in Massachusetts.* New York, 1866.

MOORE, JOSEPH G., *The Religion of the Jamaica Negroes.* Ph.D. Dissertation, Northwestern Univ., 1953.

MOORE, JOSEPH G. and SIMPSON, GEORGE E., A Comparative Study of Acculturation in Morant Bay and West Kingston, Jamaica. *Zaire,* 11: 989-1019, 1957.

MYRDAL, GUNNAR, *An American Dilemma.* New York, 1944.

NADEL, S. F., *A Black Byzantium, The Kingdom of Nupe in Nigeria.* London, 1942.

NADEL, S. F., *Nupe Religion.* London, 1954.

NKETIA, J. H., *Funeral Dirges of the Akan People.* Achimota (Gold Coast), 1955.

ORTIZ, FERNANDO, *Cuban Counterpoint.* New York, 1947.

ORTIZ, FERNANDO, *Los Instrumentos de la Música Afrocubana.* (5 vols.), Habana, 1952-55.

OWEN, NICHOLAS, *Journal of a Slave-dealer.* Boston, 1930.

PARRINDER, E. G., *African Traditional Religion.* London, 1954.

PARRISH, LYDIA, *Slave Songs of the Georgia Sea Islands.* New York, 1942.

PRICE, THOMAS J., *Saints and Spirits: Differential Acculturation in Colombian Negro Communities.* Ph.D. Dissertation, Northwestern University, 1955.

PROUDFOOT, MARY, *Britain and the United States in the Caribbean.* London, 1954.

RADCLIFFE-BROWN, A. R. and FORDE, D. (eds.), *African Systems of Kinship and Marriage.* London, 1950.

REID, V. S., *New Day.* New York, 1949.

RHETT, B. S., *200 Years of Charleston Cooking.* New York, 1934.

RIBEIRO, RENE, "On the Amaziado Relationship and other Aspects of the Family in Recife (Brazil)." *Amer. Soc. Rev.,* 10: 44-51, 1945.

RIBEIRO, RENÉ, "Cultos Afrobrasileiros de Recife." *Bol. do. Inst. Joaquim Nabuco,* Recife, 1952.

RUBIN, VERA (ed.), *Caribbean Studies: a Symposium,* Kingston, 1957.

SCHLOSSER, KATESA, *Propheten in Afrika.* Braunschweig, 1949.

SIMPSON, GEORGE E., "The Vodun Service in Northern Haiti." *Amer. Anth.,* 42: 236-54, 1940.

SIMPSON, GEORGE E., "Haiti's Social Structure." *Amer. Soc. Rev.,* 6: 640-49, 1941.

SIMPSON, GEORGE E., "Sexual and Familial Institutions in Northern Haiti." *Amer. Anth.,* 44: 655-74, 1942.

SIMPSON, GEORGE E., "The Belief System of Haitian Vodun." *Amer. Anth.,* 47: 35-59, 1945.

SIMPSON, GEORGE E., "The Ras Tafari Movement in Jamaica: a Study of Race and Class Conflict." *Social Forces,* 34: 167-70, 1955.

SIMPSON, GEORGE E., "Political Cultism in West Kingston, Jamaica." *Social and Economic Studies,* 4: 133-49, 1955.

SIMPSON, GEORGE E., "Jamaican Revivalist Cults," *Social and Economic Studies,* 5: 321-441, 1956.

SIMPSON, GEORGE E. and YINGER, J. MILTON, *Racial and Cultural Minorities.* (2nd ed.) New York, 1958.

SMITH, MARY, *Baba of Karo, a Woman of the Moslem Hausa.* London, 1954.

SMITH, M. G., "Caribbean Affairs: A Framework for Caribbean Studies." Mona (Jamaica), n. d.

SMITH, RAYMOND, *The Negro Family in British Guiana.* Jamaica, 1956.

STEWARD, AUSTIN, *Twenty-two Years a Slave.* Rochester, 1857.

SUNDKLER, BENGT G. M., *Bantu Phophets in South Africa.* London, 1948.

TAX, SOL (ed.), *Acculturation in the Americas,* Chicago, 1952, pp. 48-63, 143-218.

TAYLOR, DOUGLAS M., "The Black Carib of British Honduras." *Viking Fund Pub. in Anth.,* No. 17, New York, 1951.

THOBY-MARCELIN, PHILIPPE, *Canapé Vert.* New York, 1944.

TURNER, LORENZO, *Africanisms in the Gullah Dialect.* Chicago, 1949.

TUTUOLA, AMOS, *The Palm-Wine Drinkard.* London, 1952.

VAN EVRIE, J. H., *White Supremacy and Negro Subordination.* New York, 1868.

VERGER, PIERRE, *Dieux d'Afrique.* Paris, 1954.

VERGER, PIERRE, "Notes sur le culte des orisa et vodour." *Mém. de l'Inst. Français d'Afrique Noire,* No. 51. Dakar, 1957.

WATERS, ETHEL, *His Eye is on the Sparrow.* Garden City, 1951.

WILLIAMS, ERIC, *Capitalism and Slavery.* Chapel Hill, 1944.

WILLIAMS, JAMES, *Narrative of an American Slave.* New York, 1838.

WORK, JOHN W., *American Negro Songs.* New York, 1940.

WRITERS' PROJECT, GEORGIA, *Drums and Shadows.* Athens (Georgia), 1940.

INDEX

Abnormality, attitude toward manifestations of, by United States Negroes, 255 f.

Accident, philosophical implications of deification of, by West Africans, 71

Accommodation, of slave to aspects of European culture, 110 f.

Acculturation, defined, 10
in various aspects of culture, 135 ff.
mechanisms, making for acceleration or retardation in, 140 f.

Negro, customary assumptions concerning processes of, 134 ff.
Negro, effect on, of urban and rural life, 122
need for reanalysis of historical materials on slavery for understanding of, 86
of New World Negroes, theories regarding differential character of, 110
significance of New World Negro for study of, 15 ff.
type of, found in West Africa, 69

Adjustment to slavery, of Negroes, hypothesis of analyzed, 293

Adoption, as means of enlarging family, among African and New World Negroes, 187 f.

Aesthetic life, in West African-Congo area, 84

Africa, as "badge of shame" to United States Negroes, psychological effect of, 32

Africa, belief in low state of cultures in, as element in myth of Negro past, 2

Africa, belief in poorer stock of enslaved, as element in myth of Negro past, 1

Africa, East, exportation from Congo ports, of slaves from, 37
proportion of slaves brought to United States from, 47-48

Africa, interior of, hypothesis of as source of slaves, 38 f.

Africa, project for ethnographic studies needed in, 327 ff.

Africa, source for study of West African cultures, 77, 306

Africa, West, belief and ritual in, compared with New World Negro religion, 213 f.
coastal area of, as major source of slaves, 38 f.
complexity of cultures of, 60
importance of funeral in, 198
projects for linguistic study in, 337
recent research on native cultures of, 17 f.
reflection of marriage patterns in New World, 167 ff.
role of indirection in, 157

African background, neglect of, by students of Negro family organization, 181

African continent, belief in derivation of Negroes from all parts of, 1 f.

African culture, contrasted to European, 21 f.
descriptions of by early travelers, as sources employed by students of United States Negro, 56
importance of, in study of Negro life in United States, 28-30
resilience of, under contact, 18 f.

African elements, in New World Negro music, 265
in structure and functioning of New World Negro family, 181 f.
neglect of, in explanation of United States Negro religion, 223 f.

African origins, of New World Negroes, study of, 7 ff.

African religious beliefs, tenaciousness of in New World explained by enslavement of priests, 107 f.

African survivals, research in as acculturation study, 10 f.

African traits, unfavorable, assumed presence of in American Negro behavior, 25

351

Folk beliefs, African, among United States Negroes, 236 ff.

Folk remedies, types of practitioners administering, 239-241

Folklore, in West African-Congo area, 85

New World Negro, study of, 272 ff.

unity of, in Old World cultural province, 18 f.

Forde, C. D., cited, 57, 77, 161, 306, 307, 315

Fortes, M., cited, 77, 306

Fortes, M. and S. L., cited, 77, 306

Fortier, A., cited, 285, 324

Fraternal organizations, Negro, in United States, religious emphasis on, as African survival, 166

Frazier, E. F., cited, 126, 170, 210, 312, 315, 318

quoted, 3, 31 f., 110, 135 f., 173 f., 176, 177, 178, 183, 185, 199, 202, 317

French West Indies, sources of slaves imported into, 48

Freyre, G., cited, 16, 301

"Frizzled" hen, survival of African beliefs concerning, in New World, 237

Frustration, Negro religion in United States conceived as compensation for, 207

Fuller, S., quoted, 49 f.

Funeral, delayed, as African survival among United States Negroes, 202-204

importance of, in New World Negro societies as African survival, 197 ff.

West African, role and function of, in ancestral cult, 63

Gaboon, as source of slaves, 36 f.

project for ethnographic study in, 330

Gabriel, revolt led by, 98

Gaines, F. P., quoted, 116

Gaines, W. J., cited, 208

Garrison, L., cited, 262, 268, 322, 323

"Gaspar River" camp meeting of 1830, described, 229

Gaston-Martin, cited, 43, 305

Gender, expression of, in West African and New World Negro languages, 289

Georgia, proportions of masters and slaves in, 119 f.

Ghosts, African elements in beliefs of United States Negroes concerning, 258 ff.

Giddings, J. R., cited, 99, 310

Gist, N., cited, 167, 315

Glidden, C. R., cited, 20, 302

Gods, adoption of by West Africans, significance of for study of New World Negro, 72

relationship of, to ancestors in West Africa, 63

Gold Coast, bibliography of, 17, 302

culture of, outlined, 61 ff.

pidgin dialect of, correspondences to New World Negro speech in, 285 f.

project for ethnographic study in, 329

slaving operations in, 35

Gonzales, A. E., quoted, 255, 277

Good, and evil, absence of dichotomy between, in Negro thought, 242 f.

Grandpré, cited, 37, 304

Grant, F. C. F., cited, 289, 325

Great gods, role of, in West African religion, 70

Greenberg, Joseph, field work of, among Hausa and Maguzawa, 17, 302

Greenberg, J. H., cited, 220, 319

Griaule, M., cited, 77, 269, 307

Group, local, as unit of West African political structure, 65 ff.

Guiana, British, origins of slaves imported into, 48

Guiana, project for ethnographic studies of Negro cultures in, 331

Guinea Coast, as principal area of slaving, 36

Gullah Islands, correspondences with taki-taki in speech of Negroes in, 283

forms of animal tales in, 275

Habits, industrial, African, retained in United States, 146 f.

Hair dressing, African styles of, among United States Negroes, 147-149

Haiti, assimilation of European and African marriage patterns in, 168

proportion of masters and slaves in, 121

slave revolts in, 92 f.

study of peasant cultures of, 17

syncretism between African and Catholic beliefs in, 220

tribal origins of slaves in, 44

Hall, J. A., cited, 244, 320